FOURTH EDITION
Technically—Write!
Communicating in a Technological Era

RON S. BLICQ

Senior Consultant
RGI International

PRENTICE HALL, Englewood Cliffs, New Jersey 07632

Library of Congress Cataloging-in-Publication Data

BLICQ, RON (date)
  Technically—write! : communicating in a technological era / Ron
S. Blicq.—4th ed.
    p.   cm.
  Includes index.
  ISBN 0-13-904434-5
  1. Technical writing. I. Title.
T11.B62   1993
808'.0666—dc20   92-37453
                                           CIP

Acquisitions editors: Phil Miller
    and Alison Reeves
Editorial production / supervision
    and interior design: F. Hubert
Prepress buyer: Herb Klein
Manufacturing buyer: Bob Anderson
Cover design: Patricia Kelly

*The names of people and organizations are imaginary
and no reference to real persons is implied or intended.*

 © 1993, 1986, 1981, 1972 by Prentice-Hall, Inc.
A Simon & Schuster Company
Englewood Cliffs, New Jersey 07632

Printed in the United States of America

10 9 8 7 6 5 4 3

ISBN   0-13-904434-5

PRENTICE-HALL INTERNATIONAL (UK) LIMITED, *London*
PRENTICE-HALL OF AUSTRALIA PTY. LIMITED, *Sydney*
PRENTICE-HALL CANADA INC., *Toronto*
PRENTICE-HALL HISPANOAMERICANA, S.A., *Mexico*
PRENTICE-HALL OF INDIA PRIVATE LIMITED, *New Delhi*
PRENTICE-HALL OF JAPAN, INC., *Tokyo*
SIMON & SCHUSTER ASIA PTE. LTD., *Singapore*
EDITORA PRENTICE-HALL DO BRAZIL, LTDA., *Rio de Janeiro*

Contents

CHAPTER 3

Technical Correspondence 40

CHAPTER 4

Short Informal Reports 75

CHAPTER 5

Longer Informal and Semiformal Reports **112**

CHAPTER 6

Formal Reports **140**

CHAPTER 7

Other Technical Documents **213**

CHAPTER 8

Illustrating Technical Documents **242**

CHAPTER 11
The Technique of Technical Writing

CHAPTER 12
Glossary of Technical Usage

Index

Marking Control Chart

Preface

This book presents all those aspects of technical communication that you, as a technician, technologist, engineer, scientist, or technical manager, are most likely to encounter in industry. It introduces you to the employees of two technically oriented companies, to the type of work they do, and to typical situations that call for them to communicate with clients, suppliers, and each other.

Since the earlier editions of *Technically—Write!*, changes have occurred in the way that correspondence and reports are written at both companies (H. L. Winman and Associates in Cleveland, Ohio, and Macro Engineering Inc in Phoenix, Arizona—see Chapter 2), and these changes are reflected in this fourth edition. Whereas previously Anna King—the companies' technical editor—edited reports typed from handwritten drafts, now she edits them on line (on her computer screen), calling them up from files keystroked by the engineers and engineering technicians at their individual computer terminals in both cities. To prepare you for this comparatively new writing environment, Chapter 1 suggests techniques you can use to outline and write directly on a computer.

Other changes in the fourth edition include the introduction of the "pyramid" method of writing early in the book (also in Chapter 1), greater emphasis on writing letters and memorandums (Chapter 3), examples of reports dealing with the environment (Chapter 5), the application of current electronic communications technology to small businesses (Chapter 6), and a broader range of resumes for you to consider when you start the job-seeking process (Chapter 10).

In preparing this edition, I received invaluable suggestions from Prentice Hall reviewers, and I extend my sincere thanks to them for their help and guidance. They were Rick Halstead, Terra Technical College; Linda Hinkle Maisel, Spring Garden College; Betty Cochran, Beaufort Community College; Dale Brannon, Oklahoma State University, Technical Branch; and Patricia Connors, Memphis State University.

R. S. B.

People as "Communicators"

As communicators, people are lazy and inefficient. We are equipped with a highly sophisticated communication system, yet consistently fail to use it properly. (If our predecessors had been equally lazy and failed to develop their communication "senses" to the level required to ensure survival, many species would be extinct today; indeed, it's even questionable that we would exist in our present form.) This communication system comprises a transmitter and receiver combined into a single package controlled by a computer, the brain. It accepts multiple inputs and transmits in three mediums: action, speech, and writing.

We spend much of our wakeful hours communicating, half the time as a transmitter, half as a receiver. If, as a receiver, we mentally switch off or permit ourselves to change channels while someone else is transmitting, we contribute to information loss. Similarly, if as a transmitter we permit our narrative to become disorganized, unconvincing, or simply uninteresting, we encourage frequency drift. Our listeners detune their receivers and let their computers (brains) think about the lunch that's imminent or wonder if they should rent a video tonight.

As long as a person transmits clearly, efficiently, and persuasively, the persons receiving keep their receivers "locked on" to the transmitting frequency (this applies to both written and spoken transmissions). Such conditions expedite the transfer of information, or "communication."

In direct contact, in which one person is speaking directly to another, the receiver has the opportunity to ask the transmitter to clarify vaguely presented information. But in more formal speech situations, and in all forms of written communication, the receiver no longer has this advantage. He or she cannot stop a speaker who mumbles or uses unfamiliar terminology, and ask that parts of a talk be repeated or clarified; neither can the receiver ask a writer in another city to explain an incoherent passage of a business letter.

The results of failure to communicate efficiently soon become apparent. If people fail to make themselves clear in day-to-day communication, the consequences are likely to differ from those they anticipated, as Cam Collins has discovered to his chagrin.

Cam is a junior electrical engineer at Macro Engineering Inc, and his specialty is high voltage power generation. When he first read about a recent EHV DC power conference, he wanted urgently to attend. In a memorandum to Fred Stokes, the company's chief engineer, Cam described the conference in glowing terms which he hoped would convince Fred to approve his request. This is what he wrote:

> Fred:
> The EHV DC conference described in the attached brochure is just the thing we have been looking for. Only last week you and I discussed the shortage of good technical information in this area, and now here is a conference featuring papers on many of the topics we are interested in. The cost is only $175.00 for registration, which includes a visit to the Freeling Rapids generating station. Travel and accommodation will be about $450 extra. I'm informing you of this early so you can make a decision in time for me to arrange flight bookings and accommodation.
> Cam

Fred Stokes was equally enthusiastic and wrote back:

> Cam:
> Thanks for informing me of the EHV DC conference. I certainly don't want to miss it. Please make reservations for me as suggested in your memorandum.
> Fred

Cam was the victim of his own carelessness: he had failed to communicate clearly what he wanted. There was still a chance for Cam to remedy the situation, but only at the expense of extra time and effort.

Lisa Drew, on the other hand, did not realize she had missed a golden opportunity to be first with an innovative computer technique until it was too late to do anything about it. Her story stems from an incident that occurred several years ago, when she was a recently graduated engineer employed by a manufacturer of agricultural machinery. Lisa's job was to design modifications to the machinery, and then prepare the change procedure documentation for the production department, service representatives, sales staff, and customers.

"For each modification I had to coordinate three different documents," Lisa explained to me over lunch. "First, there had to be a design change notice to send out to everyone concerned. And then there had to be an 'exploded' isometric drawing showing a clear view of every part, with each part cross-referenced to a parts list. And finally there had to be the parts list itself, with every item labeled fully and accurately."

Lisa found that cross-referring a drawing to its parts list was a tedious, time-consuming task. The isometric drawing of the part was computer-generated by the drafting department. The parts list was also keyed into a computer, but by a separate department. However, because the two computer systems were not compatible, cross-referencing had to be done manually.

"And then I hit on a technique for interfacing the two programs," Lisa explained. "It was simple, really, and I kept wondering why no one else had thought of it!"

Without telling anyone, she modified one of the computer programs and tested her idea with five different modification kits. "It worked!" she laughed. "And, best of all, I found that the cross-referencing could be done in one tenth of the time."

Lisa felt her employer should know about her idea—possibly the company could market the program or even help her copyright it. So the following day she stopped Mr. Haddon, the Engineering Manager, as they passed in the hallway, and blurted out her suggestion. This is the conversation that ensued:

Lisa	*Mr. Haddon*
Oh! Mr. Haddon! You know how long it takes to do the documentation for a new part . . . ?	
	Yes . .s . .s . . ?
The problem is in trying to interface between the graphics computer and the parts list . . .	
	(Mr. Haddon appeared to be listening politely, but internally he was growing impatient.)
. . . It has to done by hand, you see . . .	
	Doesn't the drafting department do all that?
Oh, yes! They do. I was just trying to help them . . . to speed up their work a bit.	
	You're working for the chief draftsman now?
Oh, no! It was just an idea I had—to modify one of the computer programs . . .	
	I don't remember issuing you a work order . . .
No. You didn't. I was doing it on my own . . . *(She meant she was doing it on her own time.)*	
	You mean the computer people asked you to do it?

Lisa	*Mr. Haddon*
Well—uh—no. Not exactly . . .	
	But you have been modifying one of the computer programs?
(Reluctantly) Uh-huh.	
	Without authority?
I wanted to try . . .	
	I thought I had made it quite clear to all the staff: No projects are to be undertaken without my approval! *(His tone was cold and abrupt.)* That's final! *(And he turned on his heel and continued down the hall.)*

Lisa's simple suggestion had become lost in a web of misunderstanding. By the time she was through explaining what she had been doing, she had given up trying to offer her idea to the company. And so her idea lay dormant for two years, until a major software company came out with a comparable program. Lisa knew then that perhaps there *had* been market potential for her design.

If Cam Collins and Lisa Drew had paused to consider the needs of the persons who were to receive their information, they would never have launched precipitously into discourses that omitted essential facts. Cam had only to start his memorandum with a request ("May I have your approval to attend an EHV DC conference next month?"), and Lisa with a statement of purpose ("I have designed a computer program that can save us hundreds of dollars annually. May I have a few moments to describe it to you?"), to command the attention of their department heads. Both Mr. Stokes and Mr. Haddon could then have much more effectively appraised the information.

Such circumstances occur daily. They are frustrating to those who fail to communicate their ideas, and costly when the consequences are carried into business and industry.

Bill Carr recently devised and installed a monitor unit for the remote control panel at the microwave relay station where he is the resident engineering technician. As his modification greatly improved operating methods, the head office asked him to submit an installation drawing and an accompanying description. Here is part of his description:

> Some difficulty was experienced in finding a suitable location for the monitor unit. Eventually it was mounted on a locally manufactured bracket attached to the left-hand upright of the control panel, as shown on the attached drawing.

On the strength of Bill's explicit mounting description and detailed list of hardware, the head office converted his description into an installation instruction, purchased materials, assembled 21 modification kits, and shipped them to the 21 other relay stations in the microwave link.

Within a week the 21 resident engineering technicians reported to the head office that it was impossible to mount the monitor unit as instructed, because of an adjoining control unit. No one at the head office had remembered that Bill Carr was located at site 22, the last relay station in the microwave link, where there was no need for an additional control unit.

Bill had assumed that the head office would be aware that the equipment layout at his station was unique. As he commented afterward: "No one said why I had to describe the modification, or told me what they planned to do with it."

In business and industry we *must* communicate clearly and understand fully the implications of failing to do so. A poorly worded order that results in the wrong part being supplied to a job site, a weak report that fails to motivate the reader to take the urgent action needed to avert a costly equipment breakdown, and even an inadequate job application that fails to sell an employer on the right person for a job, all increase the cost of doing business. Such mistakes and misunderstandings are wasteful of labor and resources. Many of them can be prevented by more effective communication—communication that is receiver-oriented rather than transmitter-oriented and that transmits messages using the most expeditious, economical, and efficient means at our command.

CHAPTER 1

A Technical Person's Approach to Writing

Engineering technician Dan Skinner has a report to write on an investigation he completed seven weeks ago for his employer, H. L. Winman and Associates, a consulting engineering firm you will meet in Chapter 2. He has made several halfhearted attempts to get started, but each time never seemed to be the right moment: maybe he was interrupted to resolve a circuit problem, or it was too near lunchtime, or a meeting was called, or, when nothing else interfered, he "just wasn't in the mood." And now he's up against the wire. He has neither set pen to paper nor placed his fingers on the keyboard of his personal computer. He hasn't even worked up a writing outline.

Unless Dan is one of those unusual persons who cannot produce except when under pressure, he is in danger of writing an inadequate, hastily prepared report that does not represent his true abilities. He knows he should not have left his report-writing project until the last moment, but he is human like the rest of us and constantly finds himself in situations like this. He does not realize that by leaving a writing task until it is too late to do a good job, and then frantically organizing the work, he is probably inhibiting his writing capabilities even more than necessary.

If Dan were to relax a little, instead of worrying that he has to organize himself and his writing task, he would find the physical process of writing a much more pleasant experience. But first he must change his approach.

Every technical person, from student technician through potential scientist to practicing engineer, has the ability to write clearly and logically. But this ability must be developed. Dan Skinner must first learn some basic planning and writing techniques, and practice using them until he has acquired the skill and confidence that are the trademarks of an effective writer.

SIMPLIFYING ONE'S APPROACH

Throughout this book, I will be advising you to **tell your readers right away what they most need or want to know**. This means structuring your writing so that the first paragraph (in short documents, the first *sentence*) satisfies their curiosity. Most executives and many technical readers are busy people who have time to read only essential information. By presenting the most important items first, you can help them decide whether they want to read the whole document immediately, put it aside to read later, or pass it along to a specialist in their department.

This reader-oriented style of presentation is known as the "pyramid technique." Imagine that every letter, memorandum, or report you write is shaped like a pyramid, with a small piece of essential information—the key data you want your reader to see right away—at the top, supported on a broad base of details, facts, and evidence. In most letters and short reports the pyramid has only two parts: a brief *summary* followed by the *full development*, as shown in Figure 1-1(a). In long reports an additional part—known as the *essential details*—is inserted between the summary and the full development, as in Figure 1-1(b).

Readers normally are not aware when the pyramid technique has been used to write a letter or report. They simply find the document well organized and easy to read. The key elements are right up front, in the first paragraph, and the full story containing all the technical details is in the remaining paragraphs. For example, in the opening paragraph of his letter report in Figure 1-2, Wes Hillman summarizes what Tina Mactiere most wants to know (whether the training course was a success and what results were achieved). In the remainder of the letter he fills in background details, states briefly how the course was run, reports on student participation and reaction, and suggests additional topics that could be covered in future courses.

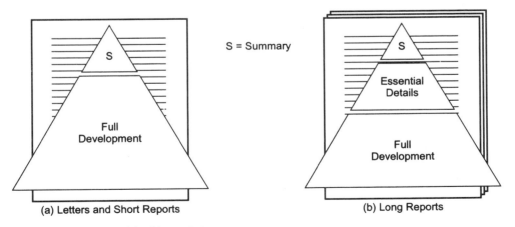

Figure 1-1. The pyramid writing technique.

RGI INTERNATIONAL

Division of
THE RONING GROUP INC

2022 South Main Drive
Montrose, OH 45287

October 18, 19xx

Tina R. Mactiere, President
Macro Engineering Inc
600 Deepdale Drive
Phoenix, AZ 85007

Dear Ms. Mactiere

<u>Results of Pilot Report Writing Course</u>

The report writing course we conducted for members of your
engineering staff was completed successfully by 14 of the 16
participants. The average mark was 63%.

This was a pilot course set up in response to an August 13,
19xx enquiry from Mr. Fred Stokes. At his request, we placed
most emphasis on providing your staff with practical experience
in writing business letters and technical reports. Attendance
was voluntary, the 16 participants having been selected at
random from 29 applicants.

Best results were achieved by participants who recognized their
writing problems before they started the course, and willingly
became actively involved in the practical work. A few said
they had expected to attend an "information" type of course,
and at first were mildly reluctant to take part in the heavy
writing program. Our comments on the work done by individual
participants are attached.

Course critiques completed by the participants indicate that
the course met their needs from a letter and report writing
viewpoint, but that they felt more emphasis could have been
placed on technical proposals and oral reporting. Perhaps such
topics could be covered in a short follow-up course.

We enjoyed developing and teaching this pilot course for your
staff, and particularly appreciated their enthusiastic
participation.

Sincerely

Wes Hillman

Wesley G. Hillman
Course Leader

enc

Figure 1-2.   A letter report written using the pyramid technique.

Every document shown in this textbook can be structured using the pyramid technique. How the pyramid is applied to letters, memorandums, reports, proposals, instructions, descriptions, and even resumes and oral presentations, is described in Chapters 3 through 7, and 9 and 10. For the moment, just remember that using the pyramid is the simplest, fastest, most effective way to plan and write any document, regardless of its length. If Dan Skinner had known about the pyramid technique, his pain would have been eased considerably.

PLANNING THE WRITING TASK

The word "planning" seems to imply that report writers must start by thoroughly *organizing* both themselves and their material. I disagree. Organizing too diligently, or too early in the writing process, for many people inhibits rather than accelerates the writing process. The key is to organize one's information in a spontaneous, creative manner, allowing one's mind to freewheel through the initial planning stages until the topics have been collected, scrutinized, sorted, grouped, and written into a logical outline that will appeal to the reader.

I recommend that at first Dan Skinner does nothing about making an outline or taking any action that smacks of organization. Instead, he should work through six simple planning stages that will not become the chore he expects. These stages are shown in Figure 1-3.

1. Gather Information

Dan's first step should be to assemble all the documents, results of tests, photographs, samples, computer data, specifications, and so on that he will need to write his report or that he will insert into it. He must gather everything he will

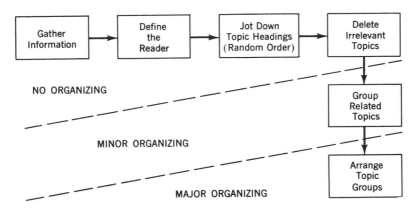

Figure 1-3. The six planning stages. In practice, these stages can overlap.

need now, because later he will not want to interrupt his writing to look for additional facts and figures.

2. Define the Reader

Next, Dan must identify who he is writing to. This is probably the most important part of his planning, for if he does not clearly identify his audience he may write an unfocused report that misses its mark. He must conjure up an image of the person or people who will read his report by asking himself six questions:

1. Who, specifically, is my reader? This may be someone he knows, in which case his task is simplified, or it may be someone he is not familiar with (such as a customer in an out-of-town firm), for whom he must imagine a persona.
2. Is he or she a technical person? Dan needs to known whether he can use or must avoid technical terms.
3. How much does the reader know about the subject I will be describing? This will give Dan a starting point, since there is no need to cover information a reader already knows.
4. What does the reader want to know or expect to be told? Dan must be able to anticipate whether the reader will be receptive or hostile to the information he has to present.
5. Will more than one person read my report? If so, Dan must repeat questions 2 through 4 for the additional readers.
6. Who is my *primary* reader? The primary reader is the person who will make a decision or take specific action after reading Dan's report. Often it is the person to whom the report is directed, but on occasion it may be one of the secondary readers. For example, a report may be addressed to a department manager, but the person who will use it or do something about it may be an engineer on the manager's staff.

Dan's inability to identify his reader(s) was one of the reasons he had difficulty getting started on his report-writing task.

3. Make Notes

Now that Dan has a reader in mind, he can start making notes, either with a pen and paper or at a computer keyboard. At this third stage he must "loosen up" enough to generate ideas spontaneously. He needs to let his mind freewheel so that it comes up with ideas and pieces of information quickly and easily. He must not stop to question the relevance of this information (that will come later); his role for the moment is purely to collect it.

Normally, at the outlining stage, a technical person will write down or key in a set of familiar or arbitrary headings, such as "Introduction," "Initial Tests," and "Material Resources," and arrange them in logical order. (These seem to be standard headings in many reports.) But I want our report writer to be different. I want Dan Skinner to free his mind of the elementary organized headings, and even to refuse to divide his subject mentally into blocks of information. Then I want him to jot down or type into the computer a series of main topics that he will discuss, writing only brief headings rather than full sentences. He must do this in random order, making no attempt to force the topics into groups (although it is quite possible that grouping will occur naturally, since many interdependent topics are likely to occur to him in logical order). The topics that he knows best will spring readily to mind; then there will be a gradual slowdown as he encounters less familiar topics.

When he stops his initial list, he should go back and examine each topic to see if it suggests less obvious topics. As additional topics come to mind, he must jot them down or key them in, still in random order, until he finds he is straining to find new ideas. This should be a signal for him to stop before he becomes too objective.

Dan must not try to decide whether each topic is relevant during this spontaneous freewheeling session. If he does, he will immediately inhibit his creativeness because he will become too logical and organized. He must list all topics, regardless of their importance and eventual position in the final report.

At the end of this session Dan's list should look like Figure 1-4. He can now take a break, knowing that the first details of his report have been committed to paper. What he may not yet realize is that he has almost painlessly produced his first outline.

4. Delete Irrelevent Topics

The fourth stage calls for Dan to examine his list of headings with a critical eye, dividing them into headings that bear directly on the subject and those that introduce topics of only marginal interest. His knowledge of the reader—identified in stage two—will help him decide whether each topic is really necessary so he can delete irrelevant topics, as has been done in Figure 1-5.

5. Group the Topics

The headings that remain should be grouped into "topic areas" that will be discussed together. This Dan can do simply by coding related topics with the same symbol or letter. In Figure 1-5, letter (A) identifies one group of related topics, letter (B) another group, and so on.

6. Organize the Topics

Only now should Dan take his first major organizational step, which is to arrange the groups of information in the most suitable order and at the same time sort out the order of the headings within each group. He must consider three factors: which order of presentation will be most interesting, which will be most logical, and which will be simplest to understand. The result will become his final writing plan or report outline, which may be handwritten or obtained as a computer printout, as in Figure 1-6. (If Dan has keyed in his outline, he may prefer to call it onto the computer screen whenever he wants to refer to it rather than work from a printed outline.)

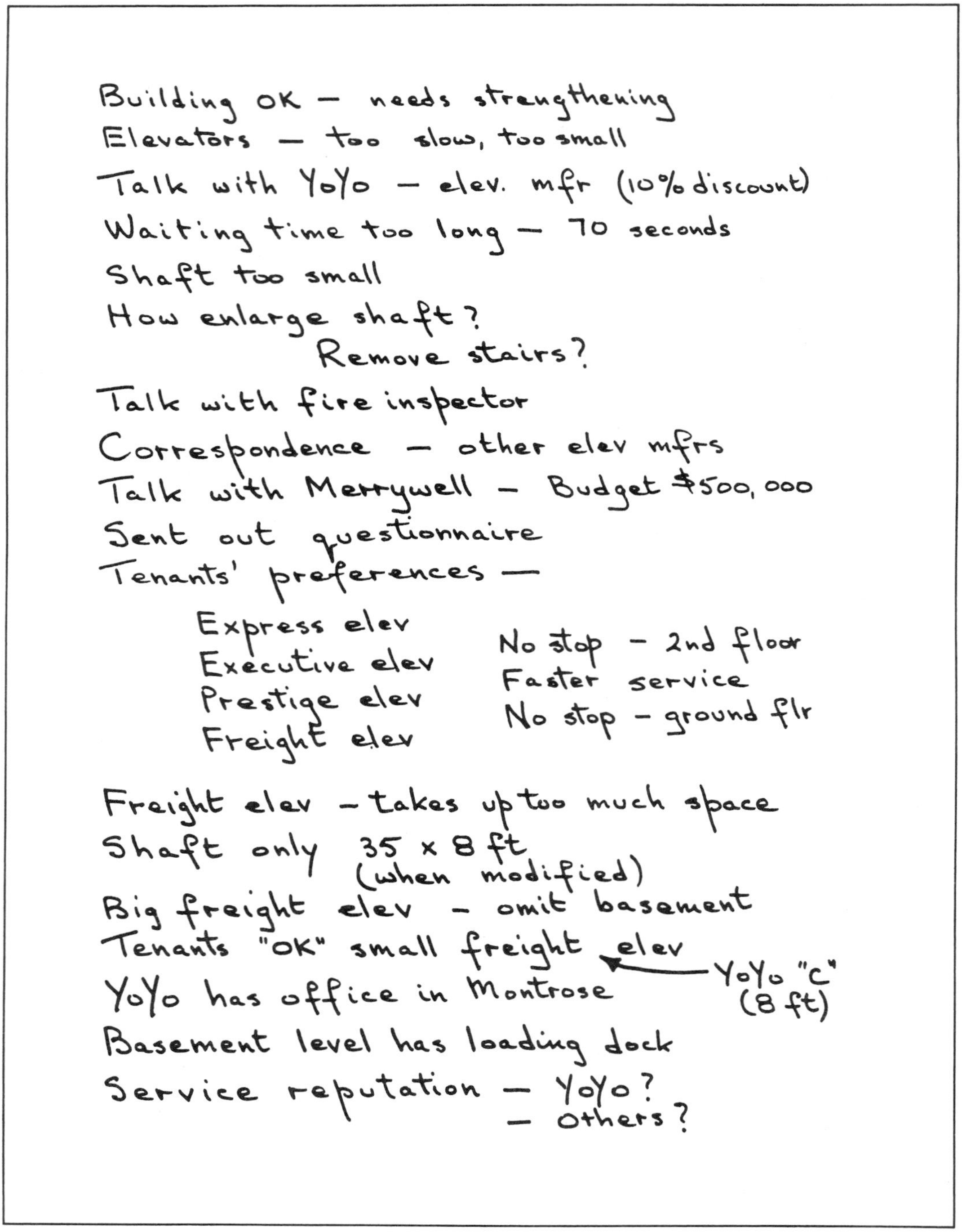

Building OK — needs strengthening
Elevators — too slow, too small
Talk with YoYo — elev. mfr (10% discount)
Waiting time too long — 70 seconds
Shaft too small
How enlarge shaft?
            Remove stairs?
Talk with fire inspector
Correspondence — other elev mfrs
Talk with Merrywell — Budget $500,000
Sent out questionnaire
Tenants' preferences —

    Express elev   No stop — 2nd floor
    Executive elev   Faster service
    Prestige elev   No stop — ground flr
    Freight elev

Freight elev — takes up too much space
Shaft only 35 x 8 ft
            (when modified)
Big freight elev — omit basement
Tenants "OK" small freight elev
YoYo has office in Montrose   ← YoYo "C"
Basement level has loading dock   (8 ft)
Service reputation — YoYo?
                    — others?

Figure 1-4. Initial list of topic headings, jotted down in random order.

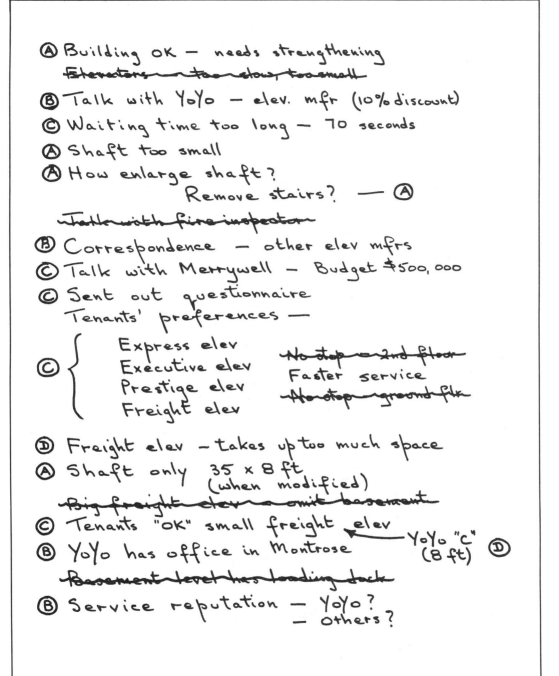

(A) Building OK — needs strengthening

~~Elevators — too slow, too small~~

(B) Talk with YoYo — elev. mfr (10% discount)

(C) Waiting time too long — 70 seconds

(A) Shaft too small

(A) How enlarge shaft?

    Remove stairs? — (A)

~~Talk with fire inspector~~

(B) Correspondence — other elev mfrs

(C) Talk with Merrywell — Budget $500,000

(C) Sent out questionnaire

Tenants' preferences —

(C) { Express elev   ~~No stop — 2nd floor~~

  Executive elev   Faster service

  Prestige elev   ~~No stop — ground flr~~

  Freight elev

(D) Freight elev — takes up too much space

(A) Shaft only 35 × 8 ft

    (when modified)

~~Big freight elev — omit basement~~

(C) Tenants "OK" small freight elev ← YoYo "C"

(B) YoYo has office in Montrose   (8 ft) (D)

~~Basement level has loading dock~~

(B) Service reputation — YoYo?

       — Others?

Figure 1-5. The same list of topic headings, but with irrelevant topics deleted and remaining topics coded into subject groups (A—structural implications; B—elevator manufacturers; C—tenants' preferences; D—freight elevator).

```
Building condition:

        OK - needs strengthening (shaft area)
        Existing elev shaft too small
        Remove adjoining staircase
        Shaft size now 35 x 8 ft

Tenants' needs:

        Sent out questionnaire
        Identified 5 major requests
        Requests we must meet:
                Cut waiting time: 32 sec max
                Handle freight up to 7 ft 6 in.
        Requests we should try to meet:
                Express elev to top 4 floors
                Deluxe models (for prestige)
                Private elev (for executives)

Budget: $500,000 max

Elevator manufacturers:

        Researched 3
        Only YoYo Company offers discount
        Only YoYo has Montrose office
```

Figure 1-6. Topic headings rearranged into a writing outline, typed at a computer terminal and printed by a dot-matrix line printer. This was part of the writing plan for the formal report on elevator selection in Chapter 6. Compare it with the final product on pages 192 and 193.

And now a final comment about outlining: If you have already developed an outlining method of your own that works well for you, or you are using outlining software successfully, then I suggest you continue as you have been doing. The outlining method I suggest here is for people who are seeking a simpler, more creative way to develop outlines than they are currently using.

WRITING THE FIRST DRAFT

As I sit at my desk, the heading "Writing the First Draft" at the top of my computer screen, I find I am experiencing exactly the same problem that every writer encounters from time to time: an inability to find the right words—*any* words—that can be strung together to make coherent sentences and paragraphs. The ideas are there, circling around inside my skull, and the outline is there, so I cannot excuse myself by saying I have not prepared adequately. What, then, is wrong?

The answer is simple. Ten minutes ago the telephone rang and a neighbor announced he would shortly bring over a block "Neighborhood Watch" plan for me to sign. I paused to switch on the coffee, for I know that Jack will expect a cup while we converse, and now I can hear the percolator grumbling away in the distance. I cannot concentrate when I know my continuity of thought is so soon to be broken.

Continuity is the key to getting one's writing done. In my case, this means writing at fairly long sittings during which I *know* I will not be disturbed. I must be out of reach of the telephone, visiting friends, and children wanting to say good night, so that I can write continuously. Only when I have reached a logical break in the writing, or have temporarily exhausted an easy flow of words, can I afford to stop and enjoy that cup of coffee.

It is no easier to find a quiet place to write in the business world. The average technical person who tries to write a report in a large office cannot simply ignore the surroundings. A conversation taking place a few desks away will interfere with one's creative thought processes. And even a co-worker collecting money for the pool on that night's NHL game between the New York Rangers and the L. A. Kings will interrupt writing continuity.

The problem of finding a quiet place to write can be hard to solve, particularly now that many people write their reports at a computer keyboard and so cannot move away from their desks (unless they are fortunate enough to own a *portable* computer). In business, I recommend a short walk around the premises to find a hidden corner, or an office which is temporarily unoccupied. Then perhaps, if you are lucky, you can just "disappear" for a while. For technical students, who frequently have to work in a tiny writing area in a crowded classroom, or in a roomful of computer terminals, conditions are seldom ideal. Outlining in the classroom, followed by writing in the seclusion of a library cubicle, is a possible alternative.

If you plan to key your own letter or report into a computer, before you start you have to consider aspects which normally do not concern you when you write with a pen or pencil—items that a typist normally looks after. For example, you have to key specific instructions into the computer to define:

- Whether you will print the report in 10 or 12 point type (i.e. with 10 or 12 characters to the lineal inch).
- The number of lines you want on a page, and the width of your planned typing lines (the number of characters or letters you want in a line).
- The width of margin you want on either side of the text.
- Whether you want the right margin to be justified (straight) or ragged.

- Where you want the page numbers to be positioned (top or bottom of the page, and either centered or to one side of the page); on most systems page numbers are printed automatically, but you can select where they are to appear.
- The line spacing you want (single or double), and how many blank lines you want between paragraphs (normally one or two).
- Whether the first line of each paragraph is to be indented or set "flush" with the left margin; and, if indented, how many spaces the indention is to be.
- For long words at the end of a line, whether you or the computer is to decide where the word is to be hyphenated (you can also elect no hyphenation).
- The levels of headings you will use and whether they are to be capitalized, underlined, or centered. (See Figure 11-2 of Chapter 11—page 324—for guidelines.)

How you enter these instructions, and whether some are set up as "defaults," will depend on the word-processing program you use. (A default is a condition, such as the position of page numbers, which is set automatically. If you want to change the condition for the report you are writing, before you start keying in the report you have to instruct the computer to make the change. You do this by keying in a predetermined symbol, which differs according to the word-processing program in use.)

When Dan Skinner has found a peaceful location out of reach of the telephone, his friends, and even his boss, he can make a fresh start. But this is exactly where a second difficulty may occur. Equipped with an outline on his left and a pad of paper or a keyboard in front of him, he may find that he does not know where to begin. Or he may tackle the task enthusiastically, determined to write a really effective introduction, only to find that everything he writes sounds trite, unrealistic, or downright silly. Discouraged after many false starts, he sweeps the growing pile of discarded paper into the wastebasket or erases what he has typed from the computer's memory.

I have frequently advised technical people who have encountered this "no start" block that the best place for them to start writing is at paragraph two, or even somewhere in the middle. For example, if Dan finds that a particular part of his project interests him more than other parts, he should write about that part first. His interest and familiarity with the subject will help him to write those first few words, and keep him going once he has started. The most important thing is to *start* writing, to put any first words down, even if they are not the right words or ideas, and to let them lead naturally into the next group of ideas.

This is where continuity becomes essential. A writer who has found a quiet place to work, and has allowed sufficient time to write a sizable chunk of text, simply must not interrupt the writing process to correct a minor point of construction. If you are to maintain continuity, you must not strive to write perfect grammar, find exactly the right word, insert perfect punctuation, or construct effective sentences and paragraphs of just the right length. That can be done later, during revision. The important thing is to keep building on that rought draft (no matter how "rough" it is), so that when you do stop for a break you know you have assembled some words that you can work up into a presentable document.

If, as he writes, Dan cannot find exactly the word he wants, he should jot down a similar word and draw a circle around it (or, at a computer terminal, type a question mark enclosed in parentheses immediately after it) as a reminder to change the word when the first draft is finished. Quite likely, when he reviews the draft, the correct word will spring to mind. Similarly, if he is not sure how to spell a certain word, he should resist the temptation to turn to the dictionary for that will disrupt the natural flow of his writing. Again, he should draw attention to the word as a reminder that he must consult his dictionary later.

I cannot stress too strongly that writers should not correct their work as they write. Writing and revising are two entirely separate functions, and they call for different approaches; they cannot be done simultaneously without the one lessening the effectiveness of the other. Writing calls for creativeness and total immersion in the subject so that the words tumble out in a constant flow. Revision calls for lucidity and logic, which forces a writer to reason and query the suitability of the words he or she has written. The first requires exclusion of every thought but the subject; the second demands an objectivity that constantly challenges the material from the reader's point of view. Writers who try to correct their work as they write soon become frustrated, for creativity and objectivity are constantly fighting for control of their fingers.

Most technical people find that, once they have written those first few difficult sentences, their inhibitions begin to fall away and they write more naturally. Their style begins to change from the dull, stereotyped writing we expect from the average technical person, to a relaxed narrative that sounds as though the writer is telling a co-worker about the subject. As the writer becomes interested in the topic, speed builds up and he or she has difficulty writing or typing fast enough. The words may not be exactly those of the final report, but when the writer starts checking the first draft he or she will be surprised at the effectiveness of many of the sentences and paragraphs written earlier.

The length of each writing session will vary, depending on the writer's experience and the complexity of the topic. If a document is reasonably short, it should be written all at one sitting. If it is long, it should be divided into several medium-length sessions that suit the writer's particular staying power.

At the end of each session, Dan should glance back over his work, note the words he has circled or questioned, and make a few necessary changes (Figure 1-7 is a page from a typical first draft). He must not yet attempt to rewrite paragraphs and sentences for better emphasis. Such major changes must be left until later, when enough time has elapsed for him to read his work objectively. Only then can he review his work as a complete document and see the relationship among its parts. Only then can he be completely critical.

TAKING A BREAK

When the final paragraph of a long report has been written, I have to resist the temptation to start revising it immediately. There are sections that I know are weak, or passages with which I am not happy, and the desire to correct them is

---

TENANTS NEEDS

To find out what the building's tenants most needed in
elevator service, we asked each company to fill out a
questionaire (sp?). From their answers we were able to
identify 5 factors needing consideration:

1. A major problem seems to be the length of time a
    rider must wait for an elevator. Every tenant said
    we must cut out lengthy waits. A survey was carried
    out to find out how long people had to wait (during
    rush hours). This averaged out at 70 sec, more than
    twice the 32 sec established by Johnson (ref?)
    before people get impacient (sp?). From this we
    calculated we would need 3 or 4 passenger elevators.

2. At first it seemed we would be forced to include a
    full-size freight elevator in our plan. Two
    companies (which?) both carry large but light
    displays up to their floors, but both later agreed
    they could hinge them, and if they did this they
    would need only 7 ft 6 in. width (maximum). They
    also said they did not need a freight elevator all
    the time, mainly between 10:00 a.m. and 3:30 p.m.

Figure 1-7. Part of the author's first draft, typed at a computer terminal and
then printed by a line printer. Note that the author has not stopped
to hunt up minor details. Several revisions were made between this
draft and the final product (see pages 192 and 193).

strong. But I know it is too soon. I must take my pages, staple them together, and
set them aside while I tackle a task that is completely unrelated.

Immediate reading without a suitable waiting period encourages writers
to look at their work through rose-tinted glasses. Sentences which normally they
would recognize as weak or too wordy appear to contain words of wisdom. Gross
inaccuracies which under normal circumstances they would pounce upon go

unnoticed. Paragraphs that may not be understood by a reader new to the subject, to their writers seem abundantly clear. Their familiarity with their work blinds them to its weaknesses.

The only remedy is to wait. If Dan has handwritten his report, he should arrange to have a double- or triple-spaced draft typed during this time. Not only will this make his work easier to read, but he may also be fortunate enough to find that the typist has corrected some of his punctuation and spelling errors. If he has keyed his report into a computer, he can run a spelling-check program and have a double-spaced copy printed out.

READING WITH A PLAN

If you have handwritten a report, always try to read from a *typed* draft so you will see the same cold, clinical type your readers will see. There is a danger in reading your own handwriting, for you may "hear" an emphasis which your readers will not notice when they read the typewritten words.

Dan Skinner's first reading should take him straight through the draft without stopping to make corrections, so that he can gain an overall impression of the report. Subsequent readings should be slower and more critical, with changes written in as he goes along. As he reads he should check for clearness, correct tone and style, and technical and grammatical accuracy.

CHECKING FOR CLEARNESS

Checking for clearness means searching for passages that are vague or ambiguous. If the following paragraph remained uncorrected, it would confuse and annoy a reader:

Muddled Paragraph   When the owners were contacted on April 15, the assistant manager, Mr. Pierson, informed the engineer that they were thinking of advertising Lot 36 for sale. He however reiterated his inability to make a definite decision by requesting this company to confirm their intentions with regard to buying the land within two months, when his boss, Mr. Davidson, general manager of the company, will have come back from a business tour in Europe. This will be June 8.

The only facts you can be sure about after reading this paragraph are that the owners of the land were contacted on April 15 and the general manager will be returning on June 8. The important information about the possible sale of Lot 36 is confusing. Probably the writer was trying to say something like this:

Revised Paragraph   The engineer spoke to the owners on April 15 to inquire if Lot 36 was for sale. He was informed by Mr. Pierson, the assistant manager, that the company was thinking of selling the lot, but that no decision would be made until after June 8, when the general manager returns from a business tour in Europe. Mr. Pierson suggested that the engineer submit a formal request to purchase the land by that date.

The more complex the topic, the more important it is to write clear paragraphs. Although the following paragraph is quite technical, it would be generally understood even by nontechnical readers:

Clear Paragraph   A sound survey confirmed that the high noise level was caused mainly by the radar equipment blower motors, with a lesser contribution from the air-conditioning equipment. Tests showed that with the radar equipment shut down the ambient noise level at the microphone positions dropped by 10 dB, whereas with the air-conditioning equipment shut down the noise level dropped by 2.5 dB. General clatter and impact noise caused by the movement of furniture and personnel also contributed to the noisy working conditions, but could not be measured other than as sudden sporadic peaks of 2 to 5 dB.

Here paragraph unity has added much to readability. The writer has made sure that:

The topic is clearly stated in the first sentence (the topic sentence).
The topic is developed adequately by the remaining sentences.
No sentence contains information that does not substantiate the topic.

If any paragraph meets these basic requirements, its writer can feel reasonably sure that the message has been conveyed clearly.

Writers who know their subject thoroughly may find it difficult to identify paragraphs that contain ambiguities. A passage that is abundantly clear to them may be meaningless or offer alternative interpretations to a reader unfamiliar with the subject. For example:

Our examination indicates that the receiver requires both repair and recalibration, whereas the transmitter needs recalibration only, and the modulator requires the same.

This sentence plants a question in the reader's mind: Does the modulator require both repair and recalibration, or only recalibration? The technician who wrote it knows, because he has been working on the equipment, but the reader will never know unless he or she cares to write or phone and ask. The technician could have clarified the message easily, simply by rearranging the information:

Our examination indicates that the receiver requires both repair and recalibration, whereas the transmitter and modulator need only recalibration.

Sometimes ambiguities are so well buried that they are surprisingly difficult to identify, as in this excerpt from a chief draftsperson's report to a department head:

> The Drafting Section will need three Model D7 drawing boards. The current price is $1175 and the supplier has indicated that his quotation is "firm" for three months. We should therefore budget accordingly.

The department head took the message at face value and inserted $1175 for drawing boards in the budget. Two months later he received an invoice for $3525. Unable by then to return the three boards, he had to overshoot his budget by $2350. This financial mismanagement was caused by the chief draftsperson, who omitted to say whether the price quoted applied to one drawing board or to three. If the word "each" had been inserted after $1175, the message would have been clear.

Many ambiguities can be sorted out by simple deduction, although it really should not be the reader's job to interpret the author's intentions. Occasionally such ambiguities provide a humorous note, as in this extract from a field trip report:

> High grass and brush around the storage tanks impeded the technicians' progress and should be cleared before they grow too dense.

So that readers will not mistakenly think it is the technicians who are growing too dense, the two thoughts in this sentence should be separated:

> High grass and brush around the storage tanks impeded the technicians' progress. This undergrowth should be cleared before it grows too dense.

CHECKING FOR CORRECT TONE AND STYLE

How do you know when your writing has the right tone? One of the most difficult aspects of technical writing is establishing a tone that is correct for the reader, suitable for the subject, and comfortable for you, the writer. If you know your subject well and have thoroughly researched your reader, you will most likely write confidently and, often, will automatically establish the correct tone. But if you try to set a tone that does not feel natural, or if you are a little uncertain about the subject and the reader, your reader will sense an unsureness in your writing. And no matter how skillfully you edit your work, the hesitancy will show up in the final sentences and paragraphs.

Finding the Best Writing Level. To check that he has set the right tone, Dan Skinner must assess whether his writing is suitable for both the subject matter and the reader. If he is writing on a specific aspect of a very technical topic and knows that his reader is an engineer with a thorough grounding in the subject, he can use all the technical terms and abbreviations that his reader will recognize. Conversely, if he is writing on the same topic for a nontechnical reader who has little or no knowledge of the subject, Dan may have to generalize rather than state specific details, explain technical terms that he would normally expect to be understood, and generally write in a more informative manner.

Writing on a technical subject for readers who do not have the same technical knowledge as yourself is not easy. You have to be much more objective when checking your work, and try to think in the same way as the nontechnical reader. You can use only those technical words which you know will be recognized, yet you must avoid oversimplified language that may irritate the reader.

For example, when engineer Rita Corrigan wrote this in a modification report, she knew her readers would be electronics technicians at radar-equipped airfields.

> We modified the M.T.I. by installing a K-59 double-decade circuit. This brightened moving targets by 12% and reduced ground clutter by 23%.

Although this statement would be readily understood by the readers for whom it was intended, to any reader not familiar with radar terminology it would mean very little. So when Rita reported on the same subject to the airport manager, she wrote this:

> We modified the radar set's Moving Target Indicator by installing a special circuit known as the K-59. This increased the brightness of responses from aircraft and decreased returns from fixed objects on the ground.

For this reader Rita has included more descriptive details and eliminated specific technical details that might not be meaningful to the airport manager. In their place she had made a general statement that aircraft responses were "increased" and ground returns "decreased." She also knew that the airport manager would be familiar with terms such as *Moving Target Indicator, responses,* and *returns.*

Now suppose that Rita also had to write to the local Chamber of Commerce to describe improvements in the airport's traffic control system. This time her readers would be entirely nontechnical, so she would have to avoid using *any* technical terms:

> We have modified the airfield radar system to improve its performance, which has helped us to differentiate more clearly between low-flying aircraft and high objects on the ground.

Keeping to the Subject. Having established that he is writing at the correct level, Dan must now check that he has kept to the subject. He must take each paragraph and ask: Is this truly relevant? Is it direct? And is it to the point?

If Dan prepared his outline using the method described earlier, and followed it closely as he wrote his report, he can be reasonably sure that most of his writing is relevant. To check that his subject development follows his planned theme, he should identify the topic sentences of some paragraphs and check them against the headings in his outline. If they follow the outline, he has kept to the main theme; if they tend to diverge from the outline or if he has difficulty in identifying them, he should read the paragraphs carefully to see whether they need to be rewritten, or possibly even eliminated. (For more information about topic sentences, see Chapter 11.)

Technical writing should always be as direct and specific as possible. Technical writers should convey just enough information for their readers to understand the subject thoroughly. Technical writing, unlike literary writing, has no room for details that are not essential to the main theme. This is readily apparent in the following descriptions of the same equipment.

Literary Description   The new cabinet has a rough-textured dove gray finish that reflects the sun's rays in varying hues. Contrary to most instruments of this type, its controls are grouped artistically in one corner, where the deep black of the knobs provides an interesting contrast with the soft gray and white background. A cover plate, hardly noticeable to the layman's inexperienced eye, conceals a cluster of unsightly adjustment screws that would otherwise mar the overall appearance of the cabinet and would nullify the esthetic appeal of its surprisingly effective design.

Technical Description   The gray cabinet is functional, with the operator's controls grouped at the top right corner where they can be grasped easily with one hand. Subsidiary controls and adjustment screws used by the maintenance crews are grouped at the bottom left corner, where they are hidden by a hinged cover plate.

Comparison of these examples shows how a technical description concentrates on details that are important to the reader (it tells *where* the controls are and why they have been so placed); it maintains an efficient, businesslike tone.

Using Simple Words.   A writer who uses unnecessary superlatives sets an unnaturally pompous tone. The engineer who writes that a design "contains ultrasophisticated circuitry" seems to be justifying the importance and complexity of his or her work rather than saying that the design has a very complex circuit. The supervisor who recommends that technician Johannes Schmitt be "given an increase in remuneration" may be understood by the company controller but will only be considered pompous by Johannes. If he had simply written that Johannes should be "given a raise," he would have been understood by both. Unnecessary use of big words, when smaller, more generally recognized, and equally effective synonyms are available, clouds technical writing and destroys the smooth flow that such writing demands.

Removing "Fat."   During the reading stage, Dan should be critical of sentences and paragraphs that seem to contain too many words. He should check that he has not inserted words of low information content, that is, phrases and expressions that add little or no information. Their removal or replacement by simpler, more descriptive words, can tighten up a sentence and add to its clarity. Low information content words and phrases are often hard to identify, because the sentences in which they appear seem to be satisfactory. Consider this sentence:

> For your information, we have tested your spectrum analyzer and are of the opinion that it needs calibration.

The words of low information content are "for your information" and "are of the opinion that." The first can be deleted, and the second replaced by "consider," so that the sentence now reads:

> We have tested your spectrum analyzer and consider it needs calibration.

The same applies to this sentence:

> If you require further information, please feel free to telephone Mr. Thompson at 489-9039.

The phrase "if you require," although not wrong, could be replaced by the single word "for", but "please feel free to" is archaic and should be eliminated. The result:

> For further information please telephone Mr. Thompson at 489-9039.

Tables 11-2 and 11-3 in Chapter 11 contain lists of low information content words and wordy expressions.

Inadvertent repetition of information can also contribute to excessive length. Although repetition can be an effective way to emphasize a point, in most cases its use is accidental. For example, Dan may write:

> We tested the modem to check its compatibility with the remote terminal. After completing the modem tests we transmitted messages at low, medium, and high baud rates. The results of the transmission tests showed . . .

If he deletes the repeated words in sentence 2 ("After completing the modem tests") and sentence 3 (". . . of the transmission tests . . ."), he can achieve a much tauter paragraph:

> We tested the modem to check its compatibility with the remote terminal, and then transmitted messages at low, medium, and high baud rates. The results showed . . .

CHECKING FOR ACCURACY

Checking accuracy means examining one's work to ensure that the information is correct and that technicalities such as grammar, punctuation, and spelling have not been overlooked.

Nothing annoys readers more than to discover that they have been presented with inaccurate information (particularly if they have been using the information for some time before they discover the error). They automatically assume that a writer knows his or her facts and has checked that they have been correctly transcribed into the report. Discovering even one error in a report can undermine the writer's credibility in the reader's eyes.

There is no way to prevent some errors when quantities and details are copied from one document to another. Therefore, Dan must carefully check that facts, figures, equations, quantities, and extracts from other documents are all copied correctly. He must do this regardless of whether he typed the report at a computer terminal or a typist typed it from his handwritten draft. Even though he may feel the typist is responsible for proofreading the report, the ultimate responsibility for checking the accuracy of the written work is his.

This is also the time for Dan to identify poor grammar, inadequate punctuation, and incorrect spelling. He must do this with care because his knowledge of the subject may blind him to obvious errors. (How many of us have inadvertently written "their" when we intended to write "there," and "too" when we meant "two"?)

Many word-processing programs now have a built-in spelling checker, yet spell-check programs are not infallible. You may use a word—particularly a technical word—which is not in the checker's memory, or you may mistype a word but accidentally form another word the checker recognizes. For example, you may plan to type "word" but accidentally type "work," which the checker would not recognize as an error. (Neither would it flag "their" and "too" as errors.) The spell-check program only reads what you have written; it does not comprehend what you have said. It examines each word and compares it to its master list: If it finds a "match," it says nothing. It cannot warn you that you have written the wrong word.

REVISING ONE'S OWN WORDS

As Dan reads his work, he should make any corrections he finds necessary. Some corrections may require only minor changes to individual sentences; others may require complete revision of whole paragraphs, and even of complete sections. If Dan is working at a computer terminal, he can make the changes himself and immediately see the improvements, either on the screen or on a printout. But if he is handwriting extensive revisions to a typed draft, he should have the draft retyped, and then he should carefully reread and revise it. This process should be repeated until a draft emerges which says exactly what he wants to say in as few words as possible.

There are two ways Dan can revise and edit at his keyboard. He can call the whole document back onto the screen, read it a screenfull at a time, and make corrections without using a paper printout. Alternatively, he can make a printout and read and correct the hard copy, returning to the keyboard and screen to make final corrections. Both methods are valid.

As he reads, Dan must continually ask himself:

Can my readers understand me?

Will the person I am writing for be able to read my report all the way through without becoming lost?

What about other readers who might also see my report: Will they understand it?

Is the focus right?

Is my report reader-oriented?

Are the important points clearly visible?

Have I summarized the main points in an opening statement which the reader will see right away?

Is my information correct?

Is it accurate?

Is it complete?

Is all of it necessary?

Is my language good?

Is it clear, definite, and unambiguous?

Are there any grammatical, punctuation, or spelling errors?

Does every paragraph have a topic sentence (preferably at the start of the paragraph)?

Have I used any big "overblown" words where simpler words would do a better job?

Are there any low information content words and phrases?

Have I kept my report as short as possible while still meeting my readers' needs and covering the topic adequately?

By now Dan's draft should be in good shape. Any further reading and revising will be final polishing, and the amount will depend on the importance of the report. If his report is for limited or in-company distribution, a standard-quality job probably will suffice. But if it is to be distributed outside the company, or submitted to an important client, then Dan will spend as much time as necessary to ensure that it conveys a good image of both himself and H. L. Winman and Associates (his employer).

REVIEWING THE FINAL DRAFT

The final step occurs when Dan feels that his report is ready for final typing. Before charging across to the typing pool, or instructing the computer to print out a letter-quality copy, he should ask himself three more questions:

1. Would I want to receive what I have written?
2. What reaction will it incur from the intended reader?
3. Is this the reaction I want?

Now he must be extremely self-critical, ready to doubt his ability to be truly objective. If his answers are at all hesitant, he should ask an independent reviewer to read his report. Ideally, this person will be technically and mentally equivalent to the eventual reader, but not too familiar with the project. The reviewer must be able to criticize constructively, reading the report completely rather than scanning it, and making notes describing possible weaknesses and ambiguities. The reviewer should take time to discuss the report with Dan and explain to him why certain sentences or paragraphs need to be reworked.

Dan Skinner will now be able to issue his report with confidence, knowing that he has fashioned a good product. The approach described here will not have made report writing a simple task for him, but it will have assisted him through the difficult conceptual stages and helped him to read and revise more efficiently. When Dan has to write his next report, he will be much less likely to put it off until it is so late that he has to do a rush job.

WORKING WITH A SECRETARY/TYPIST

In the more traditional setting, a letter- or report-writing team comprises two people: the person who writes the letter or report (the author) and the person who types it. Good communication between author and secretary is extremely important, otherwise both will become frustrated and irritated: the author, because the work is not being produced in the intended form, and the secretary, because the work repeatedly needs to be retyped. Consequently, much time is wasted unnecessarily.

You can help a typist produce your letters and reports in the form you want them if you *remember to keep the secretary in the picture*. These are details secretaries need to know:

- Whether the work is to be typed as a draft or in final form, ready for your signature. (Nothing annoys a typist more than to painstakingly prepare a business letter in its final form, and then to see you writing all over it because you *assumed* that the typist understood you only wanted a draft.)

- Whether the original or a copy of the report or letter will be sent to the addressee. If the original will be mailed, then the typist must produce a perfect document; if a copy will be mailed, then the typist can use white correction fluid. (If the typist is working at a word processor terminal, rather than on an electric or electronic typewriter, corrections are much easier to make.)
- Any "mechanical" considerations you have in mind, such as the size of the typing area (the number of lines vertically), the width of the column, the spacing between lines (single spacing, 1½ line spacing, or double spacing), and the "pitch" of the type (10 or 12 characters per inch).
- Where spaces are to be left for illustrations, with exact dimensions for each.
- The date the typing is required.

Assignments

Exercise 1.1. Describe why the pyramid method of writing will help you become a better presenter of information.

Exercise 1.2. Which do you feel is the better way for *you* to develop an outline for a report: the organized method or the "random" method? Explain why.

Exercise 1.3. (a) What are the six stages advocated for planning a report? (b) Must the stages be followed exactly in the sequence listed?

Exercise 1.4. If a report is likely to be read by several people, how would you identify which is your primary reader?

Exercise 1.5. If you had a long project report to write next month, describe where you would write it and why you would write it there.

Exercise 1.6. Is it better to write a report without stopping to "clean up" the construction along the way, or to write a page at a time and to edit it before going on to the next page? Explain why.

Exercise 1.7. What two factors will help you write more confidently, and probably help you set the right tone?

Exercise 1.8. Assume you are keying a report directly into a computer. List four instructions that you would key in to set up the appearance of the pages (after you have brought the word-processing program into the computer's memory).

Exercise 1.9. From the list of five main questions that you, as a writer, should ask yourself during the revision stage (see the boldface questions on page 26), which do you think is most important? Explain why.

Exercise 1.10. What kind of person is best equipped to do a final review of a report you have written?

CHAPTER 2

Meet "H. L. Winman and Associates"

Many of the projects you read or write about in this book are based upon events affecting two companies: H. L. Winman and Associates, a Cleveland firm of consulting engineers, and its affiliate, Macro Engineering Inc, of Phoenix, Arizona. To help you identify yourself in these working environments, this chapter contains short histories and partial organization charts of the companies, plus brief biographical sketches of the men and women who have brought them to their current position and who most influence how their engineers, scientists, technologists, and technicians write. It also describes what your role might be if you were employed by either company.

CORPORATE STRUCTURE

H. L. Winman and Associates was founded by Harvey Winman some 30 years ago. Starting with a handful of engineers working in a small office in one of the older sections of Cleveland, Harvey gradually built up his company until now it has 860 employees: 380 at the head office, 340 at the Southern affiliate in Phoenix, and 140 in smaller offices throughout the United States, Canada, and Mexico. Some are branches of H. L. Winman and Associates (there may be one in your area), while others are local engineering firms like Macro Engineering Inc that Harvey has purchased. Figure 2-1 illustrates the company's two main locations.

HEAD OFFICE: H. L. WINMAN AND ASSOCIATES

The head office of H. L. Winman and Associates is located in a modern building at 475 Reston Avenue, a major thoroughfare in the business district of

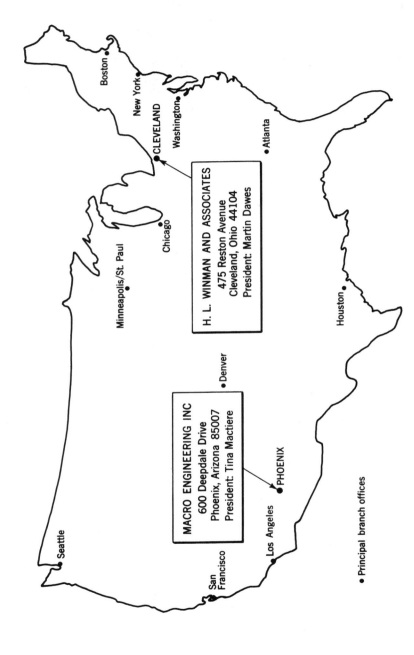

MACRO ENGINEERING INC
600 Deepdale Drive
Phoenix, Arizona 85007
President: Tina Mactiere

H. L. WINMAN AND ASSOCIATES
475 Reston Avenue
Cleveland, Ohio 44104
President: Martin Dawes

Seattle

San Francisco

Los Angeles

PHOENIX

Denver

Minneapolis/St. Paul

Chicago

CLEVELAND

Boston

New York

Washington

Atlanta

Houston

Principal branch offices

Figure 2-1. Locations of H. L. Winman and Associates' offices.

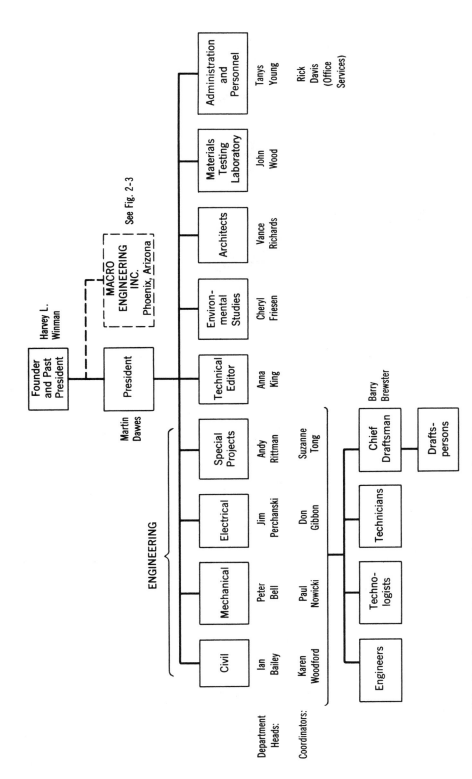

Figure 2-2. Partial organization chart—H. L. Winman and Associates.

Photo A. Harvey L. Winman.
Photo: Anthony Simmonds.

Cleveland. A partial organization chart in Figure 2-2 shows Martin Dawes as president, with nine department heads, who are responsible for the day-to-day operation of the company, reporting directly to him.

Over the years H. L. Winman and Associates has developed an excellent reputation among its clients for the management of large construction projects in remote areas of the United States and Canada. It has managed the construction of whole towns, airfields, and dams for large hydroelectric power generating stations; engineered, designed, and supervised construction of major traffic intersections, shopping complexes, bridges, and industrial parks; and designed and supervised numerous large buildings such as hotels, schools, arenas, and manufacturing plants. More recently, the company has entered the systems engineering field. Some of its current studies involve problems in air pollution, protection of the environment, and the effects of oil exploration and pipeline construction on the Alaskan arctic tundra. To handle so many diverse tasks, Harvey has built a highly adaptable, flexible staff with representation from many scientific and technical disciplines.

Although he is past retirement age, Harvey maintains an active interest in his company's business. Short, with a round face topped by a bald head with a fringe of hair, he is often seen strolling from department to department. He stops frequently to talk to his staff (most of whom he knows by first name) and to inquire about both the projects they are working on and their home lives. Well liked and thoroughly respected by all, from senior management down to the newest clerk, he is a manager of the old school, a type seldom seen in today's brusque business atmosphere.

The company's chief executive is Martin Dawes, to whom Harvey, anticipating his coming retirement, has handed full operating control. Martin is a civil engineer who developed a fine reputation for designing and constructing hydroelectric dams in remote, mountainous country. An ardent fisherman, he has often combined field projects with trips into the surrounding countryside, from which he has returned laden with salmon or trout. Before becoming president he was vice-president for four years and, before that, Head of the Civil Engineering Department for 15 years.

Both Harvey Winman and Martin Dawes write well, and they expect their staff also to be good writers. They recognize that the written word is the means that most often conveys an image of their company to customers. So, to ensure that all letters, reports, and company publications are of top quality, they have introduced a technical editor into their management team.

The editor is Anna King, who graduated with a Bachelor of Science in Electrical Engineering. She worked for the South Georgia Power and Light Company for nine years, first in the relay department and later in development engineering, where her capabilities both as engineer and report writer became evident.

Eventually Anna moved to ski country and became H. L. Winman and Associates' technical editor. "Yours will be an exacting job," Martin Dawes said on her first day. "You will be setting writing standards for the whole company. I don't want my engineers to think you are here simply to do their writing for them. Your role is to encourage them to become better writers."

SOUTHERN AFFILIATE: MACRO ENGINEERING INC

To increase his company's presence in the southern United States, Harvey Winman purchased two companies in the Phoenix area and amalgamated them to form Macro Engineering Inc (MEI). One was Robertson Engineering Company (Wayne G. Robertson, president), which had a well-established reputation for designing and custom manufacturing high-technology electronic, nucleonic, and electromechanical instruments. The other was Mactiere Microelectronics (Tina Mactiere, president), which specialized in designing computer interface systems.

When he amalgamated the two companies, Harvey combined the first three letters of Mactiere Microelectronics with the first two letters of Robertson Engineering to form Macro Engineering Inc. Tina and Wayne agreed to head the new company jointly, with Tina as President and Chief Executive Officer (CEO) and Wayne as General Manager and Marketing Manager, as shown in the company's partial organization chart in Figure 2-3.

Macro Engineering Inc has approximately 340 employees, of whom 215 are engineers, technologists, and engineering technicians. The company occupies the first two floors of the Wilshire Building, an office and light manufacturing complex in Phoenix, owned by the Wilshire Insurance Company.

Like Harvey Winman, both Tina and Wayne recognize the importance of communication in their business and particularly stress that the operating

Photo B. Tina R. Mactiere, President, Macro Engineering Inc. Photo: Maria Cerullo.

instructions and training manuals that accompany their instruments be abundantly clear. To this end, they employ a small technical writing group that can write manuals not only in English, but also in French, German, and Spanish, and occasionally have Anna King fly in from Cleveland to prepare particularly important technical proposals.

YOUR ROLE AS A TECHNICAL EMPLOYEE

As you read the remaining chapters in this book, you will be able to identify your "employer" and visualize yourself as an "employee." But you still need to know what your role will be as an engineer, technologist, or engineering technician working for either H. L. Winman and Associates or Macro Engineering Inc.

WORKING FOR A CONSULTING FIRM

Consultants provide specialist help to other companies or government organizations. Most clothing manufacturers, for example, are unlikely to have engineering help on staff, other than perhaps an electrical and a mechanical technician to deal with day-to-day maintenance. A clothing manufacturer who wants to build an extension to the plant, introduce robotics into the manufacturing operation, or correct an in-plant air pollution problem, does not have the expertise to do the work. Conceivably, the manufacturer could go directly to a

Photo C. Wayne G. Robertson, General Manager, Macro Engineering Inc. Photo: David Portigal.

builder and ask for a design and construction estimate, or to a manufacturer of robotics equipment and ask for a proposal, or to an air-conditioning company and request improved air circulation, but in each case the clothing manufacturer would be buying limited, rather biased, expertise. Often, too, a nontechnical company does not want to develop construction specifications, request and screen proposals, or coordinate and supervise contractors' work. ("That can be one big headache," most agree.) They would rather leave all these details to a consultant.

By calling in an engineering consultant, the manufacturer is in effect saying, "We have a problem and we want you to use all the resources at your command to investigate it, recommend how it can be corrected, and (sometimes) manage the corrective action for us." Because consultants are not "tied in" with particular builders, robotics manufacturers, air-conditioning installers, and so on, they have an unbiased perspective from which to view their clients' needs. Some consultants specialize in a particular field, such as civil engineering or environmental control, while others offer a fairly wide range of consulting services, as do H. L. Winman and Associates. And if a general consultant does not specialize in a particular area in which expertise is needed, the consultant will collaborate with a specialist consultant to provide the necessary services for a client.

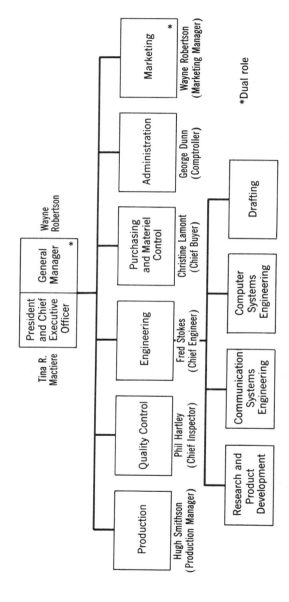

Figure 2-3. Partial organization chart showing Engineering Department of Macro Engineering Inc.

The steps in a client-consultant relationship are generally as described below and illustrated in Figure 2-4. There are two phases: the initial contact between the client and consultant, followed by a study or investigation; and the coordination of a construction or renovation contract that may evolve from the investigation.

Phase 1:

1. The client approaches the consultant, describes what is required or the problem that needs to be corrected, and, if the project is likely to be extensive, asks for a proposal and cost estimate. (If a project is small or straightforward, the client may dispense with the proposal and instruct the consultant to go ahead with the project.)
2. The consultant examines what is involved and then prepares a proposal outlining:
   - The consultant's understanding of the problem or project, and the work involved.
   - How the project will be tackled.
   - What human and material resources will be used.
   - Any reports that will be prepared.
   - The consultant's previous experience with similar projects (for the client to assess the consultant's capabilities).
   - The costs for (a) carrying out the investigation and (b) carrying out the work.

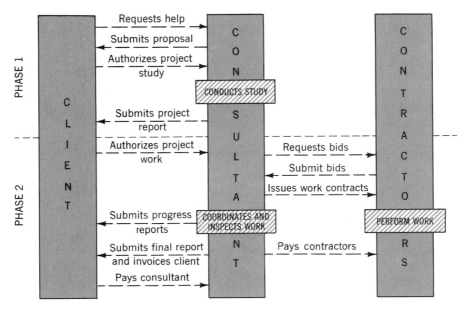

Figure 2-4. A simplified view of communication between (1) the client and consultant, and (2) the consultant and contractors.

3. If the consultant's proposal is accepted, the client writes a letter or issues a purchase order authorizing the consultant to go ahead with the project.

4. The consultant conducts the investigtion and then prepares a detailed report outlining what was found out and what corrective action needs to be taken. (If the project is lengthy, periodically the consultant will send a progress report to the client.)

Some client-consultant arrangements stop here. If, however, the client wants the consultant to manage the ensuing construction, installation, or renovation work, the relationship continues:

Phase 2:

5. The client authorizes the consultant to manage the project through to completion.
6. The consultant then:
   • Prepares a work plan and job specification.
   • Calls for bids from contractors.
   • Issues contracts to the most suitable contractors.
   • Coordinates and monitors the contractors' work.
   • Deals with problems.
   • Submits regular progress reports to the client.
   • Inspects completed work, asks for correction of unsatisfactory work, and then pays the contractors.
   • Prepares a job completion report and invoice for the whole project, and submits both to the client.

As an engineer, technologist, or engineering technician, you may be involved in any of these stages, and you will be expected to write reports at the appropriate times both for your own management and for your company's clients. Good communication is an essential component in any client-consultant relationship, because clients who are "kept in the picture" subconsciously feel that the consultant is effectively looking after their interests.

WORKING FOR AN ENGINEERING OR TECHNICAL FIRM

An engineer or engineering technician employed by a technical firm may still provide services to clients and customers. You may be a technical service representative who installs, repairs, modifies, and maintains equipment leased or sold by your company to the client. As such you will be expected to be not only an expert on the technical aspects but also an expert in interpersonal relations. A service representative is both a technical specialist and a salesperson.

As a technical service representative, you will write reports describing both routine maintenance and special conditions. After each service visit, you will have to describe what you observed about the equipment, what repairs or adjustments you made, and what components you installed or replaced. Generally, much of this information will be written onto a preprinted form that has appropriate spaces for filling in details. But when special visits are made or problems occur, you will more likely have to write a short investigation report in which you describe a problem and how you resolved it, or an inspection report in which you describe equipment condition. (For examples of such reports, see Chapter 4.)

Alternatively, you may be employed by a contractor to direct, coordinate, and supervise the work of installation teams and to provide engineering assistance when problems occur. Much of the time you will work at the job site and will keep management at the head office informed of progress by submitting reports describing the teams' progress and problems encountered. (In this capacity, you may be on the payroll of one of the contractors providing services to the consultants described earlier.)

Or you may be employed by a design company or a manufacturer, for whom you will provide engineering or technical services, which may include:

• Designing new products, processes, or procedures, or modifications to existing systems.
• Investigating manufacturing problems or assembly line "glitches."
• Proposing improved manufacturing methods.
• Installing and testing prototype equipment and components.
• Instructing operators and users of new or modified equipment.
• Inspecting work performed by assemblers and junior technicians.

In such a capacity, your reporting will have to be comprehensive, for you will be expected to describe what you have seen, to present management with an analysis or justification, and sometimes to recommend corrective action. You will have to use all the report-writing skills suggested in Chapters 4 through 7 for written reports and Chapter 9 for spoken reports.

CHAPTER 3
Technical Correspondence

When you write a personal letter to a friend or relative, you probably do not worry whether your letter is too long or contains too much information. You assume your reader will be pleased to hear from you and so launch into a general discourse, inserting comments and items of general news without concerning yourself very much about organization.

But when you write a business letter, you have to be much more in control. Your readers are busy people who want only the details that specifically concern them. Information they do not need irks them. For these people, your letters must be in focus, well-planned, brief, and clear.

FOCUS THE LETTER

Letter writing is like photography: you have to center the image in the viewfinder and then focus the image sharply before you snap the picture. But first you have to decide exactly what you want to depict; otherwise viewers will wonder why they are looking at the picture.

IDENTIFY THE MAIN MESSAGE

If you write your letters pyramid-style, you will automatically focus the reader's attention on your main message. Before you pick up a pen or place your fingers on the keyboard, fix clearly in your mind why you are writing and what you most want your reader to hear from you. Then focus sharply on this information by placing it right up front, where it will be seen immediately.

If you open a letter with background information rather than the main point, your reader will wonder why you are writing until he or she has read well into the letter. Don McKelvey's letter to Jim Connaught is a typical example of an unfocused letter:

> Dear Mr. Connaught:
>
> I refer to our purchase order No. 21438 dated April 26, 19xx, for a Vancourt microcopier model 3000, which was installed on May 14. During tests following its installation your technician discovered that some components had been damaged in transit. He ordered replacements and in a letter dated May 20 informed me that they would be shipped to us on May 27 and that he would return here to install them shortly thereafter.
>
> It is now June 10, and I have neither received the parts nor heard from your technician. I would like to know when the replacement parts will be installed and when we can expect to use the microcopier.
>
> Sincerely,
> Don McKelvey

Jim had to read more than 70 words before he discovered what Don wanted him to do. If Don had written pyramid-style, starting with a main message, Jim would have known immediately why he was reading the letter:

> Dear Mr. Connaught:
>
> We are still unable to use the Vancourt 3000 microcopier we purchased from you on April 26. Please inform me when I can expect it to be in service.

Placing the main message up front would have helped Don write a shorter explanation that would have been simpler to follow:

> The microcopier was ordered on P.O. 21438 and installed on May 14. During tests, your technician discovered that some components had been damaged in transit. He ordered replacements and in a letter dated May 20 informed me that they would be shipped to us on May 27, and that he would return here to install them. To date, I have neither received the parts nor heard from your technician.
>
> Sincerely,

Unfortunately, knowing you should open every letter with a main message is not enough in itself. You also need to know how to find exactly the right words to put at the top of the pyramid. And that is where many technical people have trouble.

FOCUS READERS' ATTENTION

To overcome this block, try using a technique recommended by Anna King, H. L. Winman and Associates' technical editor. She suggests that when you start a letter, first write these six words:

I want to tell you that . . .

And then finish the sentence with what you *most* want to tell your reader. For example:

> Dear Ms. Reynaud:
>
> *I want to tell you that* . . . the environmental data you submitted to us on October 8 will have to be substantiated if it is to be included with the Labrador study.

Then, when your sentence is complete, *delete* the first six words (the *I want to tell you that* . . . expression). What you have left will become an in-focus opening statement:

> Dear Ms. Reynaud:
>
> The environmental data you submitted to us on October 8 will have to be substantiated if it is to be included with the Labrador study.

Often you can use an opening statement formed in this way just as it stands when you remove the six "hidden" words. At other times, however, you may feel the opening statement seems a bit abrupt. If so, you can soften it by inserting one or two additional words. For example, in the letter to Ms. Reynaud, you might want to add the expression "I regret that . . .":

> Dear Ms. Reynaud:
>
> *I regret that* the environmental data you submitted to us on October 8 will have to be substantiated if it is to be included with the Labrador study.

Figure 3-1 depicts this convenient way to start a letter and concurrently create a main message. It also shows that in business letters the main message is more often referred to as the summary statement.

Here are three more examples of properly formed summary statements:

> Dear Colonel Watson:
>
> We will complete the XRS modification on June 14, eight days earlier than scheduled.

> Dear Ms. Mohammed:
>
> Your excellent paper "Export Engineering" arrived just in time to be included in the program for the Pacific Rim Conference.

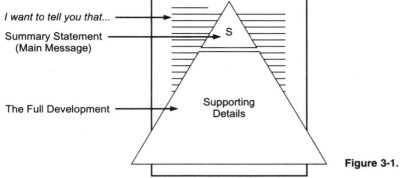

Figure 3-1. Creating a letter's summary statement.

> Dear Mr. Voorman:
> Seven defective castings were found in shipment No. 308.

You can check that the *I want to tell you that* . . . words were used to form these three opening sentences by mentally inserting them at the start of each sentence.

AVOID FALSE STARTS

If you do not use the six "hidden" words to start a letter, you may inadvertently open with an awkwardly constructed sentence that seems to be going nowhere. For example:

> Dear Mr. Corvenne:
> In answer to your enquiry of December 7 concerning erroneous read-outs you are experiencing with your Mark 17 Analyzer, and our subsequent telephone conversation of December 18, during which we tried to pinpoint the fault, we have conducted an examination into your problem.

Anna King refers to a long, rambling opening like this as "spinning one's wheels," because such a sentence does not come to grips with the topic early enough. She has prepared a list of 12 expressions (see Figure 3-2) that can easily cause you to write complicated, unfocused openings. In their place she recommends starting with the *I want to tell you that* . . . expression, which will help you focus your reader's attention on the main message. If the letter referring to the Mark 1 Analyzer had started with the six "hidden" words, it would have been much more direct:

> Dear Mr. Corvenne:
> (*I want to tell you that* . . .) The problem with your Mark 17 Analyzer seems to be in the extrapolator circuit. Following your enquiry of December 7 and your subsequent description of erroneous read-outs, we examined . . . (etc).

WHEN YOU WRITE A LETTER . . .

Never start with a word that ends in "ing":

> *Referring . . .*
> *Replying . . .*

Never start with a phrase that ends with the preposition "to":

> *With reference to . . .*
> *In answer to . . .*
> *Pursuant to . . .*

Never start with a redundant expression:

> *I am writing . . .*
> *For your information . . .*
> *This is to inform you . . .*
> *The purpose of this letter is . . .*
> *We have received your letter . . .*
> *Enclosed please find . . .*
> *Attached herewith . . .*

IN OTHER WORDS . . .

DON'T SPIN YOUR WHEELS!

Figure 3-2. Anna King's suggestion to H. L. Winman engineers.

PLAN THE LETTER

When you have identified and written the main message, you next have to select, sort, and arrange the remaining information you will convey to your reader. These details should amplify the bare message you have already presented and frequently provide evidence of its validity. For example, when Paul Shumeier wrote the following summary statement, he realized he would be presenting his reader with costly news:

> Dear Mr. Larsen:
> Tests of the environmental monitoring station at Wickens Peak show that 60% of the instruments need to be repaired and recalibrated at a cost of $5265.

He also realized that Mr. Larsen would expect the remainder of the letter to tell him why the repairs were necessary, exactly what needed to be done, and how Paul had derived the total cost. For Paul simply to list the instruments and their individual repair costs would not suffice.

If you are to provide your readers with the exact information they need, you first have to identify what questions will be foremost in their minds after they have read your main message. This means asking yourself what questions a reader is *likely* to ask, and then providing your own answers.

Paul derived his answers by asking himself six questions all based on *Who?*, *Where?*, *When?*, *Why?*, *What?*, and *How?*:

Why	(are the repairs necessary)?
What	(repairs are needed)?
How	(were the costs calculated)?
Who	(was involved)?
Where	(did this happen)?
When	(did this happen)?

Then he decided that only questions 1, 2, and 3 would provide information of immediate interest to Mr. Larsen, so he eliminated questions 4, 5, and 6.

OPENING UP THE PYRAMID

Now Paul has to insert this information into his letter, and it will become the lower part of the writer's pyramid. These are the supporting details which, as Figure 3-1 shows, are also known as the full development.

To help Paul—and you—organize a letter's full development, the lower part of the pyramid is stretched vertically and divided into three compartments known as the **background, facts**, and **outcome** (see Figure 3-3).

The **background** covers *what* has happened previously, *who* was involved, *where* and *when* the event occurred or the facts were gathered, and, sometimes, for *whom* the work was done. Paul wrote:

> Our electronics technicians examined the Wickens Peak monitoring station on May 16 and 17, in response to your May 10 request to Mr. Patrick Friesen.

The **facts** amplify the main message. They provide specific details the reader needs to fully understand the situation or to be convinced of the need to take further action. Here, Paul wrote:

> Most of the damage was caused by a tree northwest of the site that fell onto the station during a storm on April 23 and damaged parts of the roof and north and west walls. Instruments along these walls were impact-damaged and then soaked by rain. Other instruments in the station also were affected by the moisture.

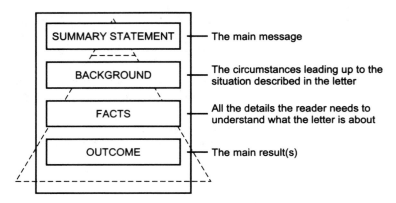

Figure 3-3. Basic writing plan for a business letter or interoffice memorandum.

> Major repairs and recalibration are required for the 16 instruments listed in attachment 1, which describes the damage and estimated repair cost for each instrument. This work will be done at our Shepperton repair depot for a total cost of $3405. Minor repairs, which can be performed on site, are necessary for the 27 instruments listed in attachment 2. These on-site repairs will cost $1860.

These two paragraphs clearly answer the *Why?*, *What?*, and *How?* questions. Note that, rather than clutter the middle of his letter with a long list, Paul placed the details in two attachments and summarized only the main points in the body of the letter. (The attachments are not shown here.)

The **outcome** describes the result or any effect the facts have had or will have. If the letter is purely informative and the reader is not expected to take any action, the outcome simply sums up the main result. Paul would have written:

> I have obtained Ms. Korton's approval to perform the repairs and a crew was sent in on May 23. They should complete their work by May 31.
> Sincerely,

But if the reader is expected to take some action, or approve somebody else taking action (often, the letter writer), then the outcome becomes a *request for action*. Since Paul wanted an answer, he wrote:

> If these repair costs are acceptable, please telephone me or fax your approval so that I can send in our repair crew.
> Sincerely,

These three parts can help you arrange the full development of any letter or memorandum into a logical, coherent sequence. Before starting, however, you have to decide whether you are writing an informative or a persuasive letter.

WRITING TO INFORM

Letters and memorandums that purely inform, with no response or action required from the reader, normally can be organized around the basic **summary statement-background-facts-outcome** writing plan shown in Figure 3-3. Kevin Toshak's memo to Tina Mactiere, in Figure 3-4, falls into this category.

Another example is a confirmation letter, in which the writer confirms previously made arrangements. In the following Macro Engineering Inc memorandum, General Manager Wayne Robertson ensures that he and Chief Buyer Christine Lamont both understand the arrangements that will evolve from a decision made at a company meeting:

	Christine:
Summary *Statement*	I am confirming that you will represent both Macro Engineering Inc and H. L. Winman and Associates at the Materials Handling conference in Houston on May
Background	15 and 16, as agreed at the Planning Meeting on March 23. At the conference you will:

MACRO ENGINEERING INC

FROM: Kevin Toshak   DATE: October 22, 19xx

TO: Tina Mactiere   SUBJECT: Monitor Installation
                                         at WRC

I have installed a TL-680 monitor unit in room 215 at   *Summary*
the Wollaston Research Center, as instructed by your   *Statement &*
memorandum of October 15.   *Background*

The unit was installed without major difficulties,
although I had to modify the equipment rack to accept   *Facts*
it as illustrated in the attached sketch. Post-
installation tests showed that the unit was accepting
signals from both the control center and the remote   *Outcome*
site.

                        Kevin T.

Figure 3-4.   An informative letter in memorandum format.

Facts
- Take part in a panel discussion on packaging electronic equipment from 10:00 to 11:15 a.m. on May 15.
- Host a wine-and-cheese reception for delegates from 5:00 to 7:00 p.m. on May 16.

Facts Janet Kominsky is making your travel and hotel reservations and the catering arrangements for the reception. Anna King will provide brochures from Cleveland, and my secretary will make up packages for you to distribute.

Outcome I'll brief you on other details before you leave.

Wayne

Although the basic writing plan for letters has four compartments (see Figure 3-3), you do not have to write exactly four paragraphs. As both Wayne's and Tina's memos show, you may combine two compartments into a single paragraph, or let one compartment be represented by several paragraphs.

WRITING TO PERSUADE

In a persuasive letter, you expect your reader either to respond to your letter or to take some form of action on reading it. Fot that reason, the **outcome** compartment of the writing plan is renamed **action,** as shown in Figure 3-5, to remind you to end a persuasive letter with an "action statement." (The "shadow" pyramid still visible in Figure 3-3 has been omitted from the writing plan in Figure 3-5. However, always remember it is still there, in the background, providing the basic structure upon which this and all future writing plans are based.) A request and a complaint are typical examples of persuasive letters.

Making a Request

Many technical people claim that placing the message at the start of a letter is not a problem until they either have to ask for something or give the reader bad news. Then they tend to lead gently up to the request or the unhappy information.

Bill Kostash is no exception. He is service manager for Mechanical Maintenance Systems Inc, and he has to write to customers to ask if they will accept a change in the preventive maintenance contracts his company has with them. He starts by writing to Ms. Bea Nguyen, the Contracts Administrator for Multiple Industries Incorporated in St. Cloud, Minnesota:

June 18, 19xx
Dear Ms. Nguyen:
I am writing with reference to our contract with you for the preventive mainte-nance services we provide on your RotoMat extruders and shapers. Under the

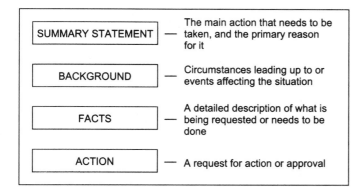

Figure 3-5. Writing plan for a persuasive letter or memorandum.

terms of the current contract (No. RE208) dated January 2, 19xx, we are required to perform monthly inspection and maintenance "... on the 15th day of each month or, if the 15th falls on a weekend or holiday, on the first working day thereafter."

Bill has got off to a bad start. Instead of opening with a summary statement, he has inserted all the background details first, with the result that Bea Nguyen does not yet know why he has written to her. He has also opened with one of the expressions Anna King lists as an awkward start in Figure 3-2. Let's see how he continues.

Our problem is that almost all of our clients ask that we perform their maintenance service between the 5th and 25th of each month, to avoid their end-of-month peak accounting periods. This in turn creates difficulty for us, in that our service technicians experience a peak workload for 20 days and then have virtually no work for 10 days.

Bea Nguyen still does not know why Bill is writing.

Consequently, to even out our workload, I am requesting your approval to shift our inspection date from the 15th to the 29th of each month. If you agree to my request, I will send our technician in to service your machines on June 29—a second time this month—rather than create a six-week period between the June and July inspections. Could you let me know by June 25 if this change of date is acceptable?
Sincerely,
William J. Kostash

Now Bea knows why Bill has written to her—but she has had to read a long way to find out.

If Bill had used the writing plan in Figure 3-5 to shape his letter, his request would have been much more effective. The revised letter is shown in Figure 3-6, in which

(1) is his **summary statement** (he states his request and what the effect will be),
(2) contains the **background** (the contract details),
(3) contains the **facts** (it describes the problem), and
(4) is the **action** statement, in which he mentions *two* actions: what he wants Bea to do (to call him) and what he will do (schedule a second visit).

Registering a Complaint

The approach is the same if you have to write a letter of complaint or ask for an adjustment. You can use the writing plan for a persuasive letter shown in Figure 3-5, inserting the following information into each compartment:

- **Summary Statement.** State the problem and say what you want the reader to do about it:

    Dear Mr. Bruyere:

    The Nabuchi 700 portable computer you recently sold me had a defective nickel-cadmium battery which had to be replaced while I was in Europe. Consequently I am requesting reimbursement of the repair expense I incurred.

    Often, it is better to generalize what action is needed in the summary and then to state exactly what has to be done in the action compartment.

- **Background.** Identify the circumstances leading up to the event, quoting specific details:

    I bought the computer and a Nubuchi 701PC international power converter from your Willows Mall store on September 4, 19xx, on Sales Invoice No. 14206A.

    If there are few background details, you may combine them with either the summary statement or the facts rather than place them in a very short independent paragraph.

- **Facts.** Describe exactly what happened, in chronological order, so that the reader will understand the reason for your complaint or request for adjustment:

    The computer worked satisfactorily for the first six weeks, but during that time I had no occasion to use it solely on battery power.

    On October 25 I left for Europe, first giving the batteries an 18-hour charge as recommended in the operating instructions. While using the computer in flight, after only 35 minutes the low battery lamp lit up and the screen warned of imminent failure. I recharged the batteries the following day, in Rheims, France, but achieved less than 25 minutes of operating time before the batteries again became fully discharged.

    As the Nabuchi line is neither sold nor serviced in France, I had to buy and install a replacement nickel cadmium battery (a Mercurio Z7S), which has since worked admirably. I enclose the defective battery, plus a copy of the sales receipt for the replacement battery I purchased from Lestrange Limitée, Rheims.

- **Action.** Identify specifically what action you want the reader to take, or that you will take.

    Please send me a check for $203.15, which at the current rate of exchange is the U.S. equivalent of the 1190 francs shown on the sales receipt.

    Sincerely,

    Suzanne Dumont, P.E.

MECHANICAL MAINTENANCE SYSTEMS LTD
2120 Cordoba Avenue
St. Paul, Minnesota 55307

June 18, 19xx

Bea Nguyen
Contracts Administrator
Multiple Industries Limited
-- Manufacturing Division
18 Commodore Bay
St. Cloud, MN 54018

Dear Ms. Nguyen

(1) I am requesting your approval to change the date of our monthly preventive maintenance visits to service your Rotomat extruders and shapers to the 29th of each month. This will help spread my technicians' workload more evenly and so provide you with better service.

(2)
(3) Our contract with you is No. RE208 dated January 2, 19xx, and it requires that we perform monthly inspection and maintenance on the 15th day of each month. Unfortunately, almost all of our clients ask that we perform their maintenance service between the 5th and the 25th. This creates a problem for us in that our service technicians experience a peak workload for 20 days and then have very little work for 10 days.

(4) Could you let me know by June 25 if you can accept the change? Then I will send a technician to your plant on June 29 for a second visit this month, rather than create a six-week space between the June and July inspections.

Sincerely

Bill Kostash

William J Kostash
Service Manager

Figure 3-6. A request letter written pyramid-style.

CREATE A CONFIDENT IMAGE

Readers react positively to letters and memorandums in which the writer conveys an image of a confident person who knows the subject well and has a firm idea of what he or she plans to do, or expects the reader to do. Such an image is conveyed by both the quality of the writing and the physical appearance of the piece of correspondence.

BE BRIEF

For technical business correspondence, brevity means writing short letters, short paragraphs, short sentences, and short words.

Short Letters

A business reader will tend to react readily to a short letter, viewing its writer as an efficient purveyor of information. In contrast, the same reader may view a long letter as "heavy going" (even before reading it) and tend to put it aside to deal with later. A short letter introduces its topic quickly, discusses it in sufficient depth, and then closes with a concluding statement, its length dictated solely by the amount of information that has to be conveyed.

I know of a company in which the managing director has ruled that no letter or memorandum may exceed one page. This is an effective way to encourage staff to be brief, and it works well for many people. But for letter writers who have more to say than they can squeeze onto a single page, the limitation can prove inhibiting. For them, I suggest borrowing a technique from report writing: Instead of placing all their information in the letter, they should change the letter into a semiformal report and then summarize the highlights— particularly the purpose and the outcome—in a one-page letter placed at the front of the report (so that the report becomes *an attachment* to the letter, as depicted in Figure 3-7).

If you use this device, in the body of the letter both refer to the attachment *and* insert a main conclusion drawn from it, as has been done here:

> During the second week we measured sound levels at various locations in the production area of the plant, at night, during the day, and on weekends. These readings (see attachment) show that a maximum of 55 dB was recorded on weekdays, and 49 dB on weekends. In both cases these peaks were recorded between 5 and 6 p.m.

Short Paragraphs

Novelists can afford to write long paragraphs because they asume they will have their readers' attention, and their readers have the time and patience to wend their way through leisurely description. But in business and industry

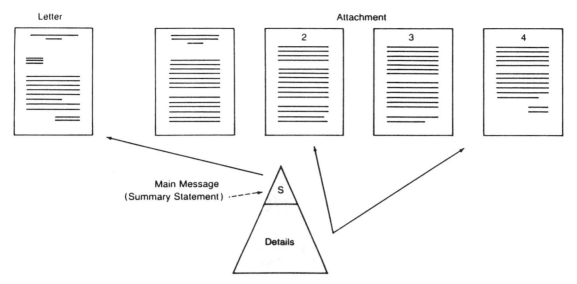

Figure 3-7. A short letter with attachments is an adaption of the pyramid method of writing. An example can be seen in Figure 5-4 (page 119).

readers are working against the clock and so want bite-size paragraphs of easy-to-digest information.

Let the first sentence of each paragraph introduce just one idea; then make sure that subsequent sentences in that paragraph develop the idea adequately and do not introduce any other ideas. In technical business writing, the first sentence of each paragraph should be a "topic sentence," so that the reader knows your main idea immediately. That is, let the first sentence of each paragraph summarize the paragraph's contents, and the remaining sentences support the first sentence by providing additional information:

> *We have tested your 15 Vancourt 801 disk drives and find that 11 require repair and recalibration.* Only minor repairs will be necessary for 6 of these drives, which will be returned to you next week. Of the 5 remaining drives, 3 require major repairs which will take approximately 20 days, and 2 are so badly damaged that repairs will cost $180 each. Since this is more than the maximum repair cost imposed by you, no work will be done on these drives.

If an idea you are developing breeds an overly long paragraph, try dividing the information into a short introductory paragraph and a series of subparagraphs, as has been done here:

My inspection of the monitoring station at Freedom Lake Narrows revealed three areas requiring attention, two immediately and one within three months:

1. The water stage manometer is recording erratic read-outs of water levels. A replacement monitor needs to be flown in immediately so that the existing unit can be returned to a repair depot for service.
2. The tubing to the bubble orifice is worn in several places and must be replaced (90 feet of ½ inch tubing will be required). This work should be done concurrently with the monitor replacement.
3. The shack's asphalt roof is wearing and will need resurfacing before freeze-up.

I am not suggesting that your letter should contain a series of small, evenly sized paragraphs, which would appear dull and stereotyped. Paragraphs should vary from quite short to medium-long to give the reader variety. How you can adjust paragraph and sentence length to suit both reader and topic, and also to place emphasis correctly, is covered in Chapter 11.

Short Sentences

If you write short, uncomplicated sentences, you will be helping your readers quickly grasp and readily understand each thought you present. Sometimes expert literary writers can successfully build sentences that develop more than one thought, but such sentences are confusing and out of place in the business world. Compare these two examples of the same information:

Complicated There has been intermittent trouble with the vacuum pumps, although the flow valves and meters seem to be recording normal output, and the 5 inch pipe to the storage tanks has twice become clogged, causing backup in the system.

Clear There has been intermittent trouble with the vacuum pumps, and twice the 5 inch pipe to the storage tank has become clogged and caused backup in the system. The flow valves and meters, however, seem to be recording normal output.

The first example is confusing because it jumps back and forth between trouble and satisfactory operation. The second example is clear because it uses two sentences to express the two different thoughts.

Short Words

Some engineers and engineering technicians feel that the technical environment in which they work, and the complex topics they have to write about, demand that they use long, complex words in their correspondence. They write "an error of considerable magnitude was perpetrated," rather than simply "we made a large error." In so doing, they make a reader's job unnecessarily difficult. Because the engineering and scientific worlds encompass many long and complex technical terms that have to be used in their original form, surround these terms with simple words to make your correspondence more readable.

BE CLEAR

A clear letter conveys information simply and effectively, so that the reader readily understands its message. To write clearly demands ingenuity and attention to detail. As a writer, you must consider not only how you will write your letters, but also how you will present them.

Create a Good Visual Impression

Experienced writers know that a nicely laid out letter impresses a reader. Subconsciously it seems to be saying "my neat appearance demonstrates that I contain quality information that is logically and clearly organized."

The appearance of a letter tells much about the writer and the company he or she represents. If a letter is sloppily arranged or contains strikeovers, visible erasures, or spelling errors, then I imagine a careless individual working in a disorganized office. But if I am presented with a neat letter, tastefully placed in the middle of the page and carefully typed with a reasonably new ribbon, then I imagine a well-organized individual working for a forward-thinking company noted for the quality of its service. I would prefer to deal with the latter company, and I will read its correspondence first.

Develop the Subject Carefully

The key to effective subject development is to present the material logically, progressing gradually from a clear, understood point to one that is more complex. This means developing and consolidating each idea for the reader's full understanding before attempting to present the next idea. The sections on paragraph unity and coherence in Chapter 11 (see pages 325 to 329) provide examples of coherent paragraphs.

Insert Headings as Signposts

In longer letters, and particularly those discussing several aspects of a situation, you can help your reader by inserting headings. Each heading must be informative, summarizing clearly what is covered in the paragraphs that follow. (Figure 11-2 on page 324 offers advice on creating and inserting headings). If, for instance, I had replaced the heading preceding this paragraph with the single word "Headings," I would not have summarized adequately what this paragraph describes.

BE DEFINITE

People who think better with their fingers on a keyboard or a pen in their hand sometimes make decisions as they write, producing indecisive letters that are irritating to read. Their writers seem to examine and discard points without really grappling with the problem. By the time they have finished a letter, they have decided what they want to say, but it has been at the readers' expense.

Decision-making does not come easily to many people. Those of us who hesitate before making a decision, who evaluate its implications from all possible angles and weigh its pros and cons, may allow our indecisiveness to creep into our writing. We hedge a little, explain too much, or try to say how or why we reached a decision before we tell our reader what the decision is. This is particularly true when we have to tell readers something unfavorable or contrary to their expectations.

The key, as before, is to use the pyramid:

1. Decide exactly what you want to say (i.e. develop your main message), and then
2. Place the main message up front (use *I want to tell you that* . . . to get started).

If you also write primarily in the active voice, you will sound even more decisive. Active verbs are strong, whereas passive verbs are weak. For example:

These passive expressions:	*Should be replaced with:*
it was our considered opinion	we consider
it is recommended that	I recommend
an investigation was made	we investigated
the outage was caused by a defective transformer	a defective transformer caused the outage

For hints on how to use the active voice, see "Emphasis" in Chapter 11 (pages 325 and 326).

USE GOOD LANGUAGE

It hardly seems necessary to tell you to use good language, but in this case I mean language that you know your readers will understand. Use only those technical terms and abbreviations they will recognize immediately. If you are in doubt, define the term or abbreviation, or replace it with a simpler expression.

CLOSE ON A STRONG NOTE

You may feel you should always end a letter with a polite closing remark, such as: *I look forward to hearing from you at your earliest convenience*, or *Thanking you in advance for your kind cooperation*. In contemporary business correspondence—and particularly in technical correspondence—such closing statements are not only outmoded but also weaken your impact on the reader. Today, you should close with a strong, definite statement.

In effect, the **outcome** part of the letter provides a natural, positive close, as illustrated by the final sentences in the letters to Mr. Larsen (page 46) and Ms. Nguyen (Figure 3-6). You should resist the temptation to add a polite but uninformative and ineffective closing remark. Simply sign off with "Regards," "Sincerely," or "Cordially."

The factors I have described so far are mostly manipulative details that can be learned. Armed with this knowledge and the basic letter formats illustrated later in this chapter, the inexperienced writer has some ground rules on the practical aspects of business letter writing. Still to be acquired is the more difficult technique of letting one's character appear in letters without letting it become too obvious.

ADOPT A PLEASANT TONE

Sincerity and tone are intangibles that defy close analysis. There is no quick and easy method that will make your letters sound sincere, nor is there a checklist that will tell you when you have imparted the right tone. Both qualities are extensions of your own personality that cannot be taught. They can only be shaped and sharpened through knowledge of yourself and which of your attributes you most need to develop.

To achieve the right tone, your correspondence should be simple and dignified, but friendly. Approach your readers on a person-to-person basis, following the five suggestions below.

KNOW YOUR READER

If you have not identified your reader properly, you may have difficulty setting the correct tone. You need to know your reader's level of technical knowledge and whether he or she is familiar with the topic you are describing. Without this focus, you may unwittingly seem condescending to a knowledgeable reader because you explain too much and use overly simple words, when the reader clearly expects to read technical terms. Conversely, you may just as easily seem overbearing to, or even overwhelm, a reader who has only limited technical knowledge if you confront him or her with heavy technical details.

Ideally, you should select just the right terminology to hold the reader's interest and perhaps offer a mild challenge. By letting readers feel they are grasping some of the complexities of a subject (often by using analogies within their range of knowledge), you can present technical information to nontechnical readers without confusing or upsetting them.

BE SINCERE

At one time it was considered good manners not to permit one's personality to creep into business correspondence. Today, business letters are much less formal and, as a result, much more effective.

Sincerity is the gift of making your readers feel you are personally interested in them and their problems. You convey this by the words you use and the way you use them. A reader would be unlikely to believe you if you came straight out and said, "I am genuinely interested in your project." The secret is to be so involved in the subject, so interested by it, that you automatically convey the ring of enthusiasm that would appear in your voice if you were talking about it.

BE HUMAN

Too many letters lack humanity. They are written from one company to another, without any indication that there is a human being at the firing end and another at the receiving end. The letters might just as well be sent from computer to computer.

Do not be afraid to use the personal pronouns, "I," "you," "he," "she," "we," and "they." Let your reader believe you are personally involved by using "I" or "we," and that you know he or she is there by using "you." Contrary to what many of us were told in school, letters may be started in the first person. If you know the reader personally, or you have corresponded with each other before, or if your topic is informal, let a personal flavor appear in your letters by using "I" and the reader's first name:

> Dear Ben:
> I read your report with interest and agree with all but one of your conclusions.

If you do not know your reader personally and are writing formally as a representative of your company, then use the first person plural and the person's surname:

> Dear Mr. Wicks:
> We read your report with interest and agree with all but one of your conclusions.

AVOID WORDS THAT ANTAGONIZE

If you use words that imply the reader is wrong, has not tried to understand, or has failed to make himself or herself understood, you will immediately place the reader on the defensive. For example, when field technician Des Tanski omitted to send motel receipts with his expense account, Andy Rittman (his supervisor) had to write to Des and ask for them. Andy wrote:

> You have failed to include motel receipts with your expense account.

This subtly antagonized Des, because the words "you have failed" seemed to imply that he was something of a failure. Andy should have written:

> Please send motel receipts to support your expense account.

Other expressions that may annoy readers or put them on the defensive are:

You have *neglected* . . . You *ought to know* . . . You seem to have *overlooked* . . . You have *not understood* . . .	*Words that make a reader feel guilty*
You *must* return the instrument . . . Your *demand* for warranty service . . . We *insist* that you . . . I am sure *you will agree* . . .	*Words that provoke a reader*
We *have to assume* . . . I *must request* . . . I *simply do not understand* your . . . *You must understand* our position . . . *Undoubtedly* you will . . .	*Words that "talk down" to a reader*

When a reader has to be corrected, the words you use should clear the air rather than electrify it. Tell readers gently if they are wrong, and demonstrate why; reiterate your point of view in clear terms, to clarify any possibility of misunderstanding; or ask for further explanation of an ambiguous statement, refraining from pointing out that the person's writing is vague.

KNOW WHEN TO STOP

When a letter is short, you may feel it looks too bare and be tempted to add an extra sentence or two to give it greater depth. If you do, you may inadvertently weaken the point you are trying to make. This is particularly true of short letters in which you have to apologize, criticize, say "thank you," or pay a compliment (i.e. "pat the reader on the back").

In all of these cases, the key is to be brief: Know clearly what you want to say, say it, and then close the letter *without repeating what you have already said.* The following writer clearly did not know when to stop:

Dear Mr. Farjeon:
I want to say how very much we appreciated the kind help you provided in overcoming a transducer problem we experienced last month. We have always received excellent service from your organization in the past, so it was only natural that we shoud turn to you again in the hour of our need. The assistance you provided in helping us to identify an improved transducer for phasing in our standby generator was overwhelming, and we would like to extend our heartfelt thanks to all concerned for their help.
Sincerely,
Paul Marchant

Paul's letter would have been much more believable if he had simply said thank you:

Dear Mr. Farjeon:
Thank you for your prompt assistance last month in identifying an improved transducer for phasing in our standby generator. Your help was very much appreciated.
Regards,
Paul Marchant

If a writer says too much when saying thank you or apologizing, the reader begins to doubt the writer's sincerity. You cannot set a realistic tone if you overstate a sentiment, if you gush, or if you overwhelm your reader with the intensity of your feelings.

USE A BUSINESSLIKE FORMAT

There are many opinions of what comprises the "correct" format for business correspondence. The examples illustrated here are those most frequently used by contemporary technical organizations under various circumstances.

LETTER STYLES

There are two letter formats: the full block and the modified block (see Figures 3-8 and 3-9). Full block is more widely used and is the format Anna King has adopted for H. L. Winman and Associates' correspondence (there are examples in the letter report in Figure 5-4 of Chapter 5 and the cover letter preceding the first report at the end of Chapter 6). Anna is also aware that letter styles are continually changing. Some companies now omit the salutation and complimentary close (such as "Sincerely"), write dates military style (day-month-year: "27 January 1993"), may omit all punctuation from names and addresses (as in "Ms Jayne K Tooke"), and even use interoffice memorandums or quick-reply letters for informal correspondence. If these trends become more firmly established, Anna will adapt H. L. Winman and Associates' letter format to reflect the changes.

The modified block style is more conservative and is used primarily by individuals for their personal correspondence and by some small businesses.

INTEROFFICE MEMORANDUM

The memorandum is a flexible document normally written on a prepared form similar to that shown in Figure 3-10. Formats vary according to the preference of individual companies, although the basic information at the head of the form is generally similar. Examples of memorandums appear in this chapter and throughout Chapters 4 and 5.

QUICK-REPLY LETTER

A quick-reply letter, like an interoffice memorandum, is handwritten or typed on a prepared form. The form has spaces for the originator to write a message and the recipient to write a reply (see Figure 3-11). It comprises three different-colored sheets printed on no-carbon-required (NCR) paper, or plain paper interleaved with carbon paper. The originator detaches and retains the middle sheet. When the recipient has written a reply, he or she detaches and keeps the third sheet and returns the front sheet to the originator.

FAX COVER SHEET

Any document sent by facsimile machine is normally preceded by a single-page fax cover sheet that identifies both addressee and sender and their respective fax and telephone numbers (see Figure 3-12). The cover sheet usually has a space for the sender to write a short explanatory note. A sender who has only a short message to send may write the message directly onto the fax cover sheet and then transmit just the single page.

ELECTRONIC MAIL

Electronic mail has virtually replaced the telegram, which was a popular and relatively fast means of communication from the 1900s through to the 1960s, and is rapidly superseding telex and similar typewriter-based direct-communication systems. The criterion for writing telegrams, telex-type messages, and intercomputer messages, however, remains the same: keep the message brief but clear. Never allow an overzealous desire for brevity to cloud your message, because it can cost more to question an obscure communication than it would have cost to write a slightly longer but clearer message in the first place.

When Mike Toller in Topeka, Kansas opened a shipment of parts from Carlson Distributors, he found the order was incomplete and contained some items he had not ordered. He made a note of the deficiencies, sat in front of his computer terminal, and keyed in this message:

> TO: CARLSON DISTRIBUTORS, NEW YORK, YOUR INV 216875 OCT 19, OUR P.O. W 1634, SHORT-SHIPPED 10 TOOLSETS MKV, 4 801 SOCKETS PLUS 2 DOZ MOD 280A LATHE BITS UNORDERED. ADVISE M. TOLLER

In New York, Christina Baird puzzled over the message on her video screen, and then keyed this reply:

> TO: M. TOLLER, CROWN MANUFACTURING, TOPEKA. YOUR MESSAGE RE OUR INV 216875 AND YOUR P.O. W1634 NOT UNDERSTOOD. PLEASE EXPLAIN. C. BAIRD.

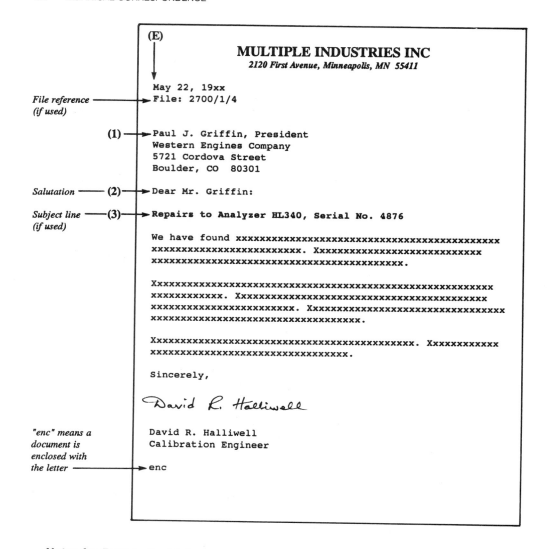

Figure 3-8. Full block letter format.

Notes for Both Letter Styles

(1) In modern correspondence the name and title of the addressee are placed ahead of the company name. If the addressee's name is not known, a suitable title (such as Purchasing Agent) should be inserted. All but essential punctuation is eliminated from the address, salutation, and signature block.

(2) Today's trend toward informality encourages writers to use first names in the salutation: "Dear Jack" instead of "Dear Mr. Sleigh."

(3) Subject lines should be *informative* (not just "Production Plan" or "Spectrum Analyzer"); they may be preceded by *Subject:*, *Ref:*, or *Re:*.

(4) The letter writer may sign for the Department Head.

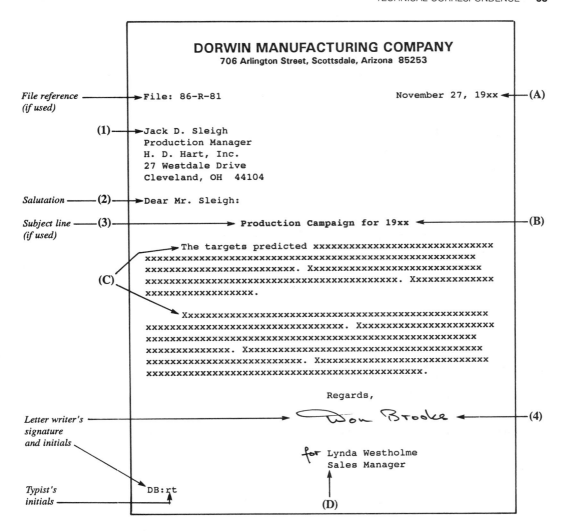

File reference ——— →File: 86-R-81   November 27, 19xx ◄—**(A)**
(if used)

(1)——→Jack D. Sleigh
Production Manager
H. D. Hart, Inc.
27 Westdale Drive
Cleveland, OH 44104

Salutation ——— **(2)**——→Dear Mr. Sleigh:

Subject line ——— **(3)**——————————→ Production Campaign for 19xx ◄——**(B)**
(if used)

→The targets predicted xxxxxxxxxxxxxxxxxxxxxxxxxxxxx
xxxxxxxxxxxxxxxxxxxxxxxxxxxxxxxxxxxxxxxxxxxxxxxx
xxxxxxxxxxxxxxxxxxxxxxxxx. Xxxxxxxxxxxxxxxxxxxxxxxxx
(C)—— xxxxxxxxxxxxxxxxxxxxxxxxxxxxxxxxxxxxxxxxx. Xxxxxxxxxxxxx
xxxxxxxxxxxxxxxxx.

    Xxxxxxxxxxxxxxxxxxxxxxxxxxxxxxxxxxxxxxxxxxxxxx
xxxxxxxxxxxxxxxxxxxxxxxxxxxxxxxxxx. Xxxxxxxxxxxxxxxxxxxxxx
xxxxxxxxxxxxxxxxxxxxxxxxxxxxxxxxxxxxxxxxxxxxxxxx
xxxxxxxxxxxxxx. Xxxxxxxxxxxxxxxxxxxxxxxxxxxxxxxxx
xxxxxxxxxxxxxxxxxxxxxxx. Xxxxxxxxxxxxxxxxxxxxxxxxx
xxxxxxxxxxxxxxxxxxxxxxxxxxxxxxxxxxxxxxxxxx.

                    Regards,

Letter writer's ———————————→ *Don Brooke* ◄——**(4)**
signature
and initials
                  for Lynda Westholme
                      Sales Manager
                           ↑
Typist's   DB:rt
initials ———   ↑
                      (D)

Notes for Modified Block Letter (Figure 3-9)

(A) If a file number is used, the date is offset to the right; if there is no file number, the date starts at the centerline.

(B) The subject line (if used) should be centered and underlined, or boldface.

(C) The first word of each paragraph may be indented about five spaces, or started flush with the left-hand margin.

(D) The signature block should start at the centerline.

Notes for Full Block Letter (Figure 3-8)

(E) Every line starts against the left margin (simple for typist to set up). The parts otherwise are the same as for the modified block letter.

Figure 3-9. Modified block letter format.

H L WINMAN AND ASSOCIATES

INTER-OFFICE MEMORANDUM

(1) To: ....Andy Rittman.................................   From: ....Rick Davis.....................

....   Subject: .....Early Mailing of Pay..... **(4)**

Date: ....December 3, 19xx...................   ....Checks for Field Staff..........

(2)
(3)
I need to know who you will have on field assignments during the week before Christmas so that I can mail their pay checks early. Please provide me with a list of names, plus their anticipated mailing addresses, by December 8.

Checks will be mailed on Tuesday, December 14. I suggest you inform your field staff of the proposed early mailing.

Rick **(5)**

Notes for Memorandum

(1) The informality of an interoffice memorandum means titles of individuals (such as Office Manager and Senior Project Engineer) may be omitted.

(2) No salutation or identification is necessary. The writer can jump straight into the subject.

(3) Paragraphs and sentences are developed properly. The informality of the memorandum is *not* an invitation to omit words so that sentences seem like extracts from telegrams.

(4) The subject line should offer the reader some information; a subject entry such as "Paychecks" would be insufficient.

(5) The writer's initials are sufficient to finish the memorandum (although some organizations repeat the name in type beneath the initials). Some people prefer to write their initials beside their name, on the "From" line, instead of signing at the foot of the memo.

Figure 3-10. Interoffice memorandum.

Mike was surprised; he thought his message was crystal clear. So he again sat at his terminal and wrote:

TO: C. BAIRD, CARLSON DISTRIBUTORS, NEW YORK. MESSAGE QUITE CLEAR: YOU SHORT-SHIPPED 10 TOOLSETS TYPE MKV AND 4 NO. 801 SOCKETS. YOU ALSO SHIPPED 2 DOZ MODEL 280A LATHE BITS WE DID NOT ORDER. PLEASE SHIP MISSING ITEMS AND ADVISE HOW YOU WANT THE BITS RETURNED. M. TOLLER.

TIME-SAVER LETTER

This form is designed to make it easy for you to reply. Simply note your answer in the space provided. Keep one copy for your files and return the other to us. Thank you.

MACRO ENGINEERING INC
600 Deepdale Drive
Phoenix, AZ
85007

REPLY

DATE   May 10, 19xx

Elmer Dubienski, Purchasing Manager
Morgan Electronics Wholesale Inc.
2160 Wavell Road
Atlanta, GA 30332

Good plan, Judy!

But please increase the air

express shipment to 8 units, to

meet our immediate commitments.

MESSAGE   DATE   May 2, 19xx

Re: Your P.O. 2837

Elmer:

    To reduce shipping costs, I plan to ship only the first 6 Microbar Environmental Control units by air express. This will be done on May 24.

Thanks,

Elmer.

    I'll then ship the remaining 24 units by rail on June 3. Is this acceptable?

Elmer Dubienski

Judy P.

Judy Paulson
Production Coordinator

®DAY-TIMERS RE-ORDER No. 23810·

SENDER - KEEP YELLOW COPY FOR YOUR FILE. MAIL WHITE AND PINK COPIES.

Figure 3-11. A quick-reply letter. When the form is folded correctly, the recipient's name will show through the window of a standard No. 10 window envelope. (Courtesy Day-Timers Inc.)

Christina replied in three words:

THANKS. WILL INVESTIGATE.

If Mike had been more explicit when he keyed in his original message, both he and Christina would have saved time.

RGI VIDEO PRODUCTIONS
Tel: (216) 488 7060
FAX: (216) 488 7294

FAX MESSAGE

TO: Microprocessor Center Inc

FAX NO.:   212 338 2191

ATTENTION: John Reeman, Ext 2207

DATE: Sept 18, 19xx   *TIME:*   14:35

No. of pages (including this sheet):   3

BRIEF MESSAGE:

Our quotation for videotaping, editing, and dubbing session 4.2.3. of the Atlanta Computer Conference onto 30 VHS videocassettes follows.

This confirms my verbal quotation of Sept 5.

Trish Kaufman

REPLY TO: RGI VIDEO PRODUCTIONS FAX: 216 488 7294

If all pages not received call 216 488 7060

Figure 3-12. A cover sheet for facsimile (fax) transmission.

In electronic mail, words must be abbreviated particularly carefully. Well-known abbreviations such as "No." and "approx" will be readily recognized. But locally coined abbreviations such as "ex" (which can mean "for example" or "coming from"), "w/o" (for "without"), and "rec" (for "received" or "recommended") should be avoided.

Assignments

Most of the letter- and memo-writing projects that follow include all the details you need to write the assignment. You are encouraged, however, to introduce additional factors if you feel they will increase the depth or scope of your letter.

PROJECT 3.1: RETURNING "USED" SOFTWARE

Three weeks ago (work out the date), you ordered Release 5.6 of *CalcuWrite*, a popular integrated database and word-processing program. (You have been using previous versions of this software in your work for four years and currently work with Release 4.2.) Today the software arrived, accompanied by invoice No. 10403 for $159.00 (the regular price for *CalcuWrite* is $695.00 but, because you were ordering an upgrade, your company has to pay only $159.00). Details of the purchase are:

- Your company is Harbison Design Group, 420 Fulham Avenue of your city.
- The software was ordered on your company's purchase order number HD2786.
- The software supplier is Metroware Inc, Suite 1216, 300 Medway Road, Seattle, WA.
- You specified on the order that the program was to be supplied on 3½ in. rather than 5¼ in. diskettes.

When you open the shipping carton, you discover that Metroware has sent you 5¼ in. diskettes. ("Oh, well," you murmur to yourself, "I can manage because my personal computer has both 3½ in. and 5¼ in. drives, plus a 40 Mbyte hard disk. I asked for the 3½ in. diskettes because the program will fit onto five of the smaller diskettes compared to nine of the larger ones.")

Then you examine the documentation that came with the program and notice that the handbook has obviously been read by someone else, because its corners are well thumbed and dog-eared. You also discover that the User Registration and Warranty Card has been filled out by someone else. (The person has written into the appropriate spaces: "Klugman, Peter F., 107–260 Oxnard Street, Portland, OR, Tel: [503] 452 6480.")

You reckon that the software was sold previously to Mr. Klugman, who subsequently returned it, and you are not happy about that. ("He could have messed up the diskettes," you mutter, "or put a virus in them.") So you pack up the software, ready to ship back to Metroware Inc, and insert a letter inside the carton in which you describe what has happened and ask for a replacement. From the literature in the box, you discern that Metroware's Customer Service Manager is Connie Vetterland.

Write the letter.

PROJECT 3.2: REVISING A LETTER

At 4:15 p.m., Norm Behouly comes to you with a problem. "I'll be on vacation tomorrow," he announces, "and I'll be away for three weeks. The trouble is, I've keyed two letters into the computer, and now the system has gone down and I can't get them out!"

Norm asks you to print them when you come in tomorrow morning and to mail them for him. He gives you two file names: SURVEY.TXT and FENCE.TXT. "You'll have to sign them for me," he adds, "and I would appreciate it if you would take the time to read them first, just in case there is a typographical error I have missed."

Now it is 9:15 on the following morning and the computer system is again operational. You key Norm's two letters onto your screen and immediately see that they need much more than just a cursory check for typographical errors.

Revise or correct each letter. Insert a full address for each recipient, including the name of your city and an appropriate zip code.

Part 1: File SURVEY.TXT

Dear Mr. Antony:

In response to your letter of June 7, 19xx, and our meeting at your residence at 960 Bidwell Street on June 14, when you showed me the plan of your Lot (Lot 271-06) and the position of the fence bordering the Lot to the south, at 964 Bidwell Street, which is Lot 271-07. You claimed there is a discrepancy between the city site plan and the physical position of the fence, and asked me to do a survey of your Lot so as to establish the correct position.

Your Lot was surveyed by me and an assistant on June 21 and while there I hammered in two markers to delineate the southeast and southwest corners. (No markers were placed on the north side because the position of that fence is not in question.) Your neighbors to the south—Mr. and Ms. Beamish—will not be happy when they find out that the fence between Lots 06 and 07 encroaches on your property. You will note from the positions of the markers that the east end of the fence is 14 inches inside your territory, but is angled toward the south so that at the west end, where it stops at the garage, it is correctly positioned.

It is assumed that you recognize that the south fence is yours, and the fence to the north is the responsibility of your neighbor to the north. Consequently you have the right to move the fence if you wish to or leave the fence where it now stands until repairs are necessary and then rebuild it in its correct position. As obviously you are aware, the fence is in good condition.

As per your request, I am writing to your neighbors today to inform them of the discrepancy and attaching our invoice.

Yours sincerely,

Part 2: File FENCE.TXT

Dear Mr. and Mrs. Beamish:

As you must have been aware, a survey of Lot 271-06 was done recently, on June 14, to determine the exact borders of the Lot at 960 Bidwell Street, to your north. While the survey was being done, markers were positioned at the southeast and southwest corners of the Lot, to establish the exact dividing line between your Lot and that of Mr. and Mrs. Antony at 960 Bidwell Street. No markers were placed at the northeast and northwest corners of the Antony's Lot.

Unfortunately the fence is incorrectly positioned between your Lot (No. 271-07) and Lot 271-06. At the southeast corner of Lot 271-06 the fence is 14 inches too far to the north and so encroaches onto your neighbor's Lot. (Actually, the fence slants toward your property as it progresses westward and at the garage end is properly positioned.)

I can only assume that you are unaware of this discrepancy, so at Mr. Antony's request I am writing to you so that you will know of the circumstances should Mr. Antony choose to reposition his fence. I am equally sure that you and the Antony's can come to an amiable agreement.

Please feel free to contact me at your convenience if you need more information concerning this matter.

I remain, yours truly,

Part 3: File GARAGE.DFT

Norms calls you from the airport: "I forgot to tell you," he says. "There's a third file—GARAGE.DFT. It's some notes about the garages on the Antonys' and the Beamishes' Lots, and I think the owners should know about them. Could you write to each of them for me? It shouldn't wait until I return."

From the notes in file GARAGE.DFT, you gather that:

1. The two garages are parallel to each other, and the space between the adjacent walls is only 17.5 inches.
2. There is a pile of lumber stacked between the garages to a height of 47 inches.
3. City bylaw 216, subparagraph 2(c) stipulates that garages must be a minimum of 24 inches apart.
4. City bylaw 216, subparagraph 2(h) requires that passageways between garages must be accessible, for fire safety reasons.

You feel the homeowners could ignore the separation discrepancy for the moment but should do something about the stacked lumber (the city inspectors may never notice the too-narrow distance between the garages, but they almost certainly will notice eventually that access between the garages is blocked, and this may lead them to measure the separation distance).

Write a letter to Mr. and Ms. Antony informing them of the problem. Tell them you are sending an identical letter to the Beamishes next door.

PROJECT 3.3: CORRECTING A BILLING ERROR

Today you receive a credit card statement from WorldCard, covering last month's purchases. There are eight debit entries, three personal and five for expenses incurred during a business trip you made to Wapiti Paper Mill between the 8th and 12th. (You are an engineering technician employed by the local branch of

H. L. Winman and Associates, and you went to the mill to investigate and rectify a problem in the process control system.) The five business expenses are:

Item	Date	Vendor/Location	Control No.	$
3	08	River Motel, Burntwood Lake	0134652	52.80
4	09	Burntwood Auto Service	0147162	272.41
5	10	Wapiti Autos	0203916	34.17
6	12	Wapiti Inn	0205771	256.50
7	12	Burntwood Auto Service	0211606	28.10

Item 4 puzzles you. You know you purchased gasoline three times and stayed one night on the road in a motel and three nights at another motel near the mill. But you could not have bought $272 of gasoline (your car's tank would not hold that much!).

Fortunately, you always keep a travel log, and in it you recorded these entries:

9th:	19.05 gal	@	$1.43/gal
10th:	24.58 gal	@	$1.39/gal
12th:	19.65 gal	@	$1.43/gal

You do not have the credit card vouchers because you attached them to the expense account you handed in to Branch Manager Vern Rogers on the 19th, and he has sent them on to the head office in Cleveland. But from your records you can work out what the error is and can guess that it occurred during data entry at WorldCard's Data Center in New York.

Write to the manager of customer accounts at the credit card company, inform him or her of the error, and ask for an adjustment. WorldCard's address is: Suite 2160, 441 Ninth Avenue, New York, NY 10001.

PROJECT 3.4: LETTER OF THANKS

Last night you attended a talk delivered by Ms. Tina Mactiere to the local chapter of the Inter-State Engineering Association (ISEA). Today you have to write a letter of thanks to Ms. Mactiere, expressing your and the ISEA chapter's appreciation. (You are the chapter's Technical Program Coordinator, and you arranged for Ms. Mactiere to give the talk.) Some details you may need are:

1. You are employed by Haarstrup Consultants Inc at 212 Broad Avenue of your city, where your company president Gavin Haarstrup encourages his technical staff to participate in ISEA activities.
2. Tina Mactiere is President and Chief Executive Officer of Macro Engineering Inc (see Chapter 2).

3. Her talk was given in the Palliser Room of the Chelmsford Hotel. The event was the Annual General Meeting (AGM) of the local ISEA Chapter. The program included a formal dinner at 6:30 p.m., Tina Mactiere's address at 8:15 p.m., and the AGM at 9:15 p.m. The affair concluded at 10:15 p.m.

4. Tina's talk was titled: "Look After the P's and Q's." Her main thrust was that technical people are so concerned about keeping abreast of new technology that they omit other aspects which are an essential part of their professional development. She cited, for example, the need for scientists, engineers, and technicians to attend courses or seminars in supervisory management, interpersonal relations, and oral and written communication—topics she referred to as "people skills."

5. Tina proved to be a dynamic speaker. She used slides and a three-minute humorous videotape that neatly underscored the points she was making.

6. There were numerous questions from the audience after her talk, and a strong round of applause.

7. Many people came up to you after the AGM and congratulated you on your choice of speaker and remarked on the appropriateness of her topic.

8. Seventy-six ISEA members attended the dinner and meeting.

PROJECT 3.5: REQUEST FOR REPLACEMENT PARTS

You are an engineering assistant employed by H. L. Winman and Associates, and you are engaged in a lake-level measurement program for the state of Wisconsin. At a critical moment in the program, your Hektik Model 370 Water Stage Manometer breaks down. You take it apart and identify that it needs a replacement spring and drive assembly. This is the third time that the fault has occurred in the past six months, and each time the thread on the drive shaft has stripped. You previously purchased spare spring and drive assemblies from the manufacturer: Hektik Company, 44 Homer Street, in Pennstown, Nebraska. The dates were May 10 and August 23, and each time the cost for the assembly was $218.50.

Today you send a fax ordering a replacement spring and drive assembly against purchase order No. 26019.

Write to the Hektik company to complain about the repeated failures (you may attribute the cause to any condition you wish, if you feel you need to point out the cause) and to request that this time the replacement assembly be supplied free of charge.

PROJECT 3.6: RESPONDING TO AN ANGRY CUSTOMER

You are an engineering techician employed by a manufacturer of technical equipment at 435 Manor Road of your city. (You may name the company yourself or assume it is the local branch of H. L. Winman and Associates.) You are the supervisor of the Service and Repair Department, and three service technicians report to you. Some of the service work is done in your plant, but more is done at customers' plants since most of the products your company manufactures are heavy industrial equipment.

Today you receive an exceedingly angry letter from Mr. Frank Rudge, owner-manager of Rudge Machine Shop at 1340 Railton Avenue of your city. In his letter, which is dated two days ago—Mr. Rudge claims that the Model 440 Neostat one of your technicians repaired in Mr. Rudge's plant three months ago does not work properly. He says the machine was not used for the first two months after it had been repaired, and now that his employees have started using it the Neostat is tearing the work instead of making a straight, clean cut.

You drive to Rudge Machine Shop to examine the Neostat. Immediately you notice that the cutting bars are out of alignment and the cutting edges are nicked and scarred. This seems to indicate that the machine has been used to cut hard metals rather than the plastics, fiberglass, and similar nonferrous materials for which it was designed. Worse still, you notice that the seal your technician placed over the lock of the box containing the relay mechanism has been broken, which means that one of Mr. Rudge's employees has been fiddling around inside, adjusting the relays. You talk briefly with the area supervisor—Daniel Fournier— but he claims no knowledge of the problem, except that ". . . the machine doesn't work properly." You ask to see the technician's repair report of three months ago, but the supervisor cannot find it.

Mr. Rudge is not in the plant during your visit, which is fortunate because you want to check the carbon copy of your technician's repair report before you reply to Mr. Rudge's complaint.

Back at the office, you dig out the report (No. B4152) and note that Wally Devries, who serviced the equipment, wrote on it:

1. Adjusted relay cut-in point, took up slack in linkage, and tested operation of cutting bars. Clearance: 0.015 in. (within specs). All satisfactory.
2. Oiled/greased all bearings and checked cutting edges: no pits or breaks.
3. Sealed relay control box (seal No. B21).

The foot of the repair report is signed "Daniel Fournier," who acknowledged completion of the repair job.

Normally your company's repair warranty for industrial equipment is six months. But the repair report contains several printed conditions affecting the warranty. Clause 3 says:

3. Breakage of seals over operating components automatically voids this warranty.

Now write a reply to Mr. Rudge's letter of complaint.

PROJECT 3.7: REQUEST TO ATTEND A COURSE

Assume that today is the second Monday of the current month, and that for the past four weeks you have been on a field assignment to San Antonio, Texas, where you have been conducting a modification program for Inter-State Telephones (IST). You have been assisted by two competent technicians (Ted McCourt and Lauren Freedman), and you are now four days ahead of schedule. The task is to be completed by the 3rd of next month.

Today you receive a folder from the University of Texas at Austin advertising a one-week course. Details are:

Course title:	Managing in a Technological Environment
Course dates:	Monday the 20th to Friday the 24th inclusive (of this month)
Type of course:	Maximum immersion: 8 a.m. to 5 p.m. daily plus 7 to 10 p.m. Wednesday evening; approximately 20 hours of home assignments
Cost:	$495; includes materials, books, and lunches, with a guest speaker from industry at each lunch
Registration:	No later than noon Wednesday the 15th; telephone registrations accepted
Participants:	Limited to 16

You are impressed by the technical standard of the course described in the folder and wish to attend. (Because of previous field assignments you missed a similar Extension Department evening course offered at your local university last winter. Your company sponsored four engineers to attend that course, for which the fee was $165 each.)

Write an interoffice memorandum to your department head, Dennis Collonni, in which you:

1. Describe the course (convince him it is a good one).
2. Ask if you can attend.
3. Ask if the company will pay the tuition fee, plus travel and lodging costs.
4. Ask to be spared from the IST task for one week (be convincing).
5. Ask for a quick reply (because time is short).

Assume that your department head can give technical approval for you to attend but must go to the department manager for financial approval. Also assume that you have a rental car for the IST project, which you can use to drive to Austin, and that the hotel in Austin will cost $95 per night.

PROJECT 3.8: A FAULTY HOME ENTERTAINMENT CENTER

Assume that you were recently in Waverly (1100 miles from home), where you visited friends Martin and Joan Tong. Martin gave Joan a PAM 98 "Home Entertainment Center" last Christmas, and you are impressed by its tone, appearance, and features. Martin tells you privately that he bought it from Craven's Discount Center at 1837 Kelly Street in Waverly, and offers to go with you if you are interested in buying one.

You are, but you are disappointed to discover that Craven's has sold all its PAM 98 entertainment centers, that no more are on order, and that no one else in Waverly carries them. However, Harry Craven, the store owner, has a suggestion: there is a demonstrator which he could sell to you at 5% off his regular discount price. You test it, and it seems okay. Martin suggests a 15% price reduction would be more normal for a demonstrator, but Mr. Craven won't budge; he adds, however, that he'll give it a good check-over if you'll leave it with him. You agree. Two days later you pick it up, pay $460.25 for it, and receive Craven's invoice No. 5603 stamped "Paid."

But when you arrive home, you find that the CD player of the PAM 98 does not work. You also discover that there is no local service center for the PAM line, so you take the entertainment center to Modern TV and Radio at 280 Waltham Avenue. When you pick it up the following day, store manager Jim Williams hands you a circuit board with several bent and twisted pins.

"There's your problem," Jim says. "Craven's in Waverly must have replaced this board—you can tell it's one of theirs because the name CRAVEN is stamped on it." He explains that whoever inserted the board did not align the pins properly and bent them by forcing it into its socket.

You pay $83.50 for the repair job on Modern TV and Radio's invoice No. 1796, and take both the PAM 98 and the ruined circuit board with you.

Write to Harry Craven, tell him what has happened, and ask for a refund of $ . . . (you decide how much). You may assume you attach copies of the two invoices to your letter.

Note: The PAM 98 is made by VICOM in Korea. It contains an AM/FM stereo radio receiver, a CD player, a dual-cassette tapedeck, and two eight-inch speakers.

PROJECT 3.9: ACKNOWLEDGING A COLLEGE AWARD

Assume that you are in the second year of the course you are enrolled in, and that three weeks ago the head of the department came to you and announced that you have been selected to be this year's winner of the Inter-State Engineering Association (ISEA) scholarship for "proficiency in computer studies." Yesterday you attended an awards luncheon attended by other scholarship winners and representatives of the firms donating the scholarships. You sat next to Calvin Wycks, vice-president of the local chapter of ISEA, who presented the award to you.

Today, you write to ISEA to thank the Association for the award. Use these details:

1. Address your letter to Marjorie McIvor, ISEA's local president.
2. ISEA's address is 1830 Mapleton Crescent of your city.
3. The award is a wall plaque inscribed with your name and a check for $500.

CHAPTER 4
Short Informal Reports

When you hear that someone has just finished writing a technical report, you may imagine a nicely bound formal document, tastefully typed and printed. In some cases you would be correct, but most of the time you would be wrong. Far fewer formal reports are issued than informal reports, which reach their readers as letters and memorandums. This chapter describes the short informal reports you are likely to write as a technologist, engineer, or engineering technician.

APPEARANCE

The memorandum report is the least formal technical report. Normally an interoffice or interdepartmental communication, its length and tone can vary considerably. It can be very direct, it can develop its topic in great detail to present a convincing case, or it can lie anywhere in between.

The letter report, although still basically informal, can vary in formality according to its purpose, the type of reader, and the subject being discussed. Some letter reports may be as informal as a memorandum report, particularly if they are conveying information between organizations whose members know each other well or have corresponded frequently. Others may be more formal, presented as business letters conveying technical information from one company to another.

Although there are many types of informal reports, and there are variations in format, content, and writing style, all are based on the writer's pyramid shown in Figure 4-1. Each contains (1) a brief statement describing what the reader most needs to know; (2) a short introduction to the topic or problem;

SUMMARY	A brief statement of the report's main features (often written last, but always placed *first*).
INTRODUCTION	**Background:** information that "sets the scene."
DISCUSSION	**Facts:** data, details; what has been and is being done.
CONCLUSIONS	**Outcome:** results and effects; it may also suggest what needs to be done.

Figure 4-1. Basic report compartments.

(3) a discussion of the data, situation, or problem, and what has been done or could be done about it; and (4) a conclusion which sums up the results and possibly recommends what should be done next.

WRITING STYLE

The reports described in this chapter are written in a direct, informative style that is crisp and to the point. Their writers are usually describing events that have already occurred, so they write mostly in the past tense, which helps them to be consistent. They shift gear into the present or future tense only when they have to describe something which is presently occurring, outline what will happen in the future, or suggest what needs to be done. All three tenses occur in the following report from Engineering Technician Dan Skinner to Coordinator Don Gibbon:

SUMMARY The indoor/outdoor carpet we installed in station DMON-TV has corrected the noise problem but is "pilling" badly. I plan to examine the carpet with the manufacturer's representative to find the cause and suggest a remedy.

BACK- The carpet was installed in the satellite studio control room during the
GROUND night of January 8–9. Sound level readings taken at 12 different
        locations on January 10 showed that the ambient noise level had
Past Tense decreased an average of 3.6 dB. Operating staff also noticed a marked decrease in clatter caused by impact noises.

FACTS
Mainly Past Tense

At the station manager's request, I returned to the control room today and checked the carpet's condition. After only two weeks of use it has tight little balls of carpet material adhering to its surface. I called the manufacturer's representative, who said that the condition is not unusual and does not mean that the carpet is wearing quickly. He suggested that it may be caused by improper carpet-cleaning techniques and probably can be easily corrected. However, it is unsightly and our client is

Present Tense

not pleased with the carpet's appearance.

OUTCOME

Future Tense

The manufacturer's representative and I will return to the control room between midnight and 2:00 a.m. on January 31 to study the carpet-cleaning techniques used by maintenance staff. I will telephone our findings to you later in the day.

Dan has written the background and most of the facts paragraph mainly in the past tense because they deal with what has already been done. At the end of the facts he has shifted into the present tense to report how the station manager feels *now*. Then for the outcome he has jumped into the future tense to outline what he *plans* to do. This past-present-future arrangement is natural and logical; reader Don Gibbon will feel comfortable making the transitions from one tense to the next. Dan's summary even follows the same pattern.

INCIDENT REPORT

Anna King is alone in the H. L. Winman office, working late one Tuesday evening, when the telephone rings. The caller is Bob Walton, a member of the electrical engineering staff who is on a field trip to Tangwell with a second staff member. He tells Anna they have had an automobile accident near Hadashville, and some of their equipment is damaged. He wants Jim Perchanski, his department head, to send out replacement items by air express.

Anna jots down notes while Bob talks. As she plans to be out of town the following day, she types out a report of the conversation and leaves it on Jim Perchanski's desk. She prepares it as a memorandum (Figure 4-2) which tells Jim Perchanski what has happened to two members of his staff, where they are now, how soon they will be able to move on, and that one of them is injured. It also tells him that equipment is damaged and replacements are needed.

Because she will not be available to answer questions the following morning, Anna takes care to describe the situation clearly (the paragraph numbers below are keyed to the memorandum):

(1) Anna knows that a subject line must be informative; it must tell what the memorandum is about and stress its importance to the reader. If she had simply written "Transcript of Telephone Call from R. Walton," she would not have captured Jim Perchanski's attention nearly as sharply.

H L WINMAN AND ASSOCIATES

INTER-OFFICE MEMORANDUM

To: .... Jim Perchanski .... From: .... Anna King ....

.... Subject: .... Accident Report and ....

Date: .... September 18, 19xx .... Request for Spare Parts ....

(1)

Bob Walton and Pete Crandell have been involved in a highway accident which will delay their inspection of the Sledgers Control project at Tangwell. They need replacement parts shipped to them tomorrow (Wednesday, September 19).

(2)

Bob telephoned from Hadashville at 7:35 p.m. to report the accident, which occurred at 5:15 p.m. some five miles north of Hadashville. Pete has been hospitalized with a fractured left knee and a suspected concussion. Bob was unhurt. The panel van and some of their equipment were damaged.

(3)

Bob wants you to ship the following items to him by air express on Remick Airlines Wednesday evening flight to Montrose and to mark the shipment HOLD FOR PICK UP BY R WALTON SEP 20:

o   1 Spectrum analyzer, HK7741
o   1 Calibrator, Vancourt model 23R
o   24 Glass phials, 300 mm long x 50 mm dia.

(4)

He has arranged to rent a van and will drive to Montrose to pick up the items Thursday morning. He will then drive on to Tangwell and will arrive there about 4:00 p.m. He has informed site RJ-17 at Tangwell of the delay.

(5)

Bob is preparing an accident report for you. If you want to call him, he is staying at Hunter's Motel in Hadashville (Tel: 614 453 6671).

(6)

Figure 4-2. A third person incident report.

(2) This brief summary gets right to the point by immediately telling Jim Perchanski in general terms what he most needs to know:

> Why the memo was written.
> What happened.
> What action has been taken.
> What action he has to take.

(3) In this paragraph Anna tells what she knows about the accident and its effects. It serves as background to the important facts that follow.

(4) Anna knows that Jim Perchanski must act quickly to ship the replacement items, so she uses a list to help him identify them. The indented list also is an attention-getter.

(5) Instructions and movement details should be explicit, otherwise the equipment and Bob Walton may not meet at Montrose. Anna has taken care throughout to identify specific days, and once even the date, to make sure that no misunderstandings occur. To state "tomorrow" or "the day after tomorrow" would be simple but might cause Jim Perchanski to assume a wrong date, since he will be reading the memorandum one day later than it was written.

(6) In this brief closing paragraph Anna indicates what further action is being taken and where Bob Walton can be contacted if more information is needed or if a message needs to be relayed to him.

When Jim Perchanski walks in on Wednesday morning, he will know immediately what has happened and what action he has to take. He does not need to ask questions because he has been placed fully in the picture.

Anna's memo is an incident report, written pyramid style, in which

- the **summary statement** is in the paragraph identified as (2),
- the **background** is at the start of paragraph (3),
- the **facts** are in the remainder of paragraph (3) and all of paragraph (4), and
- the **outcome** is in paragraphs (5) and (6).

Bob Walton also used the report writer's pyramid when he subsequently wrote to his supervisor, from Hunter's Motel in Hadashville, to describe the accident and its effect. His report is shown in Figure 4-3. Its focus and emphasis differ from those in Anna's earlier report, but it still is an incident report with the following parts:

(A) This is his **summary**: it takes a main piece of information from each compartment that follows.

(B) This is the **background**. By clearly describing the situation (*who? where? why? when?*) Bob is helping Jim more easily understand what happened. Notice how he:

- Establishes where they were, how they happened to be there, what direction they were travelling, and who else was involved.
- Itemizes vehicles, license numbers, and drivers' names in an easy-to-read list.
- Mentions that he is enclosing a sketch (so Jim can look at it *before* he reads on).

H L WINMAN AND ASSOCIATES

INTER-OFFICE MEMORANDUM

To: .... Jim Perchanski .... From: .... Bob Walton ....

.... Subject: .... Report of Auto Accident ....

Date: .... September 19, 19xx .... at Hadashville, Ohio ....

Pete Crandell and I were involved in a multiple-vehicle accident on November 18, which resulted in injuries to Pete, damage to our panel van and some equipment, and a two-day delay in our inspection of the Sledgers Control project. Ⓐ

The accident occurred at 5:15 p.m. on Highway 44, about 2 miles northeast of Hadashville. We were traveling north in company panel van T2711, on our way to site RJ-17 at Tangwell. Pete was driving and we were approaching the intersection with Highway 201.

Other vehicles involved in the accident were: Ⓑ

o   Toyota Corolla, license AHJ 735, driven by D Varlick.
o   Ford truck, license T4851, driven by F Zabetts.
o   Pontiac Grand Am, license MD3720, driven by K Schmitt.

Positions of the vehicles and our panel van immediately before the accident are shown on the attached sketch.

As the Toyota attempted a right turn into Highway 459 it skidded into the Ford truck, which was standing at the intersection waiting to enter the highway. The impact caused the Toyota's rear end to swing into our lane, where Pete could not prevent our van from colliding with it. This in turn caused the van to slide broadside into the southbound lane, where the Pontiac approaching from the opposite direction collided with its left side. Ⓒ

Pete was taken to Hadashville hospital with a broken left knee and a suspected concussion; he is likely to be there for several days. The panel van was extensively damaged and was towed to Art's Autobody, 1330 Kirby Street, Hadashville. Some of our equipment was damaged or shaken out of calibration, so I telephoned the office on Tuesday evening and asked Anna King to prepare a list of replacements for you. Ⓓ

I have rented a replacement van from Budget. I have also telephoned the duty engineer at site RJ-17 to tell him that my inspection of the Sledgers Control project will start on Friday, November 21, two days later than planned.

Bob

Figure 4-3. A first person incident report.

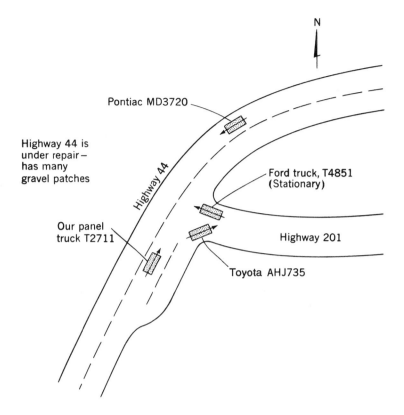

(C) Because his background information is complete, Bob's **facts** can be concise. He simply provides a chronological description of what happened from the time the Dodge started to slide until all vehicles stopped moving.

(D) In the **outcome** Bob describes the results of the accident (injuries, damage) and what he has done since (rented a van, requested replacement equipment). He closes on a strong point: What is being done about the project, which was his reason for passing through Hadashville.

Bob knows his role is to be an informative but objective (unbiased) reporter. No doubt he has an opinion of who is at fault, but to state it probably would have injected subjectivity into his report.

FIELD TRIP REPORT

After returning from a field assignment, you will be expected to write a field trip report describing what you have done. You may have been absent only a few hours, inspecting cracks in a local water reservoir; or you may have been away for two months, overhauling communications equipment at a remote defense site. Regardless of the length and complexity of the assignment, you will have to remember and transcribe many details into a logical, coherent, and factual report. To help you, carry a pocket notebook in which to jot down daily occurrences. Without such a record to rely on, you may write a disorganized report that omits many details and emphasizes the wrong parts of the project.

The simplest way to write a trip report is to answer three basic questions:

1. **Who went where?**
   Why and when did you go?
2. **What did you do?**
   What did you set out to do? Were you able to do all of it? What problems did you run into? What additional work did you do?
3. **What remains to be done?**
   What work could you not complete, and who should now complete it? When and how should it be done?

The answers to these questions provide the information for the three main writing compartments in Figure 4-1:

INTRODUCTION: Who went where? **(background)**
DISCUSSION: What did you do? **(facts)**
CONCLUSIONS: What remains to be done? **(outcome)**

Short trip reports do not need headings. A brief narrative following the **summary-background-facts-outcome** pattern carries the story:

Summary: A prototype automatic alarm has been installed at site RJ-17 for a one-month evaluation by the Roper Corporation.

Background: Dave Makepiece and I visited the site from January 15 to 17.

Facts: We completed the installation without difficulty, following installation instruction W27 throughout, and encountered no major problems. However, we omitted step 33, which called for connections to the remote control panel, because the panel has been permanently disconnected.

Outcome: The alarm will be removed by M. Tutanne on February 26, when he visits the site to discuss summer survey plans.

In practice, the very short **background** probably would be combined with either the **summary** or the **facts** to form a single paragraph.

Long trip reports require headings to help their readers identify the compartments. Typical headings might be:

- **Summary**.
- **Assignment Details** (Background).
- **Work Accomplished** and **Problems Encountered** (Facts; best treated as two separate headings).
- **Suggested Follow-up** or **Follow-up Action Required** (Outcome).

Anna King's memo shown in Figure 4-4 tells H. L. Winman and Associates' engineers how to organize their longer trip reports, describes the type of information that normally would follow each heading, and includes excerpts and sample paragraphs.

With the exception of the outcome section, trip reports should be written entirely in the past tense.

OCCASIONAL PROGRESS REPORT

Progress reports keep management aware of what its project groups are doing. Even for a short-term project, management wants to hear how the project is progressing, especially if problems are affecting its schedule. Because delays can have a marked effect on costs, management needs to know about them early.

Jack Binscarth, one of Macro Engineering Inc's senior technicians, has been assigned to Cantor Petroleums at a semiremote site in Texas, where he has to analyze oil samples. The job is expected to take five weeks, but problems develop that prevent Jack from completing the work on time. To let his chief know what is happening, he writes the brief progress report in Figure 4-5, adapting the standard background-facts-outcome arrangement into a past-present-future pattern:

(1)	**Summary**	A brief description of the overall situation.
(2)	**Progress**	The work that has been done, the problems that have been encountered, and the effect these problems have had on progress.
(3)	**Situation Now**	What is being done at present.
(4)	**Future Plans**	What will be done to complete the project, and when it will be done.

PERIODIC PROGRESS REPORT

If a project is to continue for several months, management normally will specify that progress reports be submitted at regular intervals.

A periodic progress report may be no more than a one-paragraph statement describing the progress of a simple design task, or it may be a multipage document covering many facets of a large construction project. (There are also

H L WINMAN AND ASSOCIATES

INTER-OFFICE MEMORANDUM

To: ..... Systems, Project, and ....   From: ..... Anna King ....

..... Field Engineers ....   Subject: ..... Guidelines for Writing .....

Date: ..... February 20, 19xx .... Field Trip Reports .....

A standard format is to be used for all field trip reports. These notes suggest how you can organize your information under five main headings: Summary, Assignment Details, Work Accomplished, Problems, and Follow-up Action. The headings may be omitted from very short reports.

Summary

Make your summary a short opening statement that says what was and was not accomplished, and highlights any significant outcome.

Assignment Details

Under this main heading state the purpose of the trip and include any other information the reader may want to know. If the information is too cumbersome to write as a single paragraph, use subheadings such as

o   Purpose of Trip   o   Personnel Involved
o   Background   o   Person(s) Contacted
o   Project No./Authority   o   Date(s) of Field Trip

Work Accomplished

In this section describe the work you did. Normally present it in chronological order unless more than one project is involved, in which case cover each project separately. Keep it short: don't describe at great length routine work that ran smoothly. Whenever possible refer to your work instruction or specification, and attach a copy to your report:

> The manual control was disconnected as described in steps 6 to 13 of modification instruction MI1403, enclosed as attachment 1.

Go into more detail only if you encountered difficulty, or if work was necessary beyond that anticipated by the job specification:

> At the request of the site maintenance staff, a manual control was installed in the power house as a temporary replacement for a defective GG20 control. Parts removed from the panel, together with instructions for returning the panel to its original configuration, were left with Frank Mason, the senior power house engineer.

1

Figure 4-4. How to write a long trip report.

If parts of the assignment could not be completed, identify them and explain why the work was not done:

> Test No. 46 was omitted because the RamSort equipment was being overhauled.

Problems

In contrast to the conciseness recommended for the previous section, describe problems in detail. Knowledge of problems you encountered and how you overcame them can be invaluable to the engineering or operating departments, which may be able to prevent similar problems elsewhere.

A statement that does not tell the reader what the problem was or how it was overcome is virtually useless. For example:

> Considerable time was spent in trying to mount the miniature control panel. Only by fabricating extra parts were we able to complete step 17.

If this information is to be used by the engineering or operating department, it must be more specific:

> Considerable time was spent in trying to mount the miniature control panel according to the instructions in step 17. We found that the main frame had additional equipment mounted on it, which prevented us from using most of the parts supplied. To overcome this, we fabricated a small sheetmetal extension to the main frame and mounted it with the miniature panel, as shown in attachment 2.

Follow-up Action

This section of the trip report ties up any loose ends. If any work has not been completed, draw attention to it here even though you may already have mentioned it under "Work Accomplished." Identify what needs to be done, if possible indicate how and when it should be done, and say whose responsibility it now becomes:

> The manual control mounted as a temporary replacement in the power house is to be removed when a new GG20 control panel is received. This will be done by F Mason, and we left him with instructions for doing the work.

In some cases you may direct follow-up action to someone else in your own or another department:

> The manual control panel will be removed from the power house by R Walton, who will visit the site on May 12.

If your report is very long, I suggest you insert subheadings and use a paragraph numbering system to increase its readability.

Anna King

2

Figure 4-4. *(Continued)*

MACRO ENGINEERING INC

FROM: Jack Binscarth   DATE: October 14, 19xx
        (at Cantor Petroleums)

TO: Fred Stokes   SUBJECT: Delay in Analysis of Oil
        Chief Engineer   Samples
        Head Office

My analysis of oil samples for Cantor Petroleums has been delayed by ① problems at the refinery. I now expect to complete the project on October 25, nine days later than planned.

The first problem occurred on September 23, when a strike of refinery personnel set the project back four working days. I had hoped to recover all of this lost time by working a partial overtime schedule, ② but failure of the refinery's spectrophotometer on October 13 again stopped my work. To date, I have analysed 111 samples and have 21 more to do.

The spectrophotometer is being repaired by the manufacturer, who has promised to return it to the refinery on October 19. Today I informed ③ the refinery manager of the delay, and he has agreed to an increase in the project price to offset the additional time. He will call you about this.

Providing there are no further delays I will analyze the remaining samples between October 20 and 24, and then submit my report to the ④ client the following morning. This means I should be back in the office on October 26.

Jack

Figure 4-5. An occasional progress report.

form-type progress reports, which call for simple entries of quantities consumed, yards of concrete poured, and so on, with cryptic comments.) Regardless of its size, the report should answer four main questions that the reader is likely to ask:

> Will your project be completed on schedule?
> What progress have you made?
> Have you had any problems?
> What are your plans/expectations?

To answer these questions, a periodic progress report can readily use the standard summary-background-facts-outcome arrangement:

Summary	A brief overview of the project schedule, progress made, and plans (*answers the first question*).
Background	The situation at the start of the report period.
Facts	Progress made (*answers the second question*) and problems encountered (*answers the third question*).
Outcome	Plans/expectations for the next period (*answers the last question*).

Figure 4-6 shows how survey crew chief Pat Fraser used these four compartments to write an effective progress report (the numbers in parentheses are keyed to parts of the report):

(1) The **summary** tells Civil Engineering Coordinator Karen Woodford how closely the survey project is adhering to schedule, and predicts future progress. This is the information she wants to read first.

(2) The **background** section reminds Karen of the situation at the end of the previous reporting period and predicts what Pat expected to accomplish during this period. Background should always be stated briefly.

(3) The **facts** (or **discussion**) section is broken into two parts:

> Work done during the period (3A)
>
> Problems affecting the project (3B)

Pat Fraser opens each paragraph of this compartment with a topic sentence (a summary statement) which states the main point of the paragraph in general terms:

- *Dry, clear weather . . . enabled us to progress faster than anticipated.*
- *The electrical fault in the EDM equipment . . . recurred on May 23.*
- *I have had difficulty hiring reliable people to clear brush along the route.*

Pat then describes what happened in detail, using *facts* (exact dates and milepost numbers, for example) to support each topic sentence. To prevent the report from becoming too long, Pat attaches the survey results to it and simply refers to them in the narrative. (Because of their length, they have not been printed with Figure 4-6.)

(4) In the **outcome** paragraph Pat tells Karen what the crew expects to accomplish during the forthcoming period, and even suggests when they may eventually get back on schedule. This final statement clearly supports the opening paragraph, and so brings the report to a logical close.

Other factors you should consider when writing periodic reports are:

- If a progress report is long, use headings such as these to help readers *see* your organization:

> **Adherence to Schedule** (This is your summary.)
>
> **Progress During Period** (These are your facts; background information should go at the front of the progress section.)
>
> **Problems Encountered**
>
> **Projection for Next Period** (This is the outcome.)

H L WINMAN AND ASSOCIATES

INTER-OFFICE MEMORANDUM

To: .... Karen Woodford, Coordinator .....   From: .... Pat Fraser, Survey Crew Chief

      Civil Engineering Department .....   Subject: .... Progress Report No. 4 --

Date: .... May 31, 19xx ....   Allardyce Survey Project

The Allardyce Route survey has progressed well during the May 16 to 31 period. The survey crew has regained two days, and now is only four days behind schedule. We expect to be back on schedule by June 30.   ①

Project plan AR-51 shows we should have surveyed mileposts 30 to 34 during this period. But, as stated in my May 15 report, we were six days behind schedule at the end of the previous period, having surveyed only as far as milepost 28. Consequently, we expected to survey only to milepost 32 by May 31.   ②

Dry, clear weather from May 18 to 23 enabled us to progress faster than anticipated. We reached milepost 31 on May 23, carried out a terrain analysis for the Catherine Lake diversion scheme on May 24 and 25, resumed surveying on May 26, and reached milepost 32 at 9:00 a.m. on May 29, two days earlier than expected. When we stopped work at 6:00 p.m. on May 31, we were just 300 yards short of milepost 33. Survey results are attached.   ③A

Two problems affected the project during this period:

1. The electrical fault in the EDM equipment, which delayed us several times early in the project, recurred on May 23. I had the unit repaired at Fort Wilson on May 24 and 25, while we conducted the terrain analysis, and it has since worked satisfactorily.

2. I have had difficulty hiring reliable local people to clear brush along the route. Most remain with us for only a few days and then quit, and I have had to waste time hiring replacements. This problem will continue until mid-June, when the college students we interviewed in March will join the crew.   ③B

We plan to advance to milepost 37 by June 15, which should place us only two days behind schedule. If we can maintain the same pace, I hope to make up the remaining two days during the June 16 to 30 period.   ④

P. Fraser

Figure 4-6. A periodic progress report.

- For lengthy progress or problems sections, start with a summarizing statement describing general progress; then write several subparagraphs each giving details of a particular aspect of the project. For example:

  4. Interior construction work progressed rapidly but exterior work was hampered by heavy rain:
     4.1 In the east wing, all partitions were erected, 80% of the floor tiles were laid, and 20% of the light fixtures were installed.
     4.2 In the west wing, all remaining floor tiles were laid, all light fixtures were installed, and 16 of the 24 benches were bolted down; 8 of the benches were also connected to the water supply and drains.
     4.3 Landscaping started on September 16 but had to be abandoned from September 18 to 23, when heavy rains turned the soil into a quagmire. By the end of the month only the outer areas of the parking lot had been completed.

- Be as brief as possible, particularly for routine work. Use specific information in the report narrative (*facts* rather than generalizations), and place lengthy details in an attachment. If, for example, you are reporting an extensive analysis, in your progress section you might write:

  We analyzed 142 samples, 88 (62%) of which met specifications. Results of our analyses are shown in attachment 1.

  Attachment 1 would contain several pages of tabular data (numbers, quantities, measurements) which, if included as part of the report narrative, would inhibit reading continuity.

- Describe problems, difficulties, and unusual circumstances in depth. State clearly what the problem was, how it affected your project, what measures you took to overcome it, and whether the remedial measures were successful. For example:

  Summary Statement — Juvenile vandalism has proved to be a petty but time-consuming problem. On September 3 (Labor Day) youths scaled the fence around the materials compound and stole about $170 worth of building supplies. On September 16 they started a front end loader, drove it into the excavation, then got it stuck in the mud and burned out the clutch while attempting to move it. To prevent a recurrence, from September 18 I have doubled the night watch and have had the site policed by a patrol dog. There have been no further attempts at vandalism.

  Facts

  Outcome

  Remember that management wants to hear about problems and how they were overcome. Such knowledge can be used to avert difficulties on future projects or can indicate project trends. A series of problems encountered during a design project may indicate that insufficient time has been allowed for it, that more engineering skills are needed, or that funds budgeted for the project are inadequate. While the progress section shows how much work has been done, the problems section often acts as a warning that progress may slow down unless immediate steps are taken.

- Forewarn management of any situation which, although it may not yet affect your project, may become a future problem. With such knowledge, management may be able to help you prevent a costly work stoppage or equipment breakdown. Here is a typical situation:

> 7.1 Unless the strike at Vulcan Steel Works ends shortly, it will soon curtail our construction program. Our present supply of reinforcing barmats will last until mid-October, when an alternative source of supply must be found. I have researched other suppliers but have been warned by union representatives that any attempt to obtain steel elsewhere may result in a walkout at other plants.

Where should such an entry appear in your progress report? The best position would be at the end of the facts (problems) section, immediately before the outcome.

- Number your paragraphs and subparagraphs if your report is lengthy or comprehensive (see the preceding examples); this helps you to refer to a specific part of a previous report, like this:

> The possibility of a shortage of steel mentioned in para 7.1 of my September report was averted when the strike at Vulcan Steel Works ended on October 6.

- Maintain continuity between reports. If you introduce a problem that has not been resolved in one report, then you must refer to it in your next report, even though no change may have occurred or it has been solved only a day later. You must never simply drop a problem because it no longer applies.
- If management expects you to include project cost information in your progress report, insert it in three places:

> In the **summary** (comment briefly on how closely you are adhering to projected costs).
> In the **progress** section (give more details of costs, and particularly cost implications of problems).
> In the **outcome** section (indicate future cost trends).

Costs are usually closely linked with your adherence to schedule: the more you drop behind schedule, the more likely you will have to report a cost overrun.

PROJECT COMPLETION REPORT

A project completion report is similar to an occasional progress report in that it describes work that has been done (that is, progress achieved). Normally, however, a project completion report omits any reference to present or future work, because when a job is complete no work is currently being done or is planned. Thus the summary-background-facts-outcome arrangement shown in Figure 4-1 can be adhered to fairly closely, with the **facts** compartment being relabelled **project highlights**. Sometimes an additional compartment called **exceptions** is inserted immediately after the project highlights to draw attention to deviations from the original plan for the project. These five compartment are outlined in Figure 4-7.

The project completion report written by Jack Binscarth at the end of his analysis of oil samples for Cantor Petroleums has the five writing compartments identified beside each part of the report. (See Figure 4-8; Jack's progress report for this project is shown in Figure 4-5.) Note particularly that in a short report you can combine two, or sometimes more, writing compartments into a single paragraph. In Jack's project completion report, paragraph 1 contains both the

SUMMARY	A brief statement that the project or job is complete, plus a short description of the result(s).
BACKGROUND	The cicumstances affecting the job, such as purpose, terms of reference, schedule, budget, and persons involved.
PROJECT HIGHLIGHTS	Major achievements, such as work accomplished, problems encountered (plus how they were resolved, and how they affected the project), targets met, and results obtained.
EXCEPTIONS	Variances from the project plan (if there are any), which may be work that either could not be completed or deviated from the plan; includes the reason for each exception, and its effect.
OUTCOME	Normally a closing statement that identifies any follow-up action that has to be taken, such as invoicing a client or remedying an exception.

Figure 4-7. Writing plan for a project completion report.

summary and the **background**, and paragraph 2 contains both the **project highlights** and the **exceptions**.

INSPECTION REPORT

An inspection can range from a quick check of a small building to assess its suitability as a temporary storage center, to a full-scale examination of an airline's aircraft, avionic equipment, repair facilities, and maintenance methods. In both cases the inspectors will report their findings in an inspection report. The building inspector's report will be brief: it will state that the building either is or is not suitable and will give reasons. The airline inspector's report will be lengthy: it will describe in detail the condition of every aspect of the airline's operations and list every deficiency (condition that must be corrected). In both cases the inspectors' reports can follow the summary-background-facts-outcome arrangement.

Summary	The main result(s) of the inspection (very brief); what the reader most wants to know.
Background	Why the inspection was necessary; what was being inspected; who was involved; where and when the inspection took place.

Facts What the inspection revealed (the details). There are two parts to the facts:
   A. CONDITIONS FOUND. A description of:
      • Quality (condition) of an equipment or a facility, or of work done.
      • Quantity of items examined, or of work done.
   B. DEFICIENCIES. A list of:
      • Conditions that need to be corrected.
      • Work that needs to be done (or redone).

Outcome A general statement of results, possibly with a recommendation.

MACRO ENGINEERING INC

FROM: Jack Binscarth   DATE: October 25, 19xx

TO: Fred Stokes   SUBJECT: Finalizing Cantor
                                      Petroleums' Project

My analysis of oil samples for Cantor Petroleums was   *Summary*
completed on October 24, eight days later than planned.   *Statement*
The work was done at the refinery, as requested in
Cantor Petroleums' purchase order No. 376188 dated   *Background*
September 4, 19xx, and was scheduled to start on Sep-
tember 11 and end on October 16. I was assigned to
the project under work order No. 2716.

The work plan called for me to analyze 132 oil samples
within the five-week period, but three problems caused   *Project*
me to overrun the schedule and complete four fewer   *Highlights*
analyses than specified. The delay was caused by a
strike of refinery personnel and a faulty spectrophotometer that had to be sent out for repair and
recalibration. The incomplete analyses were caused   *Exceptions*
by four contaminated samples that could not be replaced
in less than six weeks.

Russ Dienstadt, the refinery manager, agreed to a cost
overrun and has corresponded with you separately about   *Outcome*
this. He also agreed that it would be uneconomical
for me to return to analyze replacements for the four
contaminated samples. When I delivered the 128 analyses
to him on October 24, he accepted the project as being
complete.

                        Jack

Figure 4-8. A project completion report.

Kevin Doherty's building inspection report in Figure 4-9 shows how these compartments helped him shape his report into a logical, easy-to-follow document. Note particularly how:

- His **summary** (1) tells the Production Manager the *one* thing he most wants to know: can they use the building?
- The **background** (2) describes who went where, why, and when.
- Kevin has opened the **conditions** section (3) with a summarizing general statement and then supported it with facts.
- He has presented the **deficiencies** (3B) as a briefly stated list, which makes it easy to identify what has to be done, and has used active verbs to demonstrate that the actions *must* be performed.
- The recommendation in his **outcome** (4) supports his summary.

For a short inspection report like this, Kevin was correct in presenting all the conditions first and then listing all the deficiencies. But for a long report that covers many items, such an arrangement could become cumbersome. For example, if Fran Hartley followed this sequence for an inspection at Remick Airlines, the organization of the facts session would be like this:

    A. CONDITIONS FOUND:
        1. Electrical Shop
        2. Avionics Calibration Center
        3. Flammable Materials Storage
                                            (etc . . .)
    B. DEFICIENCIES:
        1. Electrical Shop
        2. Avionics Calibration Center
        3. Flammable Materials Storage
                                            (etc . . .)

The more departments Fran inspects, the longer her report becomes and the further apart each department's conditions and deficiencies sections grow.

To overcome this difficulty, Fran should treat each department as a *separate* inspection and reorganize the report so that for each department the deficiencies section immediately follows the conditions section. The organization of the whole report would then become:

Summary
Background
Facts:

    1. Electrical Shop:
        A. Conditions Found
        B. Deficiencies
    2. Avionics Calibration Center:
        A. Conditions Found
        B. Deficiencies

MACRO ENGINEERING INC

FROM: Kevin Doherty   DATE: November 5, 19xx

TO: Hugh Smithson   SUBJECT: Inspection of Carter
           Production Manager   Building

The Carter Building at the corner of River Avenue and 39th Street will make a suitable storage and assembly center for the Dennison contract. ①

Christine Lamont and I inspected the Carter Building on November 3 to assess its suitability both for storage and as a work area for 20 assemblers for 15 months. We were accompanied by Ken Wiens of Wilshire Properties. ②

We found the interior of the building to be spacious and to have good facilities, but to be unsightly. Our inspection showed that: ③

o   There are 4200 ft^2 of usable floor space (see attached building plan, supplied by Mr. Wiens); we need 2400 ft^2 for the project.

o   There are two offices, each 150 ft^2, and a large unimpeded space ideal for partitioning into a storage area and four work stations.

o   The building is structurally sound and dry, but it is very dirty and smells strongly (the previous tenant was a fertilizer distributor). ③A

o   There are numerous power outlets, newly installed with heavy-duty circuits, and the building has excellent overhead lighting.

o   Several walls are damaged and many contain obnoxious graffiti.

o   There is a new loading ramp on the north side of the building, suitable for semitrailers.

o   Washroom facilities are adequate for up to 30 people, but one toilet and two washbasins are broken.

Before we rent the building, the rental agency will have to:

1.   Clean it thoroughly.

2.   Repair damaged walls, partitions, and toilet facilities. ③B

3.   Redecorate the interior.

Ken Wiens said his firm would be willing to do this.

I recommend we rent the Carter Building from Wilshire Properties, with the provision that the deficiencies listed above must be corrected before we move in. ④

Figure 4-9. A short informal inspection report.

3. Flammable Materials Storage:
   A. Conditions Found
   B. Deficiencies (*etc. . . .*)

Outcome:

Conclusions
Recommendations

Fran's inspection report now has a much more tightly knit, logical, coherent organization.

LABORATORY REPORT

There are two kinds of laboratory reports: those written in industry to document laboratory research or tests on materials or equipment, and those written in academic institutions to record laboratory tests performed by students. The former are generally known as test reports or laboratory reports; those written by students are simply called lab reports.

   Industrial laboratory reports can describe a wide range of topics, from tests of a piece of metal to determine its tensile strength, through analysis of a sample of soil (a "drill core") to identify its composition, to checks of a microwave oven to assess whether it emits radiation. Academic lab reports can also describe many topics, but their purpose is different since they describe tests which usually are intended to help students learn something or prove a theory rather than produce a result for a client.

   Laboratory reports generally conform to a standard pattern, although emphasis differs depending on the purpose of the report and how its results will be used. Readers of industrial laboratory or test reports are usually more interested in results ("Is the enclosed sample of steel safe to use for construction of microwave towers which will be exposed to temperatures as low as $-40°C$ in a North Dakota winter?" a client may ask) than in how a test was carried out. Readers of academic lab reports are usually professors and instructors who are more likely to be interested in thoroughly documented details, from which they can assess the student report writer's understanding of the subject and what the test proved.

   A laboratory report comprises several readily identifiable compartments, each usually preceded by a heading. These compartments are described briefly here:

Part	Section Title	Contents
SUMMARY	Summary	A very brief statement of the purpose of the tests, the main findings, and what can be interpreted from them. (In short laboratory reports, the summary can be combined with the next compartment.)

Part	*Section Title*	*Contents*
BACKGROUND	**Objective**	A more detailed description of why the tests were performed, on whose authority they were conducted, and what they were expected to achieve or prove.
FACTS	**Equipment Setup**	*There are four parts here:* A description of the test setup, plus a list of equipment and materials used. A drawing of the test hook-up may be inserted here. (If a series of tests is being performed, with a different equipment setup for each test, then a separate equipment description, materials list, and illustration should be inserted immediately before each test description.)
	Test Method	A detailed, step-by-step explanation of the tests. In industrial laboratory reports the depth of explanation depends on the reader's needs: if a reader is nontechnical and likely to be interested only in results, then the test description can be condensed. For lab reports written at a college or university, however, students are expected to provide a thorough description here.
	Test Results	Usually a brief statement of the test results or the findings evolving from the tests.
	Analysis (or Interpretation)	A detailed discussion of the results or findings, their implications, and what can be interpreted from them. (The analysis section is particularly important in academic lab reports.)
OUTCOME	**Conclusions**	A brief summing-up which shows how the test results, findings, and analysis meet the objective(s) established at the start of the report.
BACKUP	**Attachments**	These are pages of supporting data such as test measurements derived during the tests, or documentation such as specifications, procedures, instructions, and drawings, which would interrupt reading continuity if placed in the report narrative (in the test method section).

The compartments described here are those most likely to be used for either an industrial laboratory report or a college/university lab report. In practice, however, emphasis and labelling of the compartments probably will differ slightly, depending on the requirements of the organization employing the report writer or, in an academic setting, the professor or instructor who will evaluate the report.

Assignments

INCIDENT REPORTS

PROJECT 4.1: ACCIDENT AT CORMORANT DAM

You are an engineering technician employed in the local branch of H. L. Winman and Associates. Currently you are supervising installation work at a remote construction project at Cormorant Dam.

The day before you left for the construction site, your branch manager (Vern Rogers) called you into his office. "I'd like you to meet Harry Vincent," he said, and introduced you to a tall, gray-haired man. "Harry is with the Department of the Environment, and he wants you to take some air pollution readings while you're at Cormorant Dam."

Mr. Vincent opened a wooden box about $14 \times 10 \times 10$ inches, with a leather shoulder strap attached to it. In the box, embedded in foam rubber, you could see a battery-powered instrument. "It's a Vancourt MK 7 Air Sampler," he explained, "and it's very delicate. Don't check it with your luggage when you fly to Cormorant Dam. Always carry it with you."

For the next hour Mr. Vincent demonstrated how to use the air sampler, and made you practice with it until he was confident you could take the twice-daily measurements he wanted.

Now it is 10 days later and you have just finished taking the late-afternoon air sample measurements. You are standing on a small platform halfway up some construction framework at Cormorant Dam and are replacing the air sampler in its box.

Suddenly there is a shout from above, followed immediately by two sharp blows, one on your hardhat and the other on your shoulder. You glimpse an 8-foot length of 4-inch square construction lumber tumble past you followed by the air sampler box, which has been knocked out of your hand. The box turns end over end until it crashes to the ground. When you retrieve it the box is misshapen and splintered and the air sampler inside it is twisted. Your hand also is throbbing badly and cannot grip anything. An examination at the medical center shows you have a dislocated shoulder, and now your arm is supported by a sling. (Fortunately, it is not your writing hand.)

Part 1. Write an incident report to Harry Vincent of the Department of the Environment. Tell him:

1. What has happened.
2. That you have shipped the damaged air sampler to him on Remick Airlines flight 751, for him to pick up at your city's airport (you enclose the airline's receipt with your report).
3. That if he wants to take any more air pollution measurements, he will have to send you another air sampler.

Harry Vincent's title is Regional Inspector and his address is: Department of the Environment, Suite 306, 444 Waltham Avenue of your city.

Part 2. Write a memorandum-form incident report to Vern Rogers. You can mention that you were absent from the construction site for 24 hours; otherwise the incident has not affected your supervision work.

PROJECT 4.2: THEFT AT WHITESHELL LAKE

You are team leader of a four-person inspection crew en route to a remote site 508 miles from your office, where construction of a nuclear power generating station is in progress. You are traveling in a panel van, and after 376 miles you and the crew agree to stop for the night. At 8:05 p.m. you pull into the Clock Inn, a small motel beside the road that skirts around Whiteshell Lake.

The following morning you are having breakfast in the motel's tiny dining room when Fran Pedersen, one of the crew, goes out to the van to fetch the road map. She returns almost immediately and gasps: "The van has been broken into!"

The four of you scramble out to the parking lot and can see right away that the window on the front passenger's door has been smashed.

"They were after the radio," Shawn Mahler observes, pointing to a gaping hope in the dash.

"Check if anything else is missing," you suggest. Already you are expecting the worst, but to your surprise you find that only two other items have been taken, one inconsequential and one important: about $6.00 from a tray in the dash (it's loose change for parking meters), and a video camera and videotapes from a storage box in the rear of the van.

You try telephoning your office, but it is too early and no one answers. So, since the motel has a fax machine, you write a memo to your manager and send it by fax. In it you describe what has happened and ask for a replacement video camera to be sent to you. Here is some additional information you draw on to write your report:

- You are driving company panel van license number HLW 279; it's a Ford.
- Your trip was authorized by Travel Order N-704, dated one week ago and signed by your manager.

- The power generating station is being constructed beside the Mooswa River, 18 miles north of the small town of Freehampton.
- The Clock Inn is 3.5 miles west of Clearwater village, on highway A1136.
- The third member of your crew is Servi Dashi.
- The video camera is a Nabichi TX200 "Portacam." You rented it from Meadows Electronics at 2120 Grassmere Road of your city. Its serial number is 21784B.
- Your manager's name is M. B. Corrigan.
- Your crew's role at the power generating station is to inspect construction work to determine progress and the quality of work being done.
- You will refer to a thick inspection procedure titled "Inspection Requirements—Project NPG-7, Mooswa River" as your working guidelines throughout the inspection. It contains detailed instructions, specifications, and tolerances that the building contractors must adhere to.
- The purpose of the video camera is to record progress visually. The videotapes you shoot will be edited and then accompany your written inspection report. The edited tape will also be shown at the Power Authority Directors' Meeting scheduled for the 22nd of next month.
- You telephone the police at Clearwater village to report the break-in and theft. They ask you to drop in and report in person. You plan to do this at the start of your drive to the construction site (which will be *after* you have sent your fax).
- In your report you ask your manager to ship you a replacement video camera by bus the day after tomorrow. One bus a day passes through Freehampton, but it stops only on request. (You will drive to Freehampton to meet the bus and will telephone your manager tomorrow to check that the video camera *will* be on that particular bus.)
- You use today's date as the date on which you write your report.

Your Assignment

1. Write the incident report (as a memorandum).
2. Prepare a fax cover page.
3. Write a letter to Meadows Electronics to explain the loss of their video camera.

PROJECT 4.3: EFFECT OF A POWER OUTAGE

H. L. Winman and Associates has been carrying out a series of extreme cold and heat tests on electronic and mechanical switches for Terrapin Control Systems of Denver, Colorado. The tests have been running for four months and will last another two months. The schedule is tight because of initial problems with measuring equipment, which delayed the start by nine days and used up any spare time the project had available.

Currently, you are testing the switches for continuous periods of from 8 to 14 hours. The tests have two parts:

1. For the first 6 hours each day you increase or decrease temperature in 2°C increments until a predetermined high or low temperature is reached. At each 2° increment you test the switches and record how they perform.

2. For the remaining 2 to 8 hours, you bake or deep-freeze the switches at the preselected temperature. No monitoring is necessary during this period (although the switches are tested at room temperature the following day).

To avoid having a technician stay throughout part 2, which on some evenings runs as late as 12:30 a.m., you have installed electrical timers in the circuits of the oven and freezer chamber. At the end of part 2 each afternoon, the timers are set to switch off at the end of the prescribed bake and deep-freeze periods.

This morning when you remove batches 87H and 84C from the oven and freezer chambers you notice that, instead of being close to room temperature, the oven is still hot and the freezer is still cold. You check the electrical timers, but both are "off." Then you notice that the electric clock on the lab wall reads only 3:39, your wristwatch reads 9:03—a difference of 5 hours and 24 minutes. You telephone the local power utility:

"Was there a power cut last night?" you ask.

"Where do you live?" a voice replies.

"I'm calling from my office," you say, and quote the address.

"Yes, there was," the voice answers. "We had a transformer blowout at Penns Vale. It affected everyone in your area."

You ask when the power cut started and ended. The voice asks you to wait a minute.

"The transformer blew out at 9:23 last night," the voice eventually announces. "And we restored power to buildings in your area at 2:47 a.m."

You thank the voice, and consult your log for the previous day's tests:

You started part 1 at 9:55 a.m.

You started part 2 at 3:55 p.m., and set the timers to run for 8 hours (they were to switch off at 11:55 p.m.).

You consider what has happened:

The continuous bake and deep-freeze periods were interrupted part way through.

The oven temperature dropped, and the freezer temperature rose, for 5 hours and 24 minutes (but to what temperature?).

The power was restored and the oven temperature again increased, and the freezer temperature decreased (but to what temperature?).

The electric timers switched off at 5:19 a.m. (after their eight hours *total* running time).

You consider the implications of the power cut:

The batches have had uncontrolled, nonstandard testing and will have to be discarded.

Yesterday's tests will have to be run again (on two new batches).
The cost is:
- Labor: 14 hours (7 hours per batch) = $280.00.
- Materials: Two complete batches at $76.00 each.
- Time: One day extra to be added to the program schedule.

Write an occurrence report to your project coordinator (J. H. Grayson).
Tell him what has happened, describe the implications, and possibly suggest what
might be done to prevent a recurrence.

PROGRESS REPORTS

PROJECT 4.4: INSTALLING A LAN

Macro Engineering Inc (MEI) in Phoenix has a contract to install a local area
network (LAN) to link the personal computers (PCs) installed in the various
offices of Multiple Industries Inc across the U.S. The project involves

- installing a model 2020 processor at each branch,
- installing an expansion card (part number 30766) in every PC at each branch,
- hard-wiring all of each branch's PCs to the 2020 processor (*hard-wiring* means "connecting by wire"),
- connecting the 2020 processor, via a modem, to a dedicated telephone line (*dedicated*
  means "sole use") which will be installed by the local telephone company, and
- testing the system.

MEI has assigned the installation work to the branch offices of H. L.
Winman and Associates and has instructed each branch to install the LAN at a
local office of Multiple Industries. Vern Rogers, manager of the H. L. Winman
branch in your city, has assigned you to install and test the LAN in the Multiple
Industries' offices at 1130 Portland Avenue of your city.

Part 1.   The following details describe your involvement with the project
from November 13 to December 11:

- You were assigned to the project on November 13.
- The project coordinator at MEI in Phoenix is Shirley Woburn.
- You visited Multiple Industries on November 20, outlined to general manager Klaus
  Wiens what you would be doing (he was aware that you would be coming), and:
  1. Counted the PCs (there were 37).
  2. Asked where the 2020 processor was to be located. Klaus suggested the stationery
     storage room, behind the switchboard, and you agreed that it would be a suitable
     location because it would be simple for the telephone company to install the
     dedicated line.

3. Measured the amount of cable you would need to connect all 37 PCs to the processor. You estimated 360 meters.

4. Arranged to do the work from December 3 to 11 (except Saturday the 8th and Sunday 9th).

- You ordered the necessary equipment and materials from MEI on November 21, and they arrived at your office on November 27. You checked them against the packing list and found everything you had ordered was present.

- You started the installation work on December 3, and in the ensuing days you:

   1. Checked with each PC user to determine exactly where he or she wanted the PC to be located (Dec 3).

   2. Installed cabling between the stationery storage room and each PC (Dec 3 to 6) until you ran out of hookup wire. ("That's strange," you muttered to yourself, "I must have underestimated the amount I would need.")

   3. Identified how many PCs still had to be hooked up: there were five (Dec 6).

   4. Telephoned Shirley Woburn at MEI and obtained approval to purchase 120 meters of No. 30-2 multistrand hookup wire locally (Dec 6).

   5. Bought and installed the remaining cable (Dec 7).

   6. Installed the 2020 processor in the stationery storage room, and connected the ends of the hookup wires to the processor (Dec 6 and 7).

   7. Installed an expansion card into a spare slot in each PC. In eight cases there were no slots and you had to modify the PC to accept the card (Dec 10). You had anticipated this during your November 20 visit, having noticed during your initial inspection that there were several older PCs, and so had brought sufficient hardware to modify up to 12 terminals.

   8. Discovered you were short six expansion cards (Dec 11). You recounted the PCs and found there were 43, not the 37 you had counted previously. ("Oh, yes!" Klaus Wiens said when you told him of the problem. "I meant to tell you: we had six PCs on back order and they came in on November 26th. Sorry about that!") You asked if any more PCs were on back order, but he said no.

   9. Telephoned Shirley Woburn and asked for six additional expansion cards (Dec 11). She said she would have to order them and would not be able to get them to you until Dec 15 at the earliest.

Your Assignment: It is now December 11, the day originally scheduled for project completion, and it is clear that you will not finish the work for at least another week. Write a memo-form progress report to Shirley Woburn at MEI in Phoenix, with a copy to Vern Rogers (your branch manager). Include a revised completion date.

Part 2. The remainder of the project continued in this sequence:

- The local telephone company came in and installed a new telephone line, and wired it to the 2020 processor, on Dec 13.

- The six expansion cards arrived on Dec 18, and you installed them on Dec 18 and 19.

- You tested the LAN for internal communication and found that each of the 43 PCs was able to work, through the 2020 processor, with the other 42 PCs (Dec 20).

- You tried communicating, through the modem, with three Multiple Industries' offices in other cities, but drew a blank each time: there was no response (Dec 21).

- You called the telephone company on Dec 21 and told supervisor Walt Friesen that the dedicated line seemed to be dead. ("Of course it is," he replied. "Were you not told? We have no spare lines at the local exchange—in fact, we are over 400 lines short and many people like yourself are waiting for the new Witney exchange to open on February 1. I've already assigned you a dedicated line at the Witney exchange, so automatic hookup will occur immediately after the exchange opens.")

Your Assignment: Since your part of the installation work is complete, you return to your office and write a memo-form project completion report for Shirley Woburn (with a copy to your branch manager). Today's date is December 21.

TRIP REPORTS AND INSPECTION REPORTS

PROJECT 4.5: CHECKING AN INSURANCE CLAIM

You are employed by the local branch of H. L. Winman and Associates. When you arrive for work this morning, branch manager Vern Rogers calls you into his office.

"We've had a call from Hugh Smithson (he's production manager for Macro Engineering Inc in Phoenix)," Vern says. "Twelve cartons of special instruments they shipped three days ago were in a semitrailer which rolled near Darlington."

He pulls out a map and points to Darlington, which is 110 highway miles from your office. "The insurance company wants someone to look over the damage with one of their adjusters to confirm how much can be repaired or salvaged."

You drive to Darlington with Fern Redovich of Meredith Insurance Corporation. She takes you to a warehouse on the other side of town, where the smashed crates tell their own story of the violence of the accident. Very little of the delicate instruments could have survived such an impact.

You examine the crates one at a time. Broken glass, tangled wire, and chipped and splintered instrument cases are jumbled together. As you check each container, the adjuster notes the numbers in his book: 10, 4, 12, 11, 6, 3, 1, 9, 8 and 2 are totally beyond repair and obviously have no salvage value. Crate number 5 surprisingly is hardly marked: somehow its movements must have been cushioned. You open it carefully and check the instruments.

"This one seems okay," you say. "I think we could do something with it."

The adjuster adds up the totals. "Not very good for us," he says. "Ten out of eleven means a heavy claim."

"Twelve," you say. "There were twelve crates."

The adjuster checks his figures again while you count crates. There are 10 smashed ones and only one good one. And the smashed ones are not so smashed that one would be unrecognizable and so not be counted.

"One is missing," you say.

"Number seven," says Fern. And the two of you check the numbers against the crates.

Plainly, crate number seven is missing. You and Fern check all the other items removed from the semitrailer, but number seven is not there.

"Looks as if it has been stolen," Fern comments. "Probably before the accident. I don't think there would have been time afterward: the police were on the scene almost right away."

You telephone Vern Rogers and tell him what has happened.

"I'll telephone Hugh Smithson in Phoenix," he says. "In the meantime I want you to write me a report and send Hugh a copy by special delivery. And be sure to notify the local police."

Information you may need to write your report:

1. The shipping company was Merryhew Express Lines of Phoenix, Arizona.
2. The waybill number was C2719.
3. Macro Engineering Inc's invoice number was R13201.
4. The 12 cartons were being shipped to Melwood Test Labs, Circular Route No. 1, Hammondtown.
5. Meredith Insurance Corporation's local address is 2800 Western Avenue, Room 414, in the same town as your office.
6. The semitrailer rolled at Berryman's Corner, two miles southeast of Darlington.
7. The crates are being held at Superior Storage, 313 Crane Street, Darlington.
8. You report the missing crate to Darlington police at 2:25 p.m. on the day you checked the crates.

PROJECT 4.6: IDENTIFYING RADIO TELEPHONE INTERFERENCE

You own a small telecommunications installation and maintenance business in the city in which you live, and occasionally you do some consulting work for the Department of Communications. Yesterday, Ray Korvan, an engineer with the Department, telephoned and asked you to look into a radio interference problem reported from the town of Moorhaven, 55 miles away.

"Several people have complained of weird sounds they are hearing on their cordless telephones," he says, "like a baby crying. We've plotted where these people live and there seems to be no distinct pattern to show where the interference is coming from. As I don't have anyone available at the moment to send over there, would you mind taking a trip up to Moorhaven to see if you can identify the source and pin it down?"

Ray faxes a purchase order to you (No. 7705061) and a sketch map of Moorhaven on which he has identified the locations of the six people who have reported interference (see Figure 4-10).

This morning you drive to Moorhaven and by 11:30 a.m. you have spoken to the six people. You summarize the results:

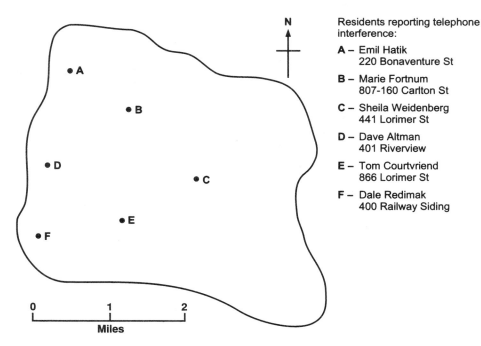

Figure 4-10. The town of Moorhaven, showing locations of residents reporting telephone interference.

• The interference is intermittent.
• It seems to occur only in the early afternoon, between about 1 and 3 p.m. (although Marie Fortnum says she has heard it once in the evening, at about 9:30 p.m. a couple of weeks ago).
• The interference does not seem to occur every day (although most of the six reported they did not always use their cordless telephones in the early afternoon).

You arrange for all six people to listen on their cordless or portable telephones between 1 and 3 p.m. this afternoon and to call one another regularly so that calls are generated on which interference can be heard. You also arrange for a friend who lives 1 mile northeast of Marie Fortnum to listen at the same time and to make calls to the others. Then you drive to the corner of Dornier St. and Wellington Ave. (1.2 miles southeast of Sheila Weidenberg's residence) and make calls on your car phone from there.

By 1:45 you have heard nothing, but the others are telling you they have heard several occurrences of what sounds like a baby crying. You telephone your friend, but he also has heard nothing. Clearly, the source of the noise must be in the western half of the town, within the area bounded by the letters A through F on the sketch map.

You drive to position E and park outside Tom Courtvriend's residence. You telephone Tom, but you still cannot hear a baby crying.

"The crying stopped seven or eight minutes ago," he explains.

Frustrated by the silence, you ask him not to speak and turn up the volume control on your telephone set. You still do not hear crying, but you *do* hear someone breathing!

You get out of your car, place a 10-inch diameter circular antenna (like a clip-on antenna for a portable television set) so that it stands vertically on your car's roof, and clip the cable end of the antenna onto the rod antenna of your portable telephone. With your left hand you hold the handset to one ear, and with your right hand you rotate the antenna clockwise. The breathing sound increases in intensity, fades, and then stops entirely. You rotate the antenna a little further and once again you can just hear the breathing. You turn the antenna back slightly and again the breathing ceases. At this "null" point you hold a magnetic compass up beside the antenna and write down the direction in which the "flat" side of the antenna is oriented. You drive farther and repeat the action twice more at locations C and B. In your notebook you now have three readings:

E—355°
C—278°
B—211°

On a town map you draw in the three bearings from the three locations, and the lines intersect to form a 1/4-inch sided triangle on a section of Radnor Street. (This process is known as triangulation and is identical to the method used by ship navigators to obtain a "fix" from three radio stations.)

You drive to Radnor Street, park, position the antenna on the roof of your car, and call Tom Courtvriend. He cannot hear the breathing, but you can (and much louder now). You rotate the antenna and stop it at the null point. The flat slide of the antenna is pointing directly toward house No. 296.

A woman answers the door.

"Do you have an electronic baby minder?" you ask.

She says she does, somewhat hesitantly, suspicious of you.

"Is it on right now?"

"Yes, it is."

You tell her that you can hear it on your car telephone, and she leads you to the kitchen where a small receiver unit rests on the countertop. You describe the problem in more detail and ask to see the transmitter.

"Not until my baby wakes up!" she says. "She sleeps most afternoons from one to three—when she stops crying! It's hard to get her to settle down."

Now you are back in your office and preparing to write a letter-form trip report to Ray Korvan (his office is at 236 Broadway Tower, in your city). Some additional information you found out is:

- The woman at 296 Radnor Street is Maisie Tillicum, and her baby's daughter is Jeanine (she is 15 months old).
- Ms. Tillicum was given the baby minder as a gift (it was sent by her sister in Plainsville, Nebraska).
- The baby minder is called "Sof-Talk" and is made by Delancy Manufacturing Inc, 200 Hugo St. North, Montrose, Ohio.
- You telephone Delancy Manufacturing and speak to Sales Manager Sharleen Reitsmar. She says that "Sof-Talk" is distributed primarily in the northeastern U.S., and the suggested retail price is $69.50. She insists the company has never had any previous complaints.
- You suspect the interference is being caused by harmonics, a poorly tuned transmitter, or a badly designed circuit that is overlapping the prescribed operating frequency.
- To give a more definitive opinion, you would have to buy a "Sof-Talk" and conduct tests, but you cannot do this without getting the Department of Communications's authorization both to buy the product and spend the additional time carrying out the tests. (So you decide to ask for approval in your report.)

Now write your trip report.

PROJECT 4.7: PROBLEM CONNECTORS AT SITE 14

H. L. Winman and Associates is management consultant to Interstate Power Company and currently is supervising the installation of parallel HV DC power transmission lines and a microwave transmitting system along a corridor between Weekaskasing Lake and Flint Narrows. The microwave transmission towers are located approximately 40 miles apart and are numbered consecutively from No. 1 at Weekaskasing Lake to No. 17 at Flint Narrows. Each tower site has a small residential community and a maintenance crew.

The maintenance crew supervisor at tower site No. 11 reported a week ago today (work out the date) that she has found nine faulty cable connectors type MT-27 and has had to replace them. She telephoned your manager (Andy Rittman), asking whether the fault had been found elsewhere. "There are so many faulty connectors," she said, "I suspect there may be more at other sites."

Andy Rittman instructed you to fly to site No. 14, the site nearest the H. L. Winman office, to investigate whether there are any other faulty connectors. "Use company test procedure TP-33 to test all the connectors," he said, "and then write me a report on your return, so I can send a copy to all site maintenance supervisors."

You flew to site No. 14 two days ago and stayed there until this morning, when you flew back to your home city. The site maintenance crew supervisor was Don Sanderson, who asked you for a copy of your report. Here are details of what you found out:

1. There are 317 type MT-27 connectors on site, with 92 in stock and 225 installed along the lines and up the tower.

2. You tested 278 of the connectors.

3. You could not test the remaining 39 connectors because they were along part of the transmission line which was powered-up throughout your visit.

4. You placed each connector under tension using test procedure TP-33.

5. 241 of the connectors were OK.

6. 37 of the connectors proved to be faulty.

7. You identified the fault as a hairline crack, which became visible when a faulty connector was placed under tension.

8. You also noticed that, although the connectors looked similar, there seemed to be two kinds of connectors on site. One batch of connectors had the letters GLA on the base. The other had the letters MVK on the base.

9. Of the 278 connectors you checked, 201 were stamped MVK, and 77 were stamped GLA.

10. All the faulty connectors had the letters GLA stamped on the base. There were no faulty connectors with the letters MVK on the base.

11. You figured that the letters must identify either different manufacturers or different batches made by the same manufacturer.

12. You instructed the maintenance crew supervisor to replace all installed GLA connectors with MVK connectors. You also told him to place all the GLA connectors in a separate box and to mark them NOT TO BE INSTALLED.

Now you have returned to your home office and you are starting to write your trip report to Andy Rittman. Write the report, telling him of your trip and your findings.

PROJECT 4.8: INSPECTING AN ACCESS ROAD

As the helicopter lifts off the tarmac at Harmonsville airport, the pilot shouts to you above the noise of the engine and the whirring blades: "I'm turning onto a heading of 072 degrees. We'll be at Sylvan Lake in 27 minutes."

You turn around to face technician Juan Martinez, seated behind you in the tiny cabin, and in turn you shout: "We'll be there at 9:34."

Juan nods and makes a note on a pad balanced on his knee.

The first time you heard of Sylvan Lake was one week ago, on September 17. Vern Rogers—your boss and manager of the local branch of H. L. Winman and Associates—dropped a map on your desk and pointed to a small lake.

"That's Sylvan Lake," he announced. "It used to be a fly-in fishing resort until the lake became overfished six years ago. Now the lodge is to be reopened by Triton Mining Corporation and used as a temporary office and living quarters while they do some test drilling nearby."

Vern told you that Triton Mining Corporation (TMC) has hired H. L. Winman and Associates to inspect the gravel road between a landing strip 4.8 miles to the east of Sylvan Lake and the fishing lodge, to assess if any road repairs need to be made before the drilling crews and rigs are flown in next spring.

At 9:31 the pilot shouts: "There's the lake!"

You look down and see a crescent-shaped lake about 2.4 miles long completely surrounded by tall trees and thick vegetation. At the southern end of the lake there is a clearing with a cluster of buildings, one considerably larger than the others, and a wooden dock jutting out into the lake.

"That's the fishing lodge," the pilot shouts. "I'll follow the road to the landing strip." The helicopter swoops down until it is flying about 30 feet above the tree tops.

The narrow road twists and turns, skirting small, weedy ponds and lakes, and is hemmed in for most of its distance by tall trees whose branches in some places meet overhead, obscuring the road from view. At one point there is a bridge where the road crosses a stream, and at another the road seems to be partly awash as it curves around the south side of a large pond.

The landing strip is simply a 2000 foot length of tarmac carved out of the brush and oriented northwest-southeast (to take advantage of the prevailing winds). You can see no buildings.

"I'll touch down here," the pilot shouts, pointing to the northwest end of the strip. "The road starts at this end."

The pilot helps you and Juan winch a four-wheel drive all-terrain vehicle down from a cavity in the helicopter's belly.

"I'll wait for you here," the pilot says, reaching into the cabin for a folding stool and a book, "to keep an eye on the chopper."

Juan starts the engine of the all-terrain vehicle, and the two of you drive to the end of the landing strip and bump down onto the road.

For the first 1.8 miles the road is rough but usable. In several places you have to skirt fallen trees that will have to be cleared away, and at the 1.1 mile point you notice a subsidence on the south side of the road. You stop and measure the area: it's about 60 yards long. In parts there is only just enough space for the 5.8 foot wide vehicle to squeeze through.

After 2.9 miles Juan stops just short of a wooden bridge over a stream. "Better inspect it first," he says.

The bridge is 13 ft 8 in. long and 14 ft 7 in. wide. It is constructed of unpainted timbers, on one of which somebody has painted in barely decipherable letters: "Sylvan Creek—No Fish." (You suspect the letters "ing" have been erased from the end of the message, probably by a disgruntled fisherman who has come away emptyhanded.)

You climb down the bank of the stream and inspect the structural timbers. "They are strong enough to support our light vehicle," you call up to Juan, "but there is some rotting of the vertical supports on the north side. They'll have to be fixed before heavier vehicles drive across."

Juan calls down to you that the planks on the bridge deck are relatively sound. "Only about 30% will have to be replaced," he adds. "That's about 75 feet of 8 inch by 2 inch lumber."

Between you, you decide that a load limit of 1.2 tons should be in effect until the repair work has been done.

At the 3.6-mile mark you encounter a 120-yard section of road that has been partly washed away where the road curves in a wide, concave arc around the south side of a moderate-sized pond. ("This is the washout we saw from the air," you comment.) The road slopes toward the water, and either subsidence or a high water level (or a combination of both) has caused the water to lap onto the road. In places the gravel is no more than 4.5 feet wide.

Juan stops the vehicle and hops out. He walks beside the water-covered part and with a long stick prods into the water. "The water has been high like this for three or four years," he says. "Except at the very edge the gravel has been completely washed away, probably by wave motion."

You negotiate the curve very slowly, the balloon tires of your vehicle providing the necessary traction and support on the slimy right side.

"This section of road will have to be completely rebuilt," you say, and make a note in the log.

The remainder of the road is in fairly good shape, and 10 minutes later you reach the fishing camp.

"Overall, the road is not in bad condition," you say to Juan, "considering that nothing has been done to it for at least six years."

Juan agrees: "Apart from the problem spots we have identified, it'll need only some grading over its entire length, and several loads of gravel."

"And gravel will be easy to find," you add. "I saw a gravel pit beside the road, almost hidden by bush, at the 3.1 mile mark. I've made a note of it."

Juan turns the vehicle around and drives back to the landing strip. The inspection has taken almost three hours.

You help winch the vehicle up into the cavity in the underside of the helicopter, and then the pilot, Juan, and you climb into the tiny cabin.

"There's something I want to show you," the pilot shouts, lifting the helicopter off the tarmac and swooping southeast along the landing strip no more than 12 feet above the ground. "While you were away I walked up to the other end of the strip. The first 900 feet can't be used."

As the helicopter swoops around the area, you see holes in the tarmac and numerous cracks through which bushes have started to grow, some 2.5 to 3 feet high.

"That effectively reduces the length of the landing strip to only 1100 feet, which is too short for many aircraft to use," the pilot adds.

You nod and make a note in your logbook. Although your assignment is to report only on the condition of the gravel road, you feel you should add a comment about the condition of the landing strip.

Now you are back at your home office, writing your report (this is September 26). You draw a sketch of the road from the landing strip to Sylvan Lake, showing the distances to the areas of fallen trees (you will have to decide where they are), the subsidence at 1.1 miles, the bridge, and the washed-out road. Some additional facts you need are:

- You drove from your office to Harmonsville, a distance of 164 miles.
- The helicopter company at Harmonsville is known as Vesuvius Helicopters Inc. The pilot's name is Chris Koshak.
- The all-terrain vehicle is manufactured by Wheels and Skids Inc of Nashville, Tennessee, and is known as "Moonwalker III."

You write your inspection report as a letter addressed to David Yanchyn at Triton Mining Corporation, 206 Hallington Street of your city. If you prefer, you may prepare your report in semiformal format and precede it with a cover letter. Include an illustration if you think it would be helpful. Indicate on the letter that Vern Rogers will receive a copy.

CHAPTER 5

Longer Informal and Semiformal Reports

The previous chapter discussed short reports that deal primarily with facts, reports in which the writer identifies the relevant details and presents them briefly and directly. This chapter describes longer reports that often deal with less tangible evidence; reports in which the writer analyzes a situation in depth before drawing a conclusion and, sometimes, making a recommendation. They may describe an investigation of a problem or unsatisfactory condition, an evaluation of alternatives to improve a situation, a study to determine the feasibility of taking certain action, or a proposal for making a change in methods or procedures. All are written in a fluent narrative style that is both persuasive and convincing; their writers have concepts or new ideas to present and they want their readers to understand their line of reasoning.

INVESTIGATION REPORT

The term "investigation report" covers any report in which the writer describes how he or she performed tests, examined data, or conducted an investigation using tangible evidence. Basically, the writer starts with known data and then analyzes and examines it so that the reader can see how the investigation was conducted and the final results were reached. The report may be issued as a letter, as an interoffice memorandum, or as a semiformal report.

Although they are not always readily identifiable, there are standard parts to a well-written investigation report that help shape the narrative and guide the reader to a full understanding of its topic. They are shown in Figure 5-1, which is an expanded version of the basic writing plan described at the start of Chapter 4. These parts are easy to recognize in long investigation reports, where headings act

SUMMARY	A brief statement of the situation or problem and what should be done about it.
INTRODUCTION	**Background** of the situation or problem
DISCUSSION	The **Facts** or **Investigation Details,** comprising:

	METHOD	How the investigation was tackled.
	FINDINGS	What the investigation revealed.
	IDEAS	Different ways the situation can be improved or the problem resolved.
	ANALYSIS	Evaluation of each idea.

CONCLUSIONS	The **Outcome**, or result of the investigation; a summing-up.
RECOMMENDATION	A positive statement advocating action.*
ATTACHMENTS	Evidence: detailed facts, figures, and statistics which support the Discussion.*

*Included only when appropriate.

Figure 5-1. Writing plan for an investigation report.

as signposts introducing each parcel of information. They are more difficult to identify in short reports which use a continuous narrative. In the two-page investigation report in Figure 5-2, the parts are identified by circled numbers:

(1) This is the **summary**.
(2) The **background** is only one sentence, which refers to the memo that invoked the investigation. Because the reader already knows the circumstances, the report writer can omit details.
(3) The **discussion** starts here, with a very brief reference to the **method**.
(4) These are the **findings**.
(5) This is the first **idea** and its **evaluation** or **analysis**.
(6) These are **ideas** 2 and 3, each followed by an **evaluation**.
(7) The **outcome** draws **conclusions** and makes a **recommendation**. (There are no **attachments**.)

DCMO--TV

MEMORANDUM

TO: Dennis Carlisle   **FROM:**   Phyllis VanDerWyck
               Operations Manager   Engineering Department

DATE: October 21, 19xx   **REF:**   Investigation of High Ambient
                                                Sound Level: Satellite Studio
                                                Control Room

I have investigated the high ambient sound level reported in the control room of our satellite studio at 21 Union Road, and have traced it to the building's air-conditioning equipment. The sound level can be reduced to an acceptable level by soundproofing the air-conditioning ducts and blowers, and by carpeting the control room.   ①

My investigation was authorized by your memo of August 28, 19xx, in which you described the audio difficulties your production crews were experiencing when programming from the satellite studio.   ②

Tests conducted with a sound level meter at various locations in the control room established that the average ambient sound level is 36 dB, with peaks of 39 dB near the west wall. This is approximately 8 to 10 dB higher than the ambient sound level measured in the control room for No. 1 studio on Westover Road.   ③

The unusually high sound level is being caused by the air-conditioning equipment, which is in an annex adjacent to the west wall of the control room. Air-conditioner rumble and blower fan noise are carried easily into the control room because the short air ducts permit little noise dissipation between the equipment and the work area. The flat hardboard surface of the west wall also acts as a sounding board and bounces the noise back into the room.   ④

I have considered three methods that we could use to reduce the ambient sound level:

1.   The most effective would be to move the air-conditioning equipment to a remote location. This, however, would require major structural alterations which would make the approach impractical except as a last resort.   ⑤

/2...

Figure 5-2. A memorandum investigation report.

<u>Dennis Carlisle</u> - page 2

2.   Alternatively, we could replace the existing blower fan assembly
      with a model TL-1 blower manufactured by the Quietaire Corporation
      of Detroit, and line the ducts with Agrafoam, a new soundproofing
      product developed by the automobile industry in Germany. Togeth-
      er, these methods would reduce the ambient sound level by about
      6 dB.

3.   We could also cover the vinyl floor tiles with indoor/outdoor
      carpet, a practice which has proved successful in Air Traffic
      Control Centers, and mount carpet on the control room's west wall,
      for a further sound level reduction of about 2.5 dB.

Since relocation of the air-conditioning equipment would be unduly
expensive, I recommend adopting methods 2 and 3. This would reduce the
overall ambient sound level by approximately 8.5 dB for a total cost of
$8080. However, I suggest we make the modifications progressively, so
that we can measure the actual sound level reduction after each step.
In this way we may be able to eliminate the final step or steps if the
earlier modifications achieve better results than anticipated. The four
steps will be:

	Modification	Expected Sound Level Reduction (dB)	Approximate Cost ($)
1.	Replace blower fan assembly	2.5	2560
2.	Line ducts with Agrafoam	3.5	2600
3.	Install floor carpet	1.5	1580
4.	Install wall carpet	1.0	1340

These modifications should provide the quieter working environment
needed by your production crews.

Figure 5-2. (*Continued*)

Every investigation report will differ, depending on the topic you are investigating and the results you obtain. Sometimes you will use all of the standard parts, sometimes you will use only some of them, and occasionally you will need to devise and insert additional parts of your own. Typical standard parts are described here.

Summary.   Start with a brief description of the whole report, stated in as few words as possible. This will give busy readers a quick understanding of the investigation from which they can learn the results and assess whether they should read the whole report.

Background.   Describe events that led up to the investigation, knowledge of which will help the reader place the report in the proper perspective. This part is often referred to as the introduction.

Investigation Details.   Describe the investigation fully, carefully organizing the details so that readers can easily follow your line of reasoning. This is the discussion. In a long report, the discussion will contain all or some of the information described here, and the parts should appear roughly in this order:

Introduce Guiding Factors. These are the requirements or limitations that controlled the direction of the investigation. They may be as diverse as a major specification stipulating definite results that must be attained, or a minor limitation such as price, size, weight, complexity, or operating speed of a recommended prototype or modification.

Outline Investigation Method. Tell readers the planned approach for the investigation so that they can understand why certain steps were taken.

Describe Equipment Used. If tests were performed that called for special instruments, describe the test setup and, if possible, include a sketch of it. In long reports this may have to be done several times, to keep details of each test setup close to its description.

Narrate Investigation Steps. Describe the course of your investigation as a series of steps, so that readers can visualize what you did. Sometimes you may be able to use a chronological description, while at other times you may want to vary the sequence to help readers better understand what was done and what was found out.

Discuss Test Results (or Investigation Results). Present the results of your investigation and discuss what effect they have had or may yet have. (Simply tabulating the results and assuming that readers will infer their implications does not suffice.)

Develop Ideas and Concepts. At some point in your narrative you may want to introduce ideas and concepts which evolve as a result of your investigation. You may introduce them periodically throughout the report to demonstrate what you had in mind before taking the next investigative step, or group them together near the end of the discussion.

Analyze Ideas. If you have developed alternative ideas (methods) for resolving a problem, evaluate them to assess which is best, using factors such as effectiveness, cost, and simplicity as your evaluation criteria. (Ensure that your readers understand what criteria you are using, and why they are important, before they read your evaluation.) You may evaluate each idea as soon as you present it, as has been done in the investigation of noise in a television studio (Figure 5-2), or you may present all the ideas first and then evaluate all of them, as Figure 5-1 suggests and Morley Wozniak does in his evaluation report later in this chapter (see Figure 5-4).

Conclusions. In a brief summing-up, draw the main conclusions that have evolved from your investigation.

Recommendation(s). State what steps need to be taken next. Recommendations are optional and must develop naturally from your discussion and conclusions.

Evidence. Attach detailed data such as calculations, cost analyses, specifications, drawings, and photographs that support the facts presented in the **investigation details** section but would interrupt reading continuity if included with it. These pages of evidence are numbered consecutively and named attachments (or, in a formal report, appendixes) and placed at the end of the report.

Some companies preface their investigation reports with a standard title and summary page similar to the H. L. Winman and Associates' design illustrated in Figure 5-3. This page saves a reader the trouble of searching for the summary and the report's identification details. Subsequent pages, which have not been included with the example, contain the report narrative, starting with the background (introduction). Reports written in this way tend to adopt a slightly more formal tone than memorandum and letter reports and are less likely to be written in the first person. Such a report sometimes may be called a form report, although the preferred name is semiformal investigation report.

EVALUATION REPORT/FEASIBILITY STUDY

Evaluation reports are similar to investigation reports, and the two names are frequently used interchangeably. Evaluation reports often start with an idea or concept their authors want to develop, prove, or disprove. They first establish guidelines to keep their report within prescribed bounds, and then research data, conduct tests, and analyze the results to determine the concept's viability. At the end of the evaluation, they draw a conclusion that the concept either is or is not feasible, or perhaps is feasible in a modified form.

The writing plan shown in Figure 5-1 can be applied to an evaluation report. Morley Wozniak's evaluation of landfill sites in Figure 5-4 follows this plan.

Morley's report is significant in that it is preceded by a one-page cover letter, thus adopting the technique suggested on page 52 and illustrated in Figure 3-7. His cover letter is similar to the executive summary that often precedes

H L WINMAN AND ASSOCIATES

INVESTIGATION REPORT

REPORT NO: 70/26   FILE REF: 53-Civ-26   DATE: March 20, 19xx

PREPARED FOR: City of Montrose, Ohio

AUTHORITY: City of Montrose letter Hwy/69/38, Nov 7, 19xx

REPORT PREPARED BY: *GWaterston*   APPROVED BY: *MDavies*

SUBJECT OR TITLE

INVESTIGATION OF STORMWATER DRAINAGE PROBLEM

Proposed Interchange at Intersection of Highways 6 and 54

SUMMARY OF INVESTIGATION

The proposed interchange to be constructed at the intersection of Highways 6 and 54, on the northern perimeter of Montrose, incorporates an underpass that will depress part of Highway 54 and some of its approach roads below the average surface level of the surrounding area. A special method for draining the stormwater from the depressed roads will have to be developed.

Two methods were investigated that could contend with the anticipated peak runoff. The standard method of direct pumping would be feasible but would demand installation of four heavy-duty pumps, plus enlargement of the 3/4-mile long drainage ditch between the interchange and Lake McKing. An alternative method of storage-pumping would allow the runoff to collect quickly in a deep storage pond that would be excavated beside the interchange; after each storm is over, the pond would be pumped slowly into the existing drainage ditch to Lake McKing.

Although both methods would be equally effective, the storage-pumping method is recommended because it would be the most economical to construct. Construction cost of a storage-pumping stormwater drainage system would be $984,000, whereas that of a direct pumping system would be $1,116,000.

Figure 5-3. Title and summary page for a semiformal investigation report.

H L WINMAN AND ASSOCIATES

PROFESSIONAL CONSULTING ENGINEERS

May 23, 1992

Robert D. Delorme, P.E.
Town Engineer
Municipal Offices
Quillicom, MI 48716

Dear Mr. Delorme

Our assessment of the three sites selected as potential landfills for the Town of Quillicom shows that each has a disadvantage or limitation. The most serious exists at Lot 18, Subdivision 5N, which is the site preferred by the Town Council. A distinct possibility exists that a landfill located here could contaminate the town's water supply.

The disadvantages of the two other sites affect only cost and convenience. Lot 47, Subdivision 6E, will be considerably more expensive to operate, while Lot 23, Subdivision 3S, will have a much lower capacity and so will have to be replaced much sooner than either of the other sites.

If the Town Council still prefers to use Lot 18, a drilling program must first be conducted to identify the soil and bedrock structure between the lot and Quillicom. Providing the boreholes show no evidence that contamination will occur, then the site would be a sound choice.

The enclosed report describes our study in detail. I will be glad to discuss it and its implications with you.

Regards

Morley Wozinak

for Vincent Hrabi
Branch Manager
H. L. Winman and Associates
Lansing, Michigan

MTW:as
enc

Figure 5-4. An evaluation report (cover letter).

H L WINMAN AND ASSOCIATES

EVALUATION OF PROPOSED LANDFILL SITES FOR
THE TOWN OF QUILLICOM, MICHIGAN

SUMMARY

Two of the three locations selected as potential landfill sites for the Town of Quillicom, both southeast of
the town, are environmentally safe. There is insufficient data to determine whether the third site, to the
north of Quillicom, poses an environmental risk.

INTRODUCTION

The Town of Quillicom in Northern Michigan currently operates a landfill 2.1 miles southeast of the town. The landfill was constructed in 1952, and since 1968 has also served the
mining community at Melody Lake, 1.7 miles to the southwest of the landfill. In a report
dated February 27, 1992, Quillicom town engineer Robert Delorme identified that the existing landfill was nearing capacity and that a new landfill must be found and operational by
April 30, 1994.

Previously, in 1989, the town had identified two sites as potential replacement landfills: Lot
18, Subdivision 5N, 2.1 miles north of Quillicom; and Lot 47, Subdivision 6E, 9.1 miles to
the southeast. The costs to set up and operate both sites were determined, and Lot 18
proved to be more economical ($2000 more to purchase and develop, but $17,000 a year
less to operate), and was favored by the Town Council. However, in a letter to the Council
dated November 15, 1991, Mr. Delorme expressed his concern that leachate from the site
could possibly contaminate the town's source of potable ground water, and recommended
that the town first carry out an environmental study.

The town subsequently engaged H. L. Winman and Associates to examine the sites and
determine both their financial viability and their environmental safety. In a letter dated
March 15, 1992, Mr. Delorme commissioned us to carry out the study, and to include a third
potential landfill site at Lot 23, Subdivision 3S, immediately adjacent to the existing landfill,
in our assessment.

STUDY PLAN

We divided our study into three components: (1) an examination of the area geology and its
ability to constrain leachate movement; (2) an examination of the physical properties of the
proposed landfill sites; and (3) an evaluation of the financial and environmental suitability of
the sites.

1

Figure 5-4.   *(Continued)*

AREA GEOLOGY AND HYDROGEOLOGY

Bedrock at Quillicom and in the area of all three proposed landfill sites is chiefly granite and gneiss lying 15 to 30 meters below the surface. A layer of till varying in thickness from 10 to 20 meters covers the bedrock, and is itself covered by 1 to 15 meters of lacustrine silts and clays.

The whole area has experienced repeated glaciation, with the most recent occurring about 20,000 years ago with the advance of the Late Wisconsonian Ice Field. The advancing ice severely scarred this granite and gneiss. When the ice began to retreat 10,000 years later, till was deposited over the region. In addition, meltwater streams below the ice field deposited vast quantities of alluvial material, which today exist as eskers.

The melting ice also created Lake Agassiz and caused silts and clays to be deposited to a depth of up to 30 meters over the entire lake bed. (In the Quillicom area these lacustrine deposits range from 5 to 15 meters deep.) Then, as the lake drained and water levels receded, streams cut into the lacustrine and till deposits. These streams eventually dried up and their channels were filled with windblown silts and sands. Today the channels are known as buried stringers and, if they are water bearing, as stringer aquifers in the weathered bedrock and eskers.

The Town of Quillicom obtains its potable water from a stringer aquifer on the surface of the weathered bedrock. Other stringer aquifers are known to exist in the area to the east and south of Quillicom, and likely exist to the west and north.

The Michigan Water Resources Board provided us with logs obtained during drillings for ground water wells in the mid-1970s, all to the south and east of Quillicom. We have plotted the locations and types of material on a topographic map, which shows that

o   the bedrock in the area slopes downward, toward the south, from the Town of Quillicom, and

o   a major 350-meter wide glacial esker starts half a mile southeast of Quillicom and continues for several miles southeast under Highway A806, to beyond the proposed landfill at Lot 47, Subdivision 6E.

As there are very few borehole records for the area north of Quillicom, we could not plot a similar map for that area.

THE PROPOSED LANDFILL SITES

The attributes of the three proposed landfill sites are discussed briefly below, and itemized in detail in the attachment. The anticipated life of each site is based on the 1992 population of Quillicom. Similarly, projected operating costs are based on 1992 prices.

Figure 5-4. (*Continued*)

Lot 23, Subdivision 3S

This narrow, 19.3-acre strip of land is immediately east of and adjacent to the existing land-fill, 2.3 miles southeast of Quillicom on Highway A806. As it is the smallest site, it will cost only $9000 to purchase and develop. Its annual operating cost will be $47,000, the same as at the present landfill, and it will have an operational life of 12 to 14 years.

Lot 47, Subdivision 6E

The largest of the three proposed sites at 88.7 acres, but also the most distant, Lot 47 is a rectangular parcel of land 9.1 miles southeast of Quillicom on Highway A806. Its combined purchase and development price will be $20,000, and its annual operating cost will be $64,000. (The high operating cost is caused primarily by the much greater distance the garbage collection vehicles will have to travel.) It will have a lifespan of almost 60 years.

Lot 18, Subdivision 5N

A roughly square, 56.2-acre parcel of land, this lot is 2.1 miles directly north of Quillicom on Highway B1017. It will cost $22,000 to purchase and develop, and $47,000 a year to operate (the same as at present). At the current fill rate, it will last for 36 to 40 years.

SITE COMPARISONS

We considered three factors when comparing the three proposed landfill sites: cost, environmental impact, and convenience.

Cost. We examined cost from two points of view: the immediate expense to purchase and develop the site, and the annual cost to operate it.

o Lot 23, adjacent to the existing landfill, offers the lowest purchase and development cost at $9000, compared with $20,000 and $22,000 for the two alternative sites.

o Lot 23 and Lot 18 (the site north of Quillicom) offer comparable operating costs at $47,000 per year, whereas Lot 47 (9 miles southeast of Quillicom) would have the highest annual operating cost at $64,000.

However, if the purchase and development costs are spread over 10 years and added to the operating costs, Lots 18 and 23 show a more comparable cost structure:

Site:	Lot 18	Lot 23	Lot 47
Annual Cost:	$49,200	$47,900	$66,000

Environmental Risk. The primary environmental risk is the effect that leachate from the landfill could have on Quillicom's source of potable water. If a landfill lies on a glacial esker, leachate from the landfill will probably contaminate ground water aquifers in the esker. If these aquifers are connected hydraulically to stringer aquifers, the stringer aquifers also probably will become contaminated.

⑥

3

Figure 5-4. (*Continued*)

o Lots 23 and 47 (and the existing landfill) lie on a major esker south of Quillicom, but offer no environmental risk because the slope of the bedrock in the area is to the south, away from the town. Consequently, even if leachate from the landfill contaminates the ground water, it will flow away from Quillicom and will not contaminate the town's water supply.

o Lot 18, however, is in the uncharted area north of Quillicom, where neither the presence of eskers nor the slope of the bedrock has been determined. Consequently it offers a potential risk that leachate from a landfill located here could contaminate the town's water supply. This will be particularly true if the slope of the bedrock south of Quillicom is the same north of the town, since then leachate will flow south, *toward* Quillicom.

Convenience. To establish convenience we considered the size of the landfill site (measured as the number of years it can be used before another site must be found) and the proximity of the site to Quillicom.

o Lot 47 is the largest site, offering close to 60 years of use, but is four times farther from Quillicom than either of the two other sites.

o Lot 18, to the north, is next largest and can provide between 36 and 40 years of use. It is a comfortable 2.1 miles from Quillicom.

o Lot 23, although the same distance as Lot 18, has a life span of only 12 to 14 years.

CONCLUSIONS

The possibility of leachate contamination of the Town of Quillicom's water supply makes Lot 18, Subdivision 5N, a doubtful choice until sufficient drilling has been done to create a profile of the strata between the lot and Quillicom.   (7)

The remaining two sites are environmentally sound but have different advantages:

o Lot 47, Subdivision 6E, provides the greatest space but will be costly to operate.

o Lot 23, Subdivision 3S, offers the lowest cost but will have only a limited life span.

RECOMMENDATIONS

We recommend that the Town of Quillicom purchases Lot 23, Subdivision 3S, and operates it as a temporary landfill from 1994 to 2004. We also recommend that the town concurrently conducts a drilling program to the north of Quillicom to determine whether Lot 18, Subdivision 5N, will be an environmentally sound site to use after the year 2004.   (8)

Morley Wozniak
Morley Wozniak, P.E.

4

Figure 5-4. (*Continued*)

ATTACHMENT

(9)

**COMPARISON OF PROPOSED LANDFILL SITES
FOR THE TOWN OF QUILLICOM**

Comparison Factor	Lot 18 Sub 5N	Lot 23 Sub 3S	Lot 47 Sub 6E
Distance from Quillicom (driving dist in miles)	2.1 N	2.4 SE	9.1 SE
Size (acres)	56.2	19.3	88.7
Life (years)	36-40	12-14	58-60
Environmental risk (the site's potential for contaminating the Quillicom water supply)	Unknown	None	None
Development costs:			
Purchase price ($)	10,000	2,000	5,000
Construction cost ($)	12,000	7,000	15,000
Ten-year cost ($/yr)	2,200	900	2,000
Operating costs ($/yr)	47,000	47,000	64,000

Figure 5-4. (*Continued*)

a formal report (executive summaries are described in Chapter 6), since it *describes and comments on* key implications drawn from the report. Its addressee (Quillicom's Town Engineer) has the option of distributing it to the Town Councilors with the report, or detaching it and replacing it with a cover letter of his own.

The parts of the writing plan shown in Figure 5-1 are identified in Morley's report by circled numbers beside the narrative. Additional comments on each part are provided here and are keyed to the numbers beside the report.

(1) Although several factors affect site selection, in his **summary** Morley addresses only environmental impact because he believes it is of overriding importance. To mention cost and space would detract from the effect he wants to create.

(2) The introduction provides **background** details leading up to the study being assigned to H. L. Winman and Associates and then to Morley.

(3) The **evaluation details** start here, with a single paragraph in which Morley outlines his **investigation method** (i.e. how he tackled the study). Note that he mentions the three components *in the same sequence* that he describes them further on in the report.

(4) These are Morley's **findings**: the results of his research. He describes the findings in detail because his readers must fully understand the geology of the area if they are to accept the conclusions he will draw later in his report. Note that he is totally objective here, reporting only facts without letting his opinions intrude.

(5) Morley presents the three possible landfill sites as his **ideas** (even though they were originally presented to him by the client—the town of Quillicom). In effect he is saying to his readers: "Now that I have described the geology of the land to you, here are three locations within the area for you to choose from." Note that he still is totally objective.

(6) In his **analysis** Morley evaluates each site (each idea) to determine its suitability. His evaluation must clearly establish the factors on which he will base his conclusions. Now he allows some subjectivity to appear in his writing (we can hear his voice behind the words).

(7) Morley's **conclusions** identify the main features affecting each site. Note that he simply offers the alternatives without saying or even implying which is preferable. This part of a report, together with the recommendations, is the outcome (sometimes referred to as the *terminal summary*).

(8) In the **recommendations** Morley states specifically what he believes the Town Council must do. He must sound definite and convincing, so he starts with "We recommend . . ." rather than the passive "It is recommended that . . .".

(9) The **attachment** brings together all the site details in an easy-to-read form and simultaneously provides readers with evidence to support what Morley says about the landfill sites in the report narrative.

Compare how Phyllis VanDerWyck and Morley Wozniak each uses the first person (see Phyllis's investigation report in Figure 5-2). Phyllis uses the informal "I" because she is writing a memorandum report to another member of the television station where she works. Morley uses the slightly more formal "we" because he knows his semiformal report will be distributed to the Quillicom Town Councilors. (Note, however, that he uses "I" in the personal cover letter to Robert Delorme that accompanies the report.)

A feasibility study also starts by introducing an idea or concept and then develops and analyzes the idea to assess whether it is technically or economically feasible. The chief difference lies in the name and application of each document. An evaluation report is generally based on an idea that is originated and evaluated within the same company; hence it is nearly always informal. A feasibility study is normally prepared at a slightly higher level: the management of company A asks company B to conduct a feasibility study for it, because company A does not have staff experienced in a specific technical field. For example, if a national wholesaler engaged solely in the distribution of dry goods were to consider purchasing an executive jet, it would seek advice from a firm of management consultants. The consultants would examine the advantages and disadvantages and publish their results in a feasibility study, which they would issue as either a letter or a formal report. Often, the differences between a feasibility study and an evaluation report are so slight that personal preference dictates which label is used for the document.

TECHNICAL PROPOSAL

There are two kinds of technical proposals: those written on a personal level between individuals, and those prepared at a company level and presented from one organization to another. The former are usually informal memorandums or letters; the latter are more often formal letters or reports.

Scientists, supervisors, technicians, engineers, and technologists write memorandums every day to their managers suggesting new ways of doing things. These memorandums become proposals if their writers fully develop their ideas. A memorandum that introduces a new idea without demonstrating fully how it can be applied is no more than a suggestion. A memorandum that introduces a new idea, discusses its advantages and disadvantages, demonstrates the effect it will have on present methods, calculates cost and time savings, and finally recommends what action needs to be taken, demonstrates that its originator has done a thorough research job before starting to write. Only then does a suggestion become a proposal.

A proposal follows the basic **summary-background-facts-outcome** arrangement suggested for reports:

1. A brief **summary** of the situation or problem, and what you propose should be done to improve or resolve it.
2. An **introduction** that defines the situation and establishes why it is undesirable or is a problem.
3. A **discussion** of how the situation or problem can be improved or resolved, covering:
   • a description of the proposed solution,
   • what will have to be done to put it into effect,

- what the proposed solution will achieve,
- what advantages will accrue,
- what disadvantages will be incurred, their likely effect, and how they can be resolved or ameliorated, and
- what the cost will be.
4. A **recommendation** for action, which should be a strong, positive statement or a direct request.

Anna King used this outline to write the proposal in Figure 5-5, in which she proposes to Martin Dawes that H. L. Winman and Associates commission a communication consultant to determine the direction the company should take as it incorporates integrated services digital network (ISDN) technology. Her proposal contains

- a **summary** (1),
- a lengthy **introduction** (2), in which she establishes what ISDN is and why the company should become involved in using the new technology,
- a **discussion** (3), in which she defines what she proposes to do and what the cost will be, and
- a **recommendation** for action (4).

When there are alternative solutions for resolving a problem or unsatisfactory condition, the **discussion** section of the proposal becomes more comprehensive because it has to describe and evaluate all the possibilities. The writing compartment is then expanded into four subsections:

The Objective: A definition of the parameters or criteria for an ideal (or good) solution.

The Proposed Solution: What it is, the result it would achieve, how it would be implemented, its cost, and its advantages and disadvantages.

Alternative Solution(s): For each: what it is, the result it would achieve, how it would be implemented, its cost, its advantages and disadvantages, and why it is not as effective as the proposed solution.

Evaluation: A brief tradeoff analysis, with reference to the **objective**, comparing the effects of: (a) adopting the proposed solution, (b) adopting an alternative solution, and (c) taking no action.

Formal technical proposals are prepared by companies seeking to impress or convince another company, or the government, of their technical capability to perform a specific task. They are normally impressive documents prepared under extreme pressure. Often they have to meet rigid format and presentation requirements spelled out in a request for proposal (generally known as an RFP) issued by the agency asking vendors to submit bids. Preparation of such proposals is beyond the scope of this book.

H L WINMAN AND ASSOCIATES

INTER-OFFICE MEMORANDUM

To: ...... Martin Dawes .... From: ...... Anna King ....

.... Subject: ...... Proposal for Feasibility Study:

Date: ...... February 26, 19xx .... Adoption of ISDN System

Within three years most high technology companies--and many that are not--will be using an Integrated Services Digital Network (ISDN) to communicate internally, nationally, and internationally. As an ISDN system needs to be adopted as part of an overall corporate strategy (it cannot be bought off-the-shelf and installed quickly), I propose that we engage TransPacific Communication Systems Inc. of Seattle to carry out a preliminary study and recommend the approach and timing that H. L. Winman and Associates should adopt.

An ISDN system uses fiberoptic cables instead of standard wire telephone lines, and transmits information through a digital rather than an analog network. It costs almost twice as much more to install than a standard telephone system, but offers significant advantages:

o   Numerous telephone conversations, pictures, graphics, computer messages, and video signals can be sent and received simultaneously over the same line, and with much greater clarity and much less information loss.

o   Computer messages can be transmitted instantaneously from one terminal to another, without a modem, with both users making changes and seeing the changes--as they are made--on their screens.

o   Fax messages can be sent 10 times more rapidly and with negligible transmission errors.

ISDN also provides:

o   Conference-call capability between a head office, branches, and staff on assignment at other ISDN-equipped locations, with the call being carried out by telephone, computer, or video, used either singly or in combination.

o   Instant access to an immense range of databanks, from airline schedules and space availability to library data stored on CD-ROM.

The cost of an ISDN also is decreasing with time, as more and more organizations invest in the system.

1

Figure 5-5. A technical proposal (page 1).

At the right moment we will have to switch to an ISDN system, so that we will not only remain competitive but also be seen to be current in the burgeoning high-technology communication field. Timing will be essential, so that we can make a rational rather than a last-minute, seat-of-the-pants decision. Our difficulty will be to know the right moment to make the changeover, how much lead time to allow, and what funds to budget.

Timing is also important because three of our branches (Indianapolis, Indiana; St. Paul, Minnesota; and Houston, Texas) will be looking for new office accommodation when their current leases expire. If the branches knew for certain that we will be changing over to ISDN, and when, they could specifically seek accommodation in buildings that have already been--or shortly will be--wired with ISDN-compatible cabling.

I am proposing that we commission TransPacific Communication Systems Inc. (TCSI) to carry out a study into H. L. Winman and Associates' ISDN requirements. TCSI has specialized in fiberoptic data transmission for six years, and developed the prototype ISDN network that is being adopted for the multi-use Washington-Alaska communication link. I have had a preliminary discussion with Anthony Symondson, TCSI's manager of development engineering, and he agrees that we should wait no longer than six months to reach a decision. He can assign an engineer to study our requirements and recommends that this be done between March 15 and April 15. The objective will be to: ③

o   Identify what our communication needs will be over the next decade.

o   Outline the advantages of implementing ISDN, and the probable effect on H. L. Winman and Associates's operations if we do not.

o   Determine which point in that period will be the optimum moment to switch over to ISDN.

o   Develop specifications for an ISDN system that will meet our particular needs.

o   Establish a schedule and a budget for incorporating the system companywide.

I have obtained a cost proposal from TCSI quoting a fee of $18,700 to carry out the study, which will include a full investigation of H. L. Winman and Associates, its branches across the country, and Macro Engineering Inc in Phoenix, all travel and related expenses, and preparation of a definitive report within 28 days of the project start date.

Tina Mactiere assures me that TCSI is a technically sound organization fully able to perform the study, and that she is willing to assign $8000 as Macro Engineering's share of the study costs. May I have your approval by March 7 to spend $10,700 as our share of the cost, so that I can authorize TCSI to start the study on March 15? ④

Anna

2

Figure 5-5. *(Continued)*

Assignments

PROJECT 5.1: RESOLVING A LANDFILL PROBLEM

You are the assistant engineer for the Town of Quillicom in Northern Michigan. Your boss is Robert D. Delorme, P.E., who is the Town Engineer.

Mr. Delorme calls you into his office today and announces: "I have a project for you. The Town Council has finally decided to do something about the landfill problem, and they want it done in a hurry."

You know about the landfill problem. The existing landfill site—at Lot 22, Subdivision 3S—is nearly full and, because the Town Engineer recognizes that a new site will not be selected before the present site reaches its capacity, in places he has authorized dumping an additional layer of garbage on top of the compacted fill.

"Before you do anything, I want you to read this," Mr. Delorme continues, placing a report in your hands. "It's a study done by H. L. Winman and Associates a while ago, and it affects what you will be doing. Take it away and read it, and then come back to me for further instructions."

You read the report (you will find it in Figure 5-4, on pages 119 to 124) and then go back to see Mr. Delorme.

"The town councilors have decided," he says a little derisively, "that before they can make a decision they need to know how much it will cost to drill a dozen boreholes north of Quillicom and analyze the results. I want you to get some quotations that I can present to them. You should also be aware that the councilors very much prefer Lot 18, Subdivision 5N, rather than either of the other locations."

"Who will plot the results of the drilling?" you ask.

"Morley Wozniak, at H. L. Winman and Associates in Lansing, Michigan. He did the previous study and wrote the report I asked you to read." Mr. Delorme hands you a map (see Figure 5-6) and a list titled *Borehole Specifications for the Area North of Quillicom*, which contains the exact positions Morley has identified where the drilling must be carried out.

"And how many quotations should I get?" you ask.

"Two, as a minimum," Mr. Delorme suggests. "Three would be better."

The following week you call on the only two drilling companies you know of in the area, one in Quillicom and one in Marquette. They give you the following quotations:

Northwest Drillers, Inc., Quillicom, Michigan	$72,760 (tax incl.)
M. J. Peabody and Company, Marquette, Michigan	$69,900 (tax extra)

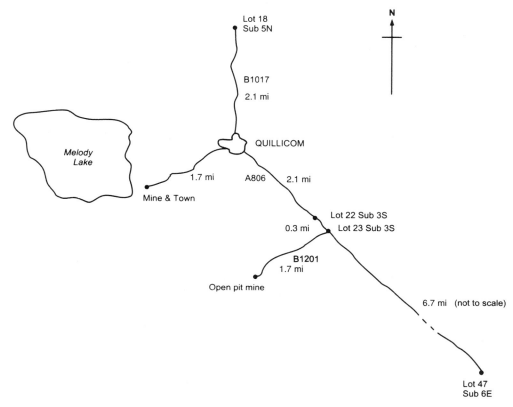

Figure 5-6. Sketch map of Quillicom area, in northern Michigan.

You had almost given up hope that you would find a third company to give you a quotation, when Mr. Delorme telephones. "Go and see Bert Knowles," he instructs you. "He's the assistant to the superintendent at Melody Lake Mine, and he has a suggestion for an alternative landfill site."

Mr. Knowles comes right to the point: "We do both open-pit and underground mining," he says. "Our open-pit mine is nearly worked out and we will finish excavating there in less than two years. The problem is that it's unsightly, and the Department of the Environment is leaning on Melody Lake Mines to do something about it. And that's where you come in." He explains that the Town of Quillicom can use the open-pit mine for a landfill, and the mine will lease it to the town for one dollar a year. "We have only two conditions: you must spread soil over the compacted garbage, and do it progressively as you go along so there will be no obnoxious smell for the people who live near the mine to contend with; and then you must seed it and plant trees."

You agree: it's something the Town would do anyway, before closing a landfill site.

Mr. Knowles drives you to the site and you stand on the lip of a shallow, roughly oval excavation varying from about 10 to 50 feet deep.

"How large is it?" you enquire.

"You'll have to talk to Inga Paullsen. She's the Mine Geologist."

When you visit Inga, she calculates the size of the excavation as 60.9 acres. "That's what it will be," she adds, "when the mining is complete. Why do you need to know?"

You describe the difficulty the Town Council is having in finding a landfill site, mention the three other sites, say the one north of the town could create an environmental problem, and explain you won't know until drilling has been completed there.

"But drilling has already been done there," Inga exclaims. "I was a junior at college and worked one summer with an exploration crew sinking boreholes north of Quillicom. We were looking for an alternative place to sink a mine shaft, but we found no ore deposits north of either Melody Lake or Quillicom. We drilled quite a few boreholes."

Inga tells you the mine does not have the records, only a report from the drilling company. She searches for it among the geology records but cannot find it. "That's strange," she mutters, "it should be here. Someone must have removed it."

The drilling company Inga worked for was Mayquill Explorations, but she says it does not exist any more. "When Ernie Mays retired he simply closed down the company. Maybe he still has all the records. You could ask him. He still lives in Quillicom."

Ernie Mays is about 70, and he lives in a bungalow at 211 Westerhill Crescent.

"I quit eight years ago," he tells you. "I sold some of my accounts to Northwest Drillers—those that were still active—and kept the remainder."

He remembers drilling for Melody Lake Mines. "We sank about 20 boreholes, all north of Quillicom, but we didn't find anything."

You ask if he remembers whether the bedrock slopes, but he shakes his head. "Not really," he says. "Nothing definite."

But he adds that he does remember there was evidence of a large sand esker running roughly south-southeast toward Quillicom.

"Do you still have the records?"

"No," he says. "The mine has them. Mr. Caldicott came to see me himself, about three years ago, and I gave them all to him."

Suddenly, everything falls into place. Frank Caldicott is not only general manager of Melody Lake Mines, but also a very influential Quillicom Town Councilor. And his youngest sister, Julie, is married to the Town Engineer—Robert Delorme, your boss.

Because you have so much new information to include, you decide to write a semiformal report of your findings. (You will have to decide whether you will include the information you now have about the previous drilling north of Quillicom and the location of the records.)

Some additional data you obtain is:

1. You are concerned about ground water contamination problems if the open-pit mine is used as a landfill, so you call Morley Wozniak at H. L. Winman and Associates in Lansing. He tells you that it will not be a problem: "Both the lake and the mining community are north of the pit, and the bedrock slopes to the south."
2. You calculate that costs to develop the open-pit mine as a landfill will be only $3000, because you can use the buildings and approach roads that are already there.
3. The open-pit mine is 2.5 miles directly south of Quillicom, but 4.1 miles by road (2.4 miles southeast along highway A806, then 1.7 miles southwest along highway B1201).
4. The annual operating cost for using the open-pit mine as a landfill will be $49,500, which is $2500 more than the cost for operating the current landfill.
5. You obtain a third drilling estimate from Quattro Drilling and Exploration Company in Houghton, Michigan, which quotes $77,600, tax included.

Before starting to write your report you visit Lansing, Michigan, on other business. On a hunch you visit the Land Titles Office and look up the surveys for the area north of Quillicom. Against Lot 18, Subdivision 5N, you find the owner listed as: *Julie Sarah Caldicott, 207 Northern Drive, Quillicom, Michigan.*

Now write your report.

PROJECT 5.2: TAKE A COMPUTER WITH YOU

You are an engineering technologist employed by H. L. Winman and Associates of Cleveland, Ohio, and you work in the Special Projects Department under the direction of Andy Rittman. You are one of five technologists who have just started working on the installation, testing, and modification of the microprocessor-controlled Westwin Environment Management System (short name: WEMS) at selected sites across the U.S., a project that will take between 2½ and 3 years to complete. The five technologists operate out of H. L. Winman and Associates' branch offices, and are:

Name	Area Responsible For	Branch Location
Quentin Dabrinski	Northeast U.S.	Boston, MA
Margaret Carriere	Southeast U.S.	Atlanta, GA
Caroline Witton	North Central U.S.	Chicago, IL
Neil Freeman	Southwest Central U.S.	Phoenix, AZ
David Cheng Ng	Pacific Coast & Alaska	San Francisco, CA

(You are to insert your name and location, in place of the person named in the list, for the area in which your college or university is located.)

The project includes some design work, detailed documentation of the work you do, preparation of maintenance and operation procedures for on-site staff (prepared to suit each particular installation), and writing of comprehensive reports that will be distributed to the companies where the WEMS systems are installed, to your own management, and to the technologists at the other sites, to keep everyone abreast of what is being done in the field. Hence, you do a lot of writing.

You want to suggest to Andy Rittman that the five technologists involved in the WEMS program should each have their own portable computer, to use when they write all the reports, procedures, and instructions that have become a major part of the project.

You research portable (laptop) computers and identify three or four that most seem to meet your needs.

Before doing any further work, you telephone Andy Rittman, tell him what you have in mind, and ask for his approval to go ahead with the study.

"I like your idea," Andy replies, "and I'm authorizing you to take the time to carry out your investigation. But before you write your report I want you to bear several things in mind."

As Andy speaks you write down his suggestions as a series of notes. He suggests that in your report you must

- establish why portable computers will be useful for people working on the WEMS project,
- identify what the company needs in a portable computer (that is, establish some guidelines or working specifications), and
- evaluate a range of computers and, within that range, demonstrate which computer most nearly fits the company's and the project's requirements.

"There are several other factors you should be aware of," Andy continues, and you jot down some more notes:

- Write a semiformal memorandum report because it will be an in-house document, but do not make the language too informal (you will be writing to management).
- Although you will be addressing your report to Andy, remember that the *real* readers will be other H. L. Winman and Associates managers who form the company's administrative committee and jointly approve capital expenditures such as this.
- Identify what the WEMS project consists of (remember that not all administrators will be familiar with it) and that it involves a lot of writing.
- Describe the types of writing that WEMS technologists do (e.g. procedures, instruction manuals, reports, installation specifications for contractors to follow, etc), and include any other writing you think will lend weight to your argument.
- Compare the ponderousness of writing on site with a pen and paper versus the simplicity of writing with a computer. Demonstrate that a person who writes manually

has to send the handwritten document in to the branch office to have it typed, and then proofread it on its return, and then send it back to the typist for corrections, before it can be issued.

- Describe the ease with which reports will be sent in to the head office by modem, at any time of the day or night, and the capability to exchange information easily between sites, also using the modem. ("You may have to explain what a modem is, for some readers," Andy adds.)
- Demonstrate that the portable computers can also be used for other projects (they won't just stop being used at the end of the WEMS project). Say that it's even possible they could be loaned to other projects on a short-term basis while the WEMS project is still in progress.
- Establish a budget for the project.

"You may even have to dispel any idea that some readers may think a laptop computer is no more than a toy," Andy finishes. "The word 'laptop' unfortunately creates the impression that it's not a professional computer."

Finally, you write down some points of your own:

1. Find three or four different computers to evaluate.
2. Identify factors to consider, such as:
   - Size (dimensions)
   - Weight
   - Number of disk drives and types (hard drive, and 3½ in. or 5¼ in. floppy diskettes)
   - Disk memory
   - CPU memory
   - Clock speed
   - Power requirements
   - Screen type and whether the screen is backlit
   - Battery type and working life between charges
   - Modem: built in or availability of a port for a modem
   - Other ports available, and whether serial, parallel, or both
   - Built-in software
   - Compatibility with other systems
   - Price
   - Options available (carrying case, cables, etc)
3. Find a convenient way to compare the computers. Compare them against the specifications or requirements, rather than against each other.
4. Make a comparison chart, to use as an appendix.

When you have assembled and documented all your data, write your report.

Note: To carry out this project, you will have to obtain data and specifications on three or four different laptop computers from local suppliers.

Portable computer technology is changing so fast that specifications for sample laptop computers are not included with these assignment instructions (because they would be current for only a very short time).

PROJECT 5.3: IDENTIFYING A POWER PLANT PROBLEM

You are an engineering assistant employed by a local consulting firm known as Machinery and Mechanical Engineers Inc. The company owner is Gerald P. Dirksen, and yesterday afternoon he called you into his office.

"I want you to drive over to Rossburn tomorrow," he announced, "to look into a technical problem in the power house at Baldur Agri-Chemicals."

(Rossburn is 43 miles from your city, has a population of 5700, and its primary industry is Baldur Agri-Chemicals, or BAC.)

"This morning," he continues, "I had a visit from Paulette Machon—she's vice-president of operations at BAC—and she is concerned because power house costs are rising just at the moment when world fertilizer prices are dropping. Apparently, this is causing the company to be uncompetitive in both national and international markets."

Mr. Dirksen repeated what Paulette Machon had told him: BAC requires a lot of hot water and steam in its manufacturing operations. However, over the past two years fuel consumption has risen by 18%, numerous breakdowns have occurred which have interfered with production, and there has been a sharp rise in production costs. Ms. Machon has visited the power house repeatedly but has never found anything that could be attributed to poor operation. In fact, the power house has always been immaculate.

Now Ms. Machon wants an independent consultant to take a look, talk to the people in the power house, and see if he or she can identify any production problems.

"Ms. Machon also hinted that the problem may not only be technical," Mr. Dirksen continued. "The present chief engineer at the BAC power house is Curt Hänness, and he is to retire in three months. BAC management has to decide whether to promote Harry Markham, the existing senior shift engineer, or to bring in a new chief engineer from outside the company. On paper, Markham is ideal for the job. He has worked in the power house for 15 years (he is now 36) and always under Hänness, so his knowledge of the plant and its operations cannot be challenged. Yet the rising costs indicate that all is not correct in the plant, and BAC management wants to be sure that the new chief engineer does not perpetuate the present conditions."

Mr. Dirksen tells you that Paulette Machon will be informing Hänness and Markham that she has engaged Machinery and Mechanical Engineers Inc to study the hot water and power generating system in their power house, and that they are to expect you.

You visit the BAC power plant in Rossburn today. During your talks to plant staff and tours of the plant, you make the following notes:

1. Housekeeping excellent—whole place shines (but is this only surface polish for impression of visitors?).
2. Maintenance logs are inadequately kept—need to be done more often. Need more detail. Equipment files not up to date and not properly filed.
3. Boiler cleaning badly neglected. Firm instructions re boiler cleaning need to be issued by head office.
4. Flow meters are of doubtful accuracy. May be overreading. Not serviced for three years. Manufacturer's service department should be contacted (these are Weston meters). Manufacturer needs to be called in to do a complete check and then recalibrate meters.
5. Overreading of meters could give false flow figures—make plant seem to produce more steam than is actually produced.
6. Good housekeeping obviously achieved by neglecting maintenance. Incorrectly placed emphasis probably caused by frequent visits from company president, who likes to bring in important visitors and impress them. Hänness likes reflected glory (so does Markham).
7. Shift engineers are responsible for maintenance of pumps and vacuum equipment. Not enough time given over to this. They seem to prefer straight replacement of whole units on failure rather than preventive maintenance. Costly method! Obviously more breakdowns: they wait for a failure before taking action. A preventive maintenance plan is needed.
8. Markham seems OK. Genial type; obviously knows his power house. Proud of it! But seems to resist change. Definitely resents suggestions. Does he lack all-round knowledge? Is he limited only to what goes on in his plant? Is he afraid of new ideas because he doesn't understand them? Young staff hinted at this: too loyal to say it outright, but I felt they were restive, hampered by his insistence that they use old techniques that are known to work but are slow. Nothing concrete was said—I just "felt" it.
9. Hänness has done a good job training Markham. Made him a carbon copy. Hänness doesn't do much now. Markham runs the show, and has for over a year. He *expects* to get the job when Hänness retires. It'll be a real blow to him if he doesn't. BAC might even lose a good company man.
10. Discussed CORLAND 200 power panel with staff. Young engineers had read about it in "Plant Maintenance"—eager to have one installed (I described the one I'd seen at Pinewood Paper Mill). But Hänness and Markham knew nothing about it—weren't interested. Are they not keeping up to date with technical magazines?

When you return to your office you tell Gerald Dirksen of your findings. He suggests you write an evaluation report for Ms. Machon and that you address both the technical problems and the personnel difficulties within the one report.

PROJECT 5.4: PROPOSAL FOR MEASURING LIGHT INTENSITY

You are an engineering technician in the Highways Department of your state, and your department supervisor is Mel Timlick. One of your responsibilities is to investigate drivers' complaints about highway conditions. Over the past two years you have received considerable mail complaining of inadequate lighting at highway intersections. At the same time you also have been receiving brochures from manufacturers describing new highway lighting products (mostly reflectors and incandescent and sodium lamps).

On a number of occasions you and Mel have discussed carrying out a lighting study but have never had sufficient funds to assign to it (the urgent need for highway repairs has always seemed to take precedence).

But now you have an idea. You are an amateur photographer, and last July 4 when you were photographing the fireworks display at a local park it occurred to you that you could use a camera to test illumination levels at highway intersections. (One of the factors inhibiting the Highways Department from carrying out lighting surveys previously has been the high cost of purchasing a digital light meter with an attached printer. It's a problem because it would have only very limited use, and therefore becomes a particularly expensive item to buy.) Your rationale is this:

- You could use inexpensive black and white high-speed 35 mm film (ASA 400 [minimum] to ASA 1000).
- You could borrow a 35 mm single lens reflex (SLR) camera from the Public Relations section of the Department of Tourism.
- You could set up a large (about 6 × 4.5 foot) board with a black and white pattern painted on it, and set it up vertically in the intersection. Then you could photograph it from different directions and at distances of 5, 10, 20, and 50 yards.
- You could paint different colored patterns or stripes on the board to see how they show up under varying lighting conditions.

When pondering further about the idea, you recognize there are additional factors you would have to consider:

1. The number of traffic intersections that would need to be tested (the majority of complaints are about 43 traffic interchanges in the state).
2. The quantity of new products to be tested (reflectors and lamps). You estimate nine manufacturers have new products that are worth evaluating.
3. The need for quality control in both film procurement and processing. All film should come from a single manufacturer and be bought from a single source at the same time (i.e. be part of the same production batch). Similarly, all exposed film should be processed and printed at a single location at the same time (i.e. using the same chemicals at the same temperature).
4. The need for similar ambient light conditions at each location (e.g. the same time of night; the absence or presence of the moon in roughly the same phase; the absence or presence of cloud cover; and the absence of traffic).
5. The need to establish a standard "shooting method." That is, by experimenting beforehand, to predetermine an optimum lens aperture setting and exposure time, such as f.16 at 1/125 second.
6. The amount of film (number of exposures) you would need at each location. Probably, two 36-exposure films would suffice.
7. The need to keep several blank exposures at the end of the film used at one site, to use at the start of the shoot at the next site (as a cross-control to establish that the new roll of film is exactly the same quality as the previous one).

8. The need to conduct a "trial run" at two locations (one well lit and one poorly lit) to ensure you are getting definable images at both ends of the range, and so iron out any bugs before you start shooting.

What would the cost be? You consider six aspects: labor; travel; new product acquisition and installation; designing and painting a pattern on a board; purchasing film; and processing and printing the film.

You work out travel costs as an average of $185 per location for 22 locations, and nothing for the remainder. This includes travel, accommodation for one night (applicable only at remote locations, which is 22 of the 43 intersections), and meals.

You obtain a quotation from Multi-Media Consultants at 2020 Whyte Avenue to make the board for $325.

You believe you could persuade product manufacturers to provide and install product samples free of charge (you think they would be eager to have their products tested, as a step toward a potential sale).

You will have to determine film cost and processing by telephoning around (to reliable, well-established film-processing houses).

You talk to Fern Kosteniuk, head of the Department of Tourism, and she agrees to lend your department an SLR camera and tripod for two months.

You telephone around and establish that the best price for a Vancourt model 2120A digital light meter with integrated printer (the standard in the light measurement trade) would be $6295. No rental agreements are available.

Now write your proposal to Mel Timlick. Remember that, if he wants approval of your idea, he will have to get the agreement of the chief engineer of the Department of Highways.

CHAPTER 6
Formal Reports

Formal reports require more careful preparation than the informal and semi-formal reports described in the previous chapters. Because they will be distributed outside the originating company, their writers must consider the impression the report will convey of the entire company. Harvey Winman recognized long ago that a well-written, esthetically pleasing report can do much to convince prospective clients that H. L. Winman and Associates should handle their business, whereas a poorly written, badly presented report can cause clients to question the company's capability. Harvey also knows that the initial impression conveyed by a report can influence a reader's readiness to plough through its technical details.

The presentation aspect must convey the originating company's "image," suit the purpose of the report, and fit the subject it describes. For instance, a report by a chemical engineer evaluating the effects of diesel fumes on the interior of bus garages would most likely be typed on standard bond paper, and its cover, if it had one, would be simple and functional. At the other end of the scale, a report by a firm of consulting engineers selecting a college site for a major metropolis might be printed professionally and bound in an artistically designed book-type folder. But regardless of the appearance of a report, its internal arrangement will be basically the same.

Formal reports are made up of several standard parts, not all of which appear in every report. Each writer uses the parts that best suit the particular subject and the intended method of presentation. There are six major and several subsidiary parts on which to draw, as shown in Table 6-1. Opinions differ throughout industry as to which is the best arrangement of these parts. The two arrangements suggested in the section on the complete formal report later in this chapter, and illustrated in the mini-reports in Figures 6-5 and 6-10, are those most frequently encountered. Knowledge of these parts and the two basic arrangements will help engineers and technicians adapt quickly to the variations in format preferred by their employers. The parts as they appear in the most widely used format are listed in Table 6-1.

MAJOR PARTS

The six major parts form the central structure of every formal report and in the traditional arrangement are known by the acronym SIDCRA (which is formed from the first letter of each part).

SUMMARY

The summary is a brief synopsis which tells readers quickly what a report is all about. Normally it appears immediately after the title page, where it can be found easily. It identifies the purpose and most important features of the report, states the main conclusion, and sometimes makes a recommendation. It does this in as few words as possible, condensing the narrative of the report to a handful of succinct sentences. It also has to be written so interestingly—so enthusiastically— that it encourages readers to read further.

The summary is considered by many to be the most important part of a report and the most difficult to write. It has to be informative, yet brief. It has to attract the reader's attention but must be written in simple, nontechnical terms. It has to be directed to the executive reader, yet it must be readily understood by almost any reader.

Table 6-1 Traditional Arrangement of Formal Report Parts

Cover or jacket
Title page
SUMMARY
Table of contents
INTRODUCTION
DISCUSSION
CONCLUSIONS
RECOMMENDATIONS
References or bibliography
APPENDIXES

NOTES:

1. MAJOR PARTS are in capital letters; subsidiary parts are in lower-case letters.

2. A cover letter or executive summary normally accompanies a formal report.

Generally, the first person in an organization who sees a report is a senior executive, who may have time to read only the summary. If the executive's interest is aroused, he or she will pass the report down to the technical staff to read in detail. But if the summary is unconvincing, the executive is more likely to put the report aside to read "later;" if this happens, the report may never be read.

Because a summary is so important, always write it last, after the remainder of your report has been written. Only then will you be fully aware of the report's highlights, main conclusions, and recommendations. To write the summary first can prove frustrating, because you will not yet have hammered out many of the finer points. You need the knowledge of a soundly developed report fixed firmly in your mind if you are to fashion an effective summary.

For a summary to be interesting, it must be informative; if it is to be informative, it must tell a story. It should have a beginning, in which it states why the project was carried out and the report written; a middle, in which it highlights the most important features of the whole report; and an end, in which it reaches a conclusion and possibly makes a recommendation. The following example illustrates how the interest is maintained in an informative summary.

Informative Summary

A specimen of steel was tested to determine whether a job lot owned by Northern Railways could be used as structural members for a short-span bridge to be built at Peele Bay in northern Alaska. The sample proved to be G40.12 structural steel, which is a good steel for general construction but subject to brittle failure at very low temperatures.

Although the steel could be used for the bridge, we consider that there is too narrow a safety margin between the $-51°C$ temperature at which failure can occur, and the $-47°C$ minimum temperature occasionally recorded at Peele Bay. A safer choice would be G40.8C structural steel, which has a minimum failure temperature of $-62°C$.

Other informative summaries preface the two formal reports at the end of this chapter and both the semiformal evaluation report in Figure 5-4 and the proposal in Figure 5-5 of Chapter 5.

Some writers prefer to write a topical summary for reports that describe history or events, or that do not draw conclusions or make recommendations. As its name implies, a topical summary simply describes the topics covered in the report without attempting to draw inferences or captivate the reader's interest.

Topical Summary

Construction of Alaska's Minnowin Point Generating Station was initiated in 1988, and first power from the 1340 MW plant is scheduled for 1994. A general description of the structures and problems peculiar to construction of this large development in an arctic climate is presented. The river diversion program, permafrost foundation conditions, and major equipments are described. The latter include the 16 propeller turbines, among the largest yet installed, each rated at 160,000 horsepower.

Because they are less results-oriented, topical summaries are *not* recommended for most formal reports.

In a formal report the summary should have a page to itself, be centered on the page, and be prefaced by the word "Summary" (or, sometimes, "Abstract"). If it is very short, it may be indented equally on both sides to form a roughly square block of information.

INTRODUCTION

The introduction begins the major narrative of the report by preparing readers for the discussion that follows. It orients them to the purpose and scope of the report and provides sufficient background information to place them mentally in the picture before they tangle with technical data. A well-written introduction contains just enough detail to lead readers quickly into the major narrative.

The length of an introduction and its depth of detail depend mostly on the reader's knowledge of the topic. If you know that the ultimate reader is technically knowledgeable, but at the same time you have to cater to the executive reader who is probably only partly technical, write the introduction (and conclusions and recommendations) in semitechnical language. This permits semitechnical executives to gain a reasonably comprehensive understanding of the report without devoting time and attention to the technical details contained in the discussion.

Most introductions contain three parts: **purpose, scope**, and **background information**. Frequently, the parts overlap, and occasionally one of them may be omitted simply because there is no reason for its inclusion. The introduction normally is a straightforward narrative of one or more consecutive paragraphs; only rarely is it divided into distinct sections preceded by headings. It always starts on a new page (normally identified as page 1 of the report) and is preceded by the report's full title. The title is followed by the single word "Introduction," which can be either a center heading or a side heading as shown in Figure 6-1.

The **purpose** explains why the project was carried out and the report is being written. It may indicate that the project has been authorized to investigate a problem and recommend a solution, or it may describe a new concept or method of work improvement that the report writer believes should be brought to the reader's attention.

The **scope** defines the parameters of the report. It describes the ground covered by the report and outlines the method of investigation used in the project. If there are limiting factors, it identifies them. For example, if 18 methods for improving packaging are investigated in a project but only four are discussed in the report, the scope indicates what factors (such as cost, delivery time, and availability of space) limited the selection. Sometimes the scope may include a short glossary of terms that need to be defined before the reader starts to read the discussion.

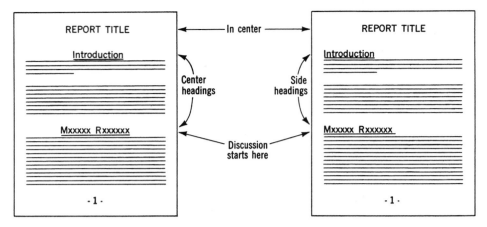

Figure 6-1. Headings used at the start of the introduction and discussion.

Background information comprises facts readers must know if they are to fully understand the discussion that follows. Facts may include descriptions of conditions or events that caused the project to be authorized, and details of previous investigations or reports on the same or a closely related subject. In a highly technical report, or when a significant time lapse has occurred between it and previous reports, background information may also provide a theory review and references to other documents. A lengthy theory review and a long list of documents are often placed in appendixes, with only a brief summary of the theory and a quick reference to the list appearing in the introduction.

The introductions shown here, plus those forming part of the two sample reports at the end of this chapter, are representative of the many ways a writer can introduce a topic.

Introduction 1

Northern Railways plans to build a short-span bridge one half mile north of Lake Peele in northern Alaska and has a job lot of steel the company wants to use for constructing the bridge. In letter NR-70/LM dated March 20, 19XX, Mr. David L. Harkness, Northern Area Manager, requested H. L. Winman and Associates to test a sample of this steel to determine its properties and to assess its suitability for use as structural members for a bridge in a very low temperature environment. *Background* *Purpose*

The test performed was the Charpy Impact Test. Two series of tests were run, one parallel to the grain of the test specimen and the other transverse to the grain, at 10° increments from +22°C down to −50°C. *Scope*

Introduction 2

H. L. Winman and Associates was commissioned by Mr. D. C. Scorobin, President and General Manager of Auto-Marts Inc. of Dallas, Texas, to select a Cleveland site for the first of a proposed chain of drive-in groceries to be built in the northeastern United States. This area was chosen since it represents an average community in which to assess customer acceptance of such a service. } *Purpose*

Aside from exterior facade and foundation details, Auto-Marts are built to a standard 80 ft by 30 ft pattern with an order window at one end of the 80 ft wall, and a delivery window at the other. Auto-Marts carry only a limited selection of household goods, but boast 90-second service from the time an order is placed to its delivery at the other end of the building. For this reason, Auto-Marts attract people hurrying home from work, the impulse buyer, and the late-night shopper, rather than the selective shopper. The store depends more on the volume of customers than on the volume of goods sold to each individual. } *Background*

The chief consideration in selecting a site must therefore be a location on the homeward-bound site of a main trunk road into a large residential area. The site must have quick and easy entry onto and exit from this road, even during peak rush hour traffic. The residential area should be occupied mainly by upper and middle class people who own cars and have money to spend. And there should be little competition from walk-in grocery stores. } *Scope*

DISCUSSION

The discussion, which normally is the longest part of a report, presents all the evidence (facts, arguments, details, data, and results of tests) that readers will need to understand the subject. The writer must organize this evidence logically to avoid confusing readers, and present it imaginatively to hold their interest.

In the discussion, a writer such as Dan Skinner (see Chapter 1) describes fully what he set out to do, how he went about it, what he actually did, and what he found out as the result of his efforts. He may consider the writing a chore and simply record his facts dutifully but unenthusiastically, or he can organize the information so interestingly that his readers feel almost as though they are reading a short story.

There is no reason why the discussion should not have a plot. It can have a beginning, in which Dan describes a problem and outlines some background events; a middle, which tells how he tackled the problem; dramatic effect, which he can use to hold readers' interest by letting them anticipate his successes, as well as share his disappointments when a path of investigation results in failure; a climax, in which he permits readers to share his pleasure as he describes how the problem was resolved; and a denouement (a literary word for final outcome), in which he ties up loose ends of information and evaluates the results. Storytelling and dramatization can be such important factors in holding reader interest that no report writer can afford to overlook them.

There are three ways you can build the discussion section of a report:

By *chronological development*—in which you present information in the order that events occurred.

By *subject development*—in which you arrange information by subjects, grouped in a predetermined order.

By *concept development*—in which you present the information as a series of ideas which reveal imaginatively and coherently how you reasoned your way to a logical conclusion.

Reports using the chronological or subject method offer less room for imaginative development than those using the concept method, mainly because they depend on a straightforward presentation of information. The concept method can be very persuasive. Identifying and describing your ideas and thought processes helps your readers organize their thoughts along the same lines.

As a report writer, you must decide early in the planning stages which method you intend to use, basing your choice on which is most suitable for the evidence you have to present. Use the following notes as a guide.

Chronological Development.   A discussion which uses the chronological method of development is simple to organize and write. It requires minimal planning: you simply arrange the major topics in the order they occurred, and eliminate irrelevant topics as you go along. You can use it for very short reports, for laboratory reports showing changes in a specimen, for progress reports showing cumulative effects or describing advances made by a project group, and for reports of investigations that cover a long time and require visits to many locations to collect evidence.

But the simplicity of the chronological method is offset by some major disadvantages. Because it reports events sequentially it tends to give emphasis to each event regardless of importance, which may cause readers to lose interest. If you read a report of five astronauts' third day in orbit, you do not want to read about every event in exact order. It may be chronologically true to report that they rose at 7:15, ate breakfast at 7:55, sighted the second stage of their rocket at 9:23, carried out metabolism tests from 9:40 to 10:50, extinguished a cabin fire at 11:02, passed directly over Houston at 11:43, and so on, until they retired for the night. But it can make deadly dull reading. Even the exciting moments of a cabin fire lose impact when they are sandwiched between routine occurrences.

When using chronological development, you must still manipulate events if you are to hold your readers' attention. You must emphasize the most interesting items by positioning them where they will be noticed, and deemphasize less important details. Note how this has been done in the following passage, which groups the previous events in descending order of interest and importance:

The highlight of the astronauts' third day in orbit was a cabin fire at 11:02. Rapid action on their part brought the fire, which was caused by a short circuit behind panel C, under control in 38 seconds. Their work for the day consisted mainly of metabolism tests and. . . .

They sighted the first stage of their rocket on three separate occasions, first at 9:23, then at . . . and . . ., when it passed directly over Houston. Their meals were similar to those of the previous day.

This narrative uses *chronological* order to describe the events on a day-to-day basis, and a modified *subject* order to describe the events for each day.

Over a five-year period H. L. Winman and Associates has been investigating the effects of salt on concrete pavement. Technical editor Anna King has suggested that the final report should have chronological development, because the investigation recorded the extent of concrete erosion at specific intervals. She wanted the engineering technician writing the report to describe how the erosion increased annually in direct relation to the amount of salt used to melt snow each year, and for the final conclusion to demonstrate the cumulative effect that salt had on the concrete.

Subject Development.   If the previous investigation had been broadened to include tests on different types of concrete pavements, or if both pure salt and various mixtures of salt and sand had been used, then the emphasis would have shifted to an analysis of erosion on different surfaces or caused by various salt/sand mixtures, rather than a direct description of the cumulative effects of pure salt. For this type of report Anna King would have suggested arranging the topics in *subject* order.

The subject order could be based on different concentrations of salt and sand. The technician would first analyze the effects of a 100% concentration of salt, then continue with salt/sand ratios of 90/10, 80/20, 70/30, and so on, describing the results obtained with each mixture. Alternatively, the technician could select the different types of pavement as the subjects, arranging them in a specific order and describing the effects of different salt/sand concentrations on each surface.

The head of the drafting department, Barry Brewster, has been investigating blueprint machines and he plans to recommend the most suitable model for installation in his department. He writes his report using the subject method. First, he tells readers that he intends to evaluate the speed, economy of operation, and usefulness of each blueprinter, and then he emphasizes the most suitable machine by discussing it either first or last. If he chooses to describe it first, he can state immediately that it is the best blueprinter, and say why. He can then discuss the remaining models in descending order of suitability, comparing each to the selected machine to show why it is less suitable. If he prefers to describe the best machine last, he can discuss the others in ascending order of suitability, stating why he has rejected each one before he starts describing the next.

Also suitable for subject development is an analysis of insulating materials being conducted by laboratory technician Rhonda Moore. In her report, she groups materials having similar basic properties or falling within a specific price bracket. Within the groups she describes the test results for each material and draws a conclusion as to its relative suitability for a specified purpose. Like Barry, she describes the materials in ascending or descending order of suitability.

The subject method of development permits Barry and Rhonda to either hide their personal preferences until almost the end of their reports or, if they prefer, let their preferences show all the way through. They have the choice of using an impersonal, objective approach, or a personal, subjective approach similar to that for the concept method. These alternative approaches are illustrated in Figure 6-2.

Rhonda prefers the primarily objective approach. She discusses each insulating material (the subject) without using any words that might make her readers think she believes the material to be a good or bad choice:

> Material A has an R-17 rating. It is a dense, brown, tightly packed fibrous substance enclosed in a continuous, one-meter wide envelope, which is supplied in 10-, 20-, and 50-meter rolls. To cover a standard 4 × 3 meter area would cost $108.00. The fire resistance of Material A is. . . .

Rhonda has only to introduce the insulating materials, and possibly outline how they were selected, before she describes them. She need not identify her selection criteria until almost the end of the discussion, after she has objectively described each material but before she subjectively compares their main features.

Barry prefers the more colorful subjective approach: he wants to persuade his readers to accept or reject each blueprinter (subject) as they read about it:

> The Image-copy is the only machine that meets all our technical requirements. It has a high speed of 25 feet per minute, an extended continuous-roll feature with automatic cutter, and both manual and automatic stacking capabilities. Although its purchase price of $5700.00 is $700.00 above budget, its high reliability (reported by users such as Multiple Industries and Apex Architectural Services) would probably cause less equipment downtime and consequent production frustrations than other models. Specific features are. . . .

Barry must establish the criteria or parameters which most affect his choice, and must convince his readers why they are important, *before* he describes each blueprinter; if he does not, readers will not readily accept his subjective statements.

Whichever approach you use for a comparative analysis, your readers should find that the discussion leads comfortably and naturally into the conclusions you eventually draw. The conclusions should never contain any surprises.

Concept Development.   By far the most interesting reports are those using the concept method of development. They need to be organized more carefully than reports using either of the previous methods, but they give the

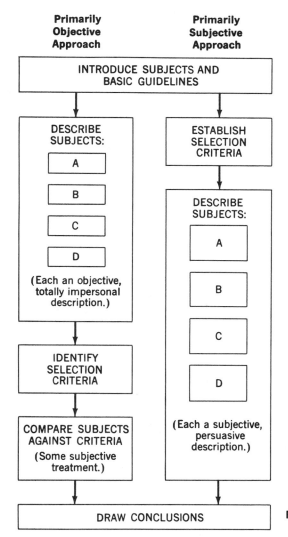

Figure 6-2. Alternative methods for a comparative analysis.

writer a tremendous opportunity to devise an imaginative arrangement of the topic—which is the best way to hold a reader's attention.

They can also be very persuasive. Because the report is organized in the order in which you reasoned your way through the investigation, your readers will much more readily appreciate the difficulties you encountered, and frequently will draw the correct conclusion even before they read it. This helps readers to feel they are personally involved in the project.

You can apply the concept approach to your reports by thinking of each project as a logical but forceful procession of ideas. If you are personally convinced that the results of your investigation are valid, and remember to explain in your report *how* you reached the results and *why* they are valid, then you will probably be using the concept method properly.

Always anticipate reader reaction. If you are presenting a concept (an idea, plan, method, or proposal) which readers are likely to accept, then use a straightforward four-step approach:

1. Describe your concept in a brief overview statement.
2. Discuss how and why your concept is valid: offer strong arguments in its favor, starting with the most important and working down to the least important.
3. Introduce negative aspects, and discuss how and why each can be overcome or is of limited importance.
4. Close with a restatement of your concept, its validity, and its usefulness.

But if you are presenting a controversial concept, or need to overcome reader bias, then modify your approach. Try to overcome objections by carefully establishing a strong case for your concept *before* you discuss it in detail.

Special Projects Department Head Andy Rittman used this approach in a report he prepared for Mark Dobrin, owner/manager of a company making extruded plastic and metal parts for the defense industry. Manufacturing costs had risen steeply over the past two years, and Mark's prices had become uncompetitive. Mark thought he should replace some of his older, less efficient equipment, so he asked H. L. Winman and Associates to evaluate his needs.

Andy Rittman quickly realized that if Mark was to avoid going completely out of business, he would have to replace much of his equipment with microprocessor-controlled machines and do it soon. Because the cost would be high, he would have to lease rather than buy the new equipment.

Here, Andy had a problem. Mark was as old-fashioned as some of his extruders and shapers, and throughout his life he had steadfastly refused to purchase anything that he could not buy outright. He was unlikely to change now.

In his report, Andy used a carefully reasoned argument to prove to Mark that he needed a lot of new equipment and that the only feasible way he could acquire it would be to lease it. Throughout, Andy wrote objectively but sincerely of his findings, hoping that the logic of his argument would swing Mark around to accepting his recommendation. Very briefly, here is the step-by-step approach Andy used:

He opened with a summary which told Mark that to avoid bankruptcy he would have to invest in a lot of expensive equipment and make extensive changes in his operating methods.

Andy then produced financial projections to prove his opening statement and discussed the productivity and profitability necessary for Mark to remain in business.

He discussed why Mark's equipment and methods were inefficient, introduced the changes Mark would have to make, established why each change was necessary, and demonstrated how each would improve productivity. (Andy referred Mark to an appendix for specific equipment descriptions, justifications, and costs.)

Andy then introduced two sets of cost figures: one for making the minimum changes necessary for Mark's business to survive, and the second for more comprehensive changes that would ensure a sound operating basis for the future. He commented that both would require capital purchases likely to be beyond Mark's financial resources.

He outlined alternative financing methods available to Mark, the implications and limitations of each, and the financial effect each would have on Mark's business. (Although he introduced leasing, Andy made no attempt to persuade Mark that he would have to lease; he let the figures speak for themselves.)

Andy concluded by summarizing the main points he had made: that new equipment *must* be acquired; that to buy even the minimum equipment was beyond Mark's financial resources; and that, of the financing methods available, leasing was the most feasible.

In his recommendation, Andy suggested that Mark should make comprehensive changes and lease the new equipment. (By then, Andy had become so involved with Mark's predicament that he wrote strongly and sincerely.)

Even though the concept method challenges a writer to fashion interesting reports, it is not always the best reporting medium. For instance, the concept method could have been used for the blueprinter report mentioned earlier, but it is doubtful whether the topic would have warranted full analysis of the author's ideas. When a topic is fairly clear-cut, there is no need to lead the reader through a lengthy "this is how I thought it out" discussion. Reserve the concept method for topics that are controversial, difficult to understand, or likely to meet reader resistance.

Whichever method you use, avoid cluttering the discussion with detailed supporting information. Unless tables, graphs, illustrations, photographs, statistics, and test results are essential for reader understanding while the report is being read, banish them to the appendix. But always refer to them in the discussion, like this:

The test results attached at Appendix C show that aircraft on a bearing of 265°T experienced considerably weaker reception than aircraft on any other bearing. This was attributed to. . . .

Readers *interested only in results* will consider that this statement tells them enough and continue reading the report. Readers interested in *knowing how the results were obtained*—who want to see the overall picture—will turn to Appendix C to find out how you went about performing the tests. They probably will not do so immediately, but will return to it when they have finished the report or have reached a suitable stopping place. In neither case are readers slowed down by supporting information which, though relevant, is not immediately essential to an understanding of the report.

Unless the discussion is very short, divide it into a series of sections each preceded by an informative heading. If a section is long or naturally subdivides into subtopics, also use subordinate headings. Major topics normally demand center or side headings, while minor topics need only side or paragraph headings, as shown in Figure 6-3 and Figure 11-4 of Chapter 11. These headings eventually form the table of contents for the report.

After each heading, start the section with an overview statement to describe what the section is about and suggest what conclusion will be drawn from it. Overview statements are miniature summaries which direct a reader's attention to the point you want to make. If the section is short, the overview statement may be a single sentence: if the section is long, it will probably be a short paragraph.

At the end of each major section, insert a concluding statement that summarizes the result of the discussion within that section. From these section conclusions you can later draw your main conclusions.

CONCLUSIONS

Conclusions briefly state the major inferences that can be drawn from the discussion. They must be based entirely on previously stated information and *never* surprise readers by introducing new material or evidence to support your argument. If there is more than one conclusion, state the main conclusion first and follow it with the remaining conclusions in decreasing order of importance. This is shown in the following examples, which present the same conclusions in both narrative and tabular form.

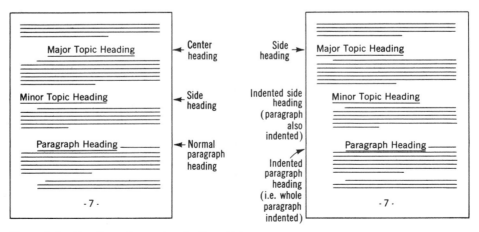

Figure 6-3. Headings show subordination of ideas.

Narrative Conclusion

The operational life of QK 3801 magnetrons can be increased 30% by installing modification XL-1 on our radar transmitters. This will save $8300 over the next 12 months, but will require that we increase magnetron run-in time from 4 to 8 hours.

Tabular Conclusion

Installation of modification XL-1 on our radar transmitters will result in:

1. A 30% increase in the operational life of QK 3801 magnetrons.
2. An $8300 saving over the next 12 months.
3. A change in operating procedures to increase magnetron run-in time from 4 to 8 hours.

Because conclusions are *opinions* (based on the evidence presented in the discussion), they must never tell the reader what to do. This must always be left to the recommendations.

RECOMMENDATIONS

Recommendations appear in a report when the discussion and conclusions indicate that further work needs to be done, or you have described several ways to resolve a problem or improve a situation and want to identify which is best. Write recommendations in strong definite terms to convince readers that the course of action you advocate is valid. Use the first person and active verbs:

I recommend that we build a five-station prototype of the Microvar system and test it operationally.

Unfortunately, in many industries, and often in government circles, using the first person is severely frowned upon. Consequently, many reports fail to contain really strong recommendations because their writers hesitate to say "I recommend . . ." and so use the indefinite "It is recommended . . .". If you feel you cannot use the personal "I," try using the plural "we" to indicate that the recommendations represent the company's viewpoint. For example:

We recommend building a five-station prototype of the Microvar system. We also recommend that you:

1. Install the prototype in Railton High School.
2. Commission a physics teacher experienced in writing programmed instruction manuals to write the first programs.
3. Test the system operationally for three months.

Because recommendations must be based solidly on the evidence presented in the discussion and conclusions, they can *never* introduce new evidence or new ideas.

APPENDIX

Related data not necessary to an immediate understanding of the discussion should be placed further back in the report, in the appendix. The data can vary from a complicated table of electrical test results to a simple photograph of a blown resistor. The appendix is a suitable place for manufacturer's specifications, graphs, comparative data, drawings, sketches, excerpts from other reports or books, cost analyses, and correspondence. There is no limit to what can be placed in the appendix, providing it is relevant and reference to it is made in the discussion.

The term "appendix" applies to only one set of data. In practice there can be many sets, each of which is called an appendix and assigned a distinguishing letter, such as A, B, or C. The sets of data are collectively referred to as either "appendixes" or "appendices"; the former term is preferred in the United States.

The importance of an appendix has no bearing on its position in the report. Whichever set of data is mentioned first in the discussion becomes Appendix A, the next set becomes Appendix B, and so on. Each appendix is considered a separate document complete in itself and is page-numbered separately, with its front page labeled 1. Examples of appendixes appear at the end of formal report 1, later in this chapter.

SUBSIDIARY PARTS

In addition to the six major parts of a report, there are several additional parts that perform more routine functions. Although I refer to these as "subsidiary," they nevertheless contribute much to the effectiveness of a report. A writer must never assume that these seemingly peripheral details will look after themselves. Because they directly affect the image conveyed of both the writer and the company, they must be prepared no less carefully than the rest of the report.

COVER

Almost every major formal report has a cover. It may be made of cardboard printed in multiple colors and bound with a dressy plastic binding, or it may be only a light cover of colored fiber material stapled on the left. The cover not only informs readers of the report's main topic but also conveys an image of the company that originated it. This "matching" of subject matter and company image plays an important part in setting the correct tone for a report.

The cover should contain the report title, the name of the originating company and, perhaps, the name of its author. The title should be set in bold letters well-balanced on the page and separated from any other information.

Choice of a title is particularly important. It should be short yet informative, implying that the report has a worthwhile story to tell. Compare these titles:

Original Vague Title

RADOME LEAKAGE

Revised Informative Title

POROSITY OF FIBERGLAS CAUSES RADOME LEAKS

Technical officers at remote radar sites would glance at the first title with only a muttered, "Looks like somebody else has the same problem we've got." But the second title would encourage them to stop and read the report. They would recognize that someone may have found the cause of (and, hopefully, a remedy for) a trouble spot that has bothered them for some time.

Note how the following titles each *tell* something about the reports they precede:

REDUCING AMBIENT NOISE IN AIR TRAFFIC
CONTROL CENTERS

EFFECTS OF DIESEL EXHAUST ON LATEX PAINTS

SALT EROSION OF CONCRETE PAVEMENTS

These titles may not attract every potential reader who comes across them, but they will certainly gain the attention of anyone interested in the topics they describe.

TITLE PAGE

The title page normally carries four main pieces of information: the report title (the same title that appears on the cover); the name of the person, company, or organization to whom the report is directed; the name of the company originating the report (sometimes with the author's name); and the date the report was completed. This page may also contain the authority or contract under which the report has been prepared, a report number, a security classification such as CONFIDENTIAL or SECRET, and a copy number (important reports given only limited distribution are sometimes assigned copy numbers to control and document their issue). All this information must be tastefully arranged on the page, as has been done in the first sample report later in this chapter.

TABLE OF CONTENTS

All but very short reports contain a Table of Contents (T of C). The T of C not only lists the report's contents but also shows how the report has been arranged. Just as a prospective book buyer will scan the contents page to find out what a book contains and whether it will be of interest, potential readers will scan a report's T of C to identify how the author has organized the work.

The pleasing arrangement for a T of C on page 173 shows that the single word CONTENTS, rather than "Table of Contents," is preferred. The contents page also contains a list of appendixes, with each identified by its full title. In long reports you may also insert a list of illustrations and their page numbers between the T of C and the list of appendixes.

REFERENCES (ENDNOTES), BIBLIOGRAPHY, AND FOOTNOTES

A report writer who refers to another document, such as a textbook, journal article, report, or correspondence, or to other persons' data or even a conversation, must identify the source of this information in the report. To avoid cluttering the report narrative with extensive cross-references, the reference details are placed in a storage area at the end of the report or at the foot of the page. This storage area is known as a List of References or a Bibliography. Specifically:

A **list of references** is the most convenient and popular way to list source documents. The references are typed as a sequentially numbered list at the end of the narrative sections of the report (usually immediately ahead of the appendix). Such numbered references are frequently referred to as *Endnotes*. For a short example, see page 183.

A **bibliography** is simply an alphabetical listing of the documents used to research and conduct a project. The documents are listed in alphabetical order of authors' surnames, and the list is placed at the end of the report narrative. A bibliography may list many more documents than are referred to in the report. An example appears on page 161.

Footnotes, which are printed at the foot of the page on which the particular reference appears, are seldom used in contemporary reports.

Preparing a List of References

References should contain specific information, arranged in this sequence:

(a) Author's name (or authors' names).
(b) Title of document (article, book, paper, report).

(c) Identification details, such as:

For a book: city and state (or country) of publication, publisher's name, and year of publication.

For a magazine article or technical paper: name of magazine or journal, volume and issue number, and date of issue.

For a report: report number, name and location of issuing organization, and date of issue.

For correspondence: name and location of issuing organization; name and location of receiving organization; and the letter's date.

For a conversation or speech: name and location of speaker's organization; name, identification, and location of listener; and the date.

(d) The page number (if applicable) on which the referenced item appears or starts.

Referencing a Book

If you are referring to information in a book authored by only one person, the entry in your list of references should contain:

(a) Author's name (in natural order: first name and/or initials, and then surname).
(b) Book title (always underlined).
(c) City of publication, publisher's name, and year of publication (all within parentheses).
(d) Page number (the first page of the referenced pages).

If it is your first reference, and you are referring to an item on page 174 of the book, your entry would look like this:

1. Laurinda K. Wicherly, <u>Fiberoptic Modes of Communication</u> (New York: The Moderate Press Inc, 1992), p. 74.

If a book has two authors, both are named:

2. David B. Shaver and John D. Williams, <u>Management Techniques for a Research Environment</u> (Boulder, Colorado: Witney Publications, 1991), p. 215.

But if there are three or more authors, only the first-named author is listed and the remaining names are replaced by "and others":

3. Donald R. Kavanagh and others, . . .

If a book is a second or subsequent edition (as this book is), the edition number is entered immediately after the book title:

4. Ron S. Blicq, <u>Technically—Write!</u> 4th ed. (Englewood Cliffs, NJ: Prentice Hall, Inc., 1993).

Some books contain sections written by several authors, each of whom is named within the book, with the whole book edited by another person. If your reference is to the whole book, then identify it by the editor's name and insert the abbreviation "ed." immediately after the name:

5. Donna R. Linwood, ed., <u>Seven Ways to Make Better Technical Presentations</u> (Portland, Oregon: Bonus Books Inc., 1992).

But if your reference is to a particular section of the book, identify it by the specific author, enclose the section title in quotation marks, underline the book title, and then name the editor:

6. Kevin G. Wilson, "Preparation: The Key to a Good Talk," <u>Seven Ways to Make Better Technical Presentations</u>, ed. Donna R. Linwood (Portland, Oregon: Bonus Books Inc., 1992), p. 71.

In examples 5 and 6, "ed." means "editor" or "edited by."

Referencing a Magazine or Journal Article

Similarly, if you are referring to an article in a magazine or journal, list these details:

(a) Author's or authors' name(s) (in natural order).
(b) Title of article (always in quotation marks, and *not* underlined).
(c) Title of journal or magazine (underlined).
(d) Volume and issue numbers (shown as two numbers separated by a colon).
(e) Journal or magazine issue date.
(f) Page on which article or excerpt starts (optional entry).

If an article is your seventh reference, it would look like this:

7. Morris Kattermole, "Electronic Conferencing Reduces Travel Costs," <u>Business Bits and Bytes</u>, 4:9, September 1992, p. 38.

If a magazine article does not show an author's name, then the entry starts with the title of the article:

8. "Selling to the EEC: Challenge of the 90's," <u>Technical Marketing</u>, 11:5, May 1991, p. 113.

Referencing a Report

To refer to a report written by yourself or another person, list this information:

(a) Author's name (or authors' names), if the author is identified on the report.
(b) Title of report (underlined).
(c) Report number, or other identification (if any).
(d) Name and location (city and state) of organization issuing report.
(e) Report date.
(f) Page number (if a specific part of the report is being referenced).

Here is an example:

9. Derek A. Lloyd, <u>Effective Communication and Its Importance in Management Consulting</u>. Report No. 61, Smyrna Development Corporation, Atlanta, Georgia, February 18, 1991).

Referencing a Letter or Memorandum

For a letter or memorandum, the entry should be like this:

10. Christine Lamont, Macro Engineering Inc, Phoenix, Arizona. Letter to Marlene Hudson, No. 7 Architectural Group, Dallas, Texas, December 12, 1992.

Referencing a Conversation or Speech

For a conversation or speech, the entry should be like this:

11. David R. Phillips, Lakeside Power and Light Company, Montrose, Ohio, in conversation with Anna King, H. L. Winman and Associates, Cleveland, Ohio, March 16, 1990.
12. Frances R. Cairns, Elwood Martens and Associates, San Diego, California, speaking to the 8th Symposium on Interactive Videodisk Technology, Chicago, Illinois, September 23, 1991.

Remember that when a magazine article, technical paper, or report is published as one of several documents bound into a volume, then it is listed within quotation marks (only the title of the volume is underlined). But if the article, technical paper, or report is published as a *separate*, stand-alone document, the quotation marks are omitted and the title of the article, paper, or report is underlined, as in entry 9.

If you use a word-processing program that allows you to print *in italics*, you may use italics instead of underlining the title of a book, magazine, or report. In effect, you are then performing the same role as a typesetter, who knows that underlined words in a handwritten or typewritten manuscript are to be typeset in italics rather than underlined. The examples in this section are shown with the words underlined; those in the sample formal report have been set in italics (see page 183).

Every entry in a list of references must have a corresponding reference to it in the discussion section of your report. At an appropriate place in the narrative, you should insert a superscript (raised) number to identify the particular reference. It should look like this:

> Earlier tests[3] showed that speeds higher than 2680 rpm were impractical.
> *(Alternatively, the superscript could go here.)*

If your report is being prepared using a word-processing program and printer that cannot create superscript numbers, state the reference number in parentheses:

> Earlier tests (3) showed that speeds . . .

If you refer to the same document several times, your list of references need show full details for that document only the first time you refer to it. Subsequent references can be shown in a shortened form containing only the author's name (or authors' names) and the page number. For example, if the first reference you make is to an item on page 48 of a book, the entry in the list of references would be:

> 1. Wayne D. Barrett, <u>Management in a Technical Domain</u> (San Francisco, California: Martin-Baisley Books, 1992), p. 48.

Now suppose that your second and third sources are other documents but for your fourth source you again refer to *Management in a Technical Domain*, this time quoting from page 159. Now you need list only the author's surname and the new page number:

> 4. Barrett, p. 159.

You would do the same for each future reference to the same document, simply changing the page number each time. (Note that the Latin terms *ibid.* and *op. cit.* are no longer used.) You can even make repeated references to several different documents by the same author by simply inserting the year of publication for the particular document between the author's name and the page number:

> 9. Barrett, 1992, p. 159.

Preparing a Bibliography

A bibliography lists not only the documents to which you make direct reference, but also many other documents which deal with the topic. The major differences between the elements of the list of references and those of the bibliography are:

- Bibliography entries are *not* numbered.
- The name of the first-named author for each entry is reversed, so that the author's surname becomes the first word in the entry. (If there is a second-named author, his or her name is *not* reversed.)

- The *first* line of each bibliography entry is extended about five computer columns or typewriter spaces to the left of all other lines (see Figure 6-4).
- The entries are arranged in alphabetical sequence of first-named authors.
- Punctuation of individual entries is significantly different, with each entry being divided into three compartments separated by periods: (1) author identification; (2) title of book or specific article; and (3) publishing details. (How the periods are positioned is shown in Figure 6-4.)
- Page numbers usually are omitted, since generally the bibliography refers to the whole document. (Reference to a specific page is made within the narrative of the report.)

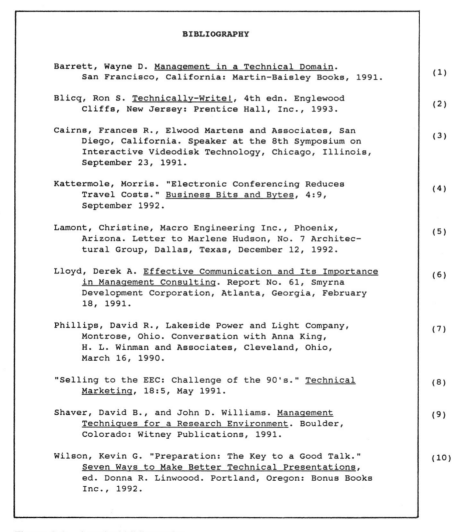

Figure 6-4. A typical bibliography.

Figure 6-4 shows how to list bibliography entries from various sources. The entries for this bibliography have been formed from some of the "reference" entries listed earlier. The number to the right of each entry is cross-referenced to the following list:

(1) Book by one author.
(2) Book by one author, fourth edition.
(3) Conference speech.
(4) Journal (i.e. magazine) article.
(5) Letter.
(6) Report.
(7) Conversation.
(8) Magazine article, with no author identification.
(9) Book by two authors.
(10) Section of book, with section written by one author and whole book edited by another.

Because a bibliography is not numbered, you cannot cross-refer directly to it simply by inserting a superscript number in the report narrative, as can be done with a list of references. The most common method is to insert a parenthetical reference in the narrative, and in it include the author's name (or authors' names) and the page number:

> Although the tests conducted in Alaska (Faversham, p. 261) showed only moderate decomposition . . .

The full descriptive listing for Faversham's book or report is carried in the bibliography.

If several publications by the same author are listed in the bibliography, then the date of the particular publication is included as a parenthetical reference to identify which document is being mentioned:

> The most significant tests were those conducted 22 miles south of Old Crow, N.W.T. (Crosby, 1991, p. 17), which showed that . . .

COVER LETTER

A cover letter is a brief letter that identifies the report and states why it is being forwarded to the addressee. The following letter accompanied the second report in this chapter (see pages 189 to 193).

> Dear Mr. Merrywell:
> We enclose our report No. 8-23, "Selecting New Elevators for the Merrywell Building," which has been prepared in response to your letter WDR/71/007 dated April 27, 19xx.

If you would like us to submit a design for the enlarged elevator shaft, or to manage the installation project on your behalf, we shall be glad to be of service.

Sincerely,

Barry V. Kingsley

H. L. WINMAN AND ASSOCIATES

Some cover letters include comments that draw attention to key factors described in the report or that evolve from it, and sometimes summarize or interpret the report's main findings. This is done in the cover letter preceding the first report in this chapter (see Figure 6-7).

EXECUTIVE SUMMARY

An executive summary is an analytical summary of the purpose of the report, its main findings and conclusions, and the author's recommendation. Unlike the normal report summary prepared for all readers, the executive summary can present detailed information on aspects of particular concern to senior executives, and often may discuss financial implications.

An executive summary can be presented in two ways: externally, as a letter pinned to the front of the report; or internally, as an integral part of the report.

If it is attached to the report like a cover letter, the executive summary allows the recipient to remove it before circulating the report to other readers. Hence, an external executive summary is written like a letter and is addressed to a specific reader—or group of readers—which permits its author to comment on aspects that are intended primarily for their rather than general readers' eyes. The executive summary that precedes Report 1 nicely serves this purpose (see Figure 6-7 on page 169).

If, however, the executive summary is bound within the report so that it will be read by everyone, its purpose becomes more general and its author simply summarizes and perhaps comments on the report's major findings. Rather than being written like a letter, the words EXECUTIVE SUMMARY are centered at the top of the page and the summary is presented like a semiformal report. Such an executive summary may be positioned immediately inside the report cover, behind the title page (in which case it replaces the normal short summary), or after the table of contents.

When an executive summary is bound inside the report, a brief cover letter may also be prepared as a transmittal document and attached to the front of the report.

An executive summary often may precede a major technical proposal, in which a company describes how it can successfully tackle a task at an economical price for the government or another company (see Chapter 5 for a more detailed description of a technical proposal). In this case, the executive summary chiefly discusses financial and legal implications, such as specifications and statements that may be open to more than one interpretation or that will have a marked effect if the proposed task is implemented.

THE COMPLETE FORMAL REPORT

THE MAIN PARTS

Two formal reports are included with this chapter, each typical of the quality of writing and presentation that the technical business industry expects of engineering, science, and computer graduates. The first report is presented in the traditional arrangement discussed so far, and the second in an alternative, pyramidal arrangement. In each case, the parts of the report remain the same but their sequence changes. Capsule descriptions of the main parts are listed in Table 6-2.

The pages preceding each report discuss the report's sequence, identify how the project was initiated and the report came to be written, and comment on both the report and the author's approach. The reports are typed single spaced with two clear lines between paragraphs, which is the style preferred by industry. In comparison, academic institutions tend to prefer double spacing throughout.

TRADITIONAL ARRANGEMENT OF REPORT PARTS

In the traditional arrangement, there is a logical flow of information: the **introduction** leads into the **discussion**, from which the writer draws **conclusions** and makes **recommendations** (the two latter parts sometimes are referred to jointly as the **terminal summary**), as illustrated in the birds-eye mini report in Figure 6-5. This arrangement is used for most technical and business reports.

Formal Report 1
Evaluating Electronic Communication Aids
for Single-Operator Business Owners
(Figures 6-7 and 6-8)

The author of this report is Craig Derwent, who is an engineering technologist at Macro Engineering Inc in Phoenix. He keystroked and edited the report on a personal computer, using *Wordstar Professional 7.0* as his word-processing software, and printed it on a Hewlett Packard III laser printer. The word-processing program automatically justified the right margin (made it vertically straight), inserted page numbers, and assembled the Contents page.

COMMENTS ON THE REPORT

To understand the circumstances leading up to the report, first read Evelyn Wollasey's letter of authorization in Figure 6-6.

The report contains all the major parts described earlier in this chapter. The comments in this section identify how Craig used these parts to develop a convincing case to present to his client.

Table 6-2

THE MAIN PARTS OF A FORMAL REPORT

Cover:	Jacket of report; contains title of report and name of originating company; its quality and use of color reflect company "image."
Title Page:	First page of report; contains title of report, name of addressee or recipient, author's name and company, date, and sometimes a report number.
Summary:	An abridged version of whole report, written in nontechnical terms; *very* short and informative; normally describes salient features of report, draws a main conclusion, and makes a recommendation; always written last, after remainder of report has been written.
Table of Contents:	Shows contents and arrangement of report; includes a list of appendixes and, sometimes, a list of illustrations.
Introduction:	Prepares reader for discussion to come; indicates purpose and scope of report, and provides background information so that reader can read discussion intelligently.
Discussion:	A narrative that provides all the details, evidence, and data needed by the reader to understand what the author was trying to do, what he or she actually did and found out, and what he or she thinks should be done next.
Conclusions:	A summary of the major conclusions or milestones reached in the discussion; conclusions are only opinions so can never advocate action.
Recommendations:	If the discussion and conclusions suggest that specific action needs to be taken, the recommendations state categorically what must be done.
References:	A list of reference documents which were used to conduct the project and which the author considers will be useful to the reader; contains sufficient information for the reader to correctly identify and order the documents.
Appendix(es):	A "storage" area at the back of the report that contains supporting data (such as charts, tables, photographs, specifications, and test results) which rightly belong in the discussion but, if included with it, would disrupt and clutter the major narrative.

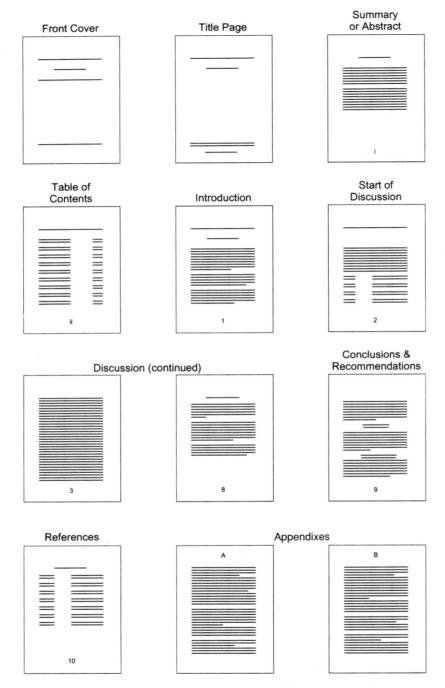

Figure 6-5. Formal report—traditional arrangement. (*These are the individual pages of Formal Report 1; to conserve space some of the discussion pages have been omitted.*)

A S O B O

THE ASSOCIATION OF SINGLE-OPERATOR BUSINESS OWNERS
Suite 2210 - 1800 Broad Avenue
Houston, Texas 77518

September 24, 1992

Tina R. Mactiere
President and Chief Executive Officer
Macro Engineering Inc
600 Deepdale Drive
Phoenix, AZ 85007

Dear Tina

The Association of Single-Operator Business Owners (ASOBO) is engaging
Macro Engineering Inc to evaluate electronic communication aids available to the Association's members, and to determine which are most
important or useful for the members to acquire. My letter confirms our
discussion and verbal agreement of September 17.

The term "Single-Operator" in the Association's name refers to businesspeople who in most cases are not only the sole owner of the business but
also the solitary employee. Examples are consultants working out of
their residences, tradespeople such as carpenters, pipefitters, and
painters who prefer to work independently rather than for a larger firm,
sales representatives, interior decorators, house agents, and "ma and
pa" shop owners. These are the people who joined forces to form ASOBO
in 1985, which now has 14,000 members across the U.S.

Many members feel they could increase their business by acquiring new
electronic technology, such as a cellular telephone, facsimile machine,
or portable computer, but in most cases their businesses are too small,
and their profit margins are too lean, to permit them to purchase or
rent more than one. To make the correct choice they need an objective
evaluation that identifies the key advantages of each equipment and how
effectively the equipment can be of assistance to them. (They cannot
rely on the opinions of vendors, who tend to be biased in favor of their
particular product.)

I would appreciate receiving your report by February 15, 1993 so that I
can distribute it to the members in March, with the Association's annual
report.

Regards

Evelyn K. Wollasey

Evelyn K Wollasey
Executive Secretary

Figure 6-6.   A letter authorizing a technical study.

Craig admitted that he had more difficulty writing this report than he has when writing many technical investigation reports. "The problem was the broad diversity of audience," he explained. "In most investigation reports I'm solving a problem for a clearly defined audience, so it's relatively easy to focus my report. There may be several ways to resolve the problem, and my audience will expect me to describe each of them, but they also expect me to make a *specific* recommendation."

He could not readily focus his report because he was writing for many diverse readers (the members of the Association of Single-Operator Business Owners—ASOBO), each pursuing a different trade or profession and each having a different need. To overcome this, he wrote the report *backwards*. This was his approach:

1. First, he made two lists: (1) a representative sample of different ASOBO members; and (2) the electronic communication aids that individual members might want to acquire.

2. Next, he divided each list into groups. The ASOBO members, he discovered, fell mainly into two groups (those who worked in one place, and those who moved around while on the job). The products and services fell into three groups (telephone-based, computer-based, and convenience-based).

3. His third step was to form a composite cross-reference table identifying who would need each product or service, and the extent of that need. This table became Appendix A (see page 184). Then, because he knew his readers would be particularly interested in prices, he developed a second table showing approximate minimum and maximum costs. This table became Appendix B.

4. Next he wrote the Introduction, to "set the scene." Here, he drew on the letter of authorization from ASOBO's executive secretary, Evelyn Wollasey, to establish his objective and criteria. By now he realized that he would be unable to write a report with clearly defined answers, so he explained this both in the introduction and in the cover letter (see Figure 6-7).

5. For the discussion, which he wrote next, Craig presented three components: a description of the individual products and services available to ASOBO members and how they were grouped; a description of the potential users and how they were grouped; and an analysis of how each product or service would suit each user group. ("But how to organize the report did not really occur to me until I had assembled the table in Appendix A," Craig added. "I would have had much more difficulty writing the discussion if I had not done that first.")

6. Now Craig wrote his conclusions and recommendations. ("The conclusions came easily," he explained, "because I had only to summarize the key outcomes from my analysis. But I found writing the recommendations more difficult, because I really wanted to tell *each* reader exactly what he or she should do!")

7. Finally, Craig wrote the summary, drawing principally on his conclusions to form it. ("The summary is longer than I usually like to write," he explained, "but I wanted to include enough detail—particularly costs—because I felt that my readers would want that.")

Remember Craig's "backwards" approach when you have to write your next long report. By documenting all the details *first*, you will find you can organize your ideas more easily and write more fluidly.

MACRO ENGINEERING INC

600 Deepdale Drive
Phoenix, AZ
85007

January 15, 1993

Evelyn K. Wollasey
Executive Secretary
Association of Single-Operator Business Owners
Suite 2210 - 1800 Broad Avenue
Houston, TX 77518

Dear Evelyn

I have completed an evaluation of electronic communication aids that would be suitable for single-operator business owners, as requested in your letter to Tina Mactiere of September 24, 1992, and enclose my report detailing the results.

In one sense my conclusions appear clearcut: most ASOBO members who remain at their place of business to do their work need simple and relatively inexpensive communication aids, whereas those who go to their customers to provide a service need more sophisticated and more expensive systems. Yet for many members such a generalization is not entirely valid, because the particular business a member is engaged in creates a different set of criteria. This means that, for each member pursuing a common profession or trade, and conducting his or her business in a particular fashion, I really should have produced a separate set of recommendations. This in turn would have created an omnibus report which, as you and I agreed when we met on December 18, would have taken too long to prepare and would have cost more than your Association had budgeted.

As a result, some of your members may be disappointed that, instead of providing definitive recommendations for each individual's situation, the report only provides guidelines they can use to assess their own needs.

I have avoided evaluating specific products so that the report will remain current for as long as possible. Because products change so frequently, to have evaluated them would rapidly "date" the report and so limit its usefulness to your members.

Sincerely

Craig M Derwent, P.E.

enc

Figure 6-7. A cover letter that also serves as an executive summary.

Comments on additional aspects of Craig's report follow, with specific references to the individual pages of the report in Figure 6-8.

- Craig's cover letter in Figure 6-7 is equivalent to an executive summary because he comments on the report's contents.
- A quick glance at the table of contents tells Craig's readers that he has divided his information into three logical units: the products and services, the users, and the analysis of needs. Immediately they will understand his approach.
- In the introduction (report page 1), the purpose is stated in paragraph 1, the background is partly in paragraph 1 but mostly in paragraph 2, and the scope is in the three "bulleted" subparagraphs and the final paragraph.
- Craig described the purpose (report page 1) in words that closely parallel those used by Evelyn Wollasey in her letter establishing project requirements for Macro Engineering Inc (see paragraph 1 in Figure 6-6). He also quoted some of the background details she used to describe vendors' biased opinions. Doing this helped ensure his report would be properly focused.
- For the discussion Craig adopted an overall "concept" arrangement (he described what equipment is available, identified who the users are, and then analyzed who is likely to need what). But *within* each major descriptive section he switched to the "by subject" arrangement to describe the equipment and users in detail.
- Craig wrote entirely in the first person plural. ("When I am presenting the results of a study I have done personally, and am offering *my* opinions, I prefer to write in the first person singular," he explained. "But in this case I did not know any of my readers personally, and none of them would know me, so I felt it would be more appropriate—and more acceptable from their point of view—for the evaluation to appear to come from the company.")
- The illustration on page 3 of the report probably is unnecessary. Craig admitted that he inserted it because he felt that, without an illustration, the seven pages of uninterrupted narrative might appear overly formidable to some readers.
- Craig also commented that, for appearance's sake, he would have liked the bottom line of type on each page to end at the same level. When setting his page parameters, he designated 53 lines of type for his page length, yet some pages had to end at about line 50 or 51. On report page 4, for example, he could not bring some of the text from the next page forward because he would have ended the page with either a heading standing alone or only the first line of type from the next paragraph. Neither would have been acceptable.
- The introduction—and the conclusions and recommendations—are assigned pages to themselves (pages 1 and 9).
- Every reference listed on page 10 is keyed to an appropriate point in the report narrative (see, for example, the superscript number at the end of paragraph 3 on page 4).
- The appendixes are not page-numbered because each consists of only one page.

PYRAMIDAL ARRANGEMENT OF REPORT PARTS

In recent years, more and more report writers have altered the organization of their reports so that they more effectively meet their readers' needs. The pyramidal arrangement brings the conclusions and recommendations forward,

MACRO ENGINEERING INC

EVALUATING ELECTRONIC COMMUNICATION AIDS
FOR SINGLE-OPERATOR BUSINESS OWNERS

Prepared for:

The Association of Single-Operator Business Owners

Prepared by:

Craig M Derwent, P.E.

Report 9303   January 13, 1993

Figure 6-8. Formal Report 1: Traditional Arrangement.

SUMMARY

A survey has been conducted to identify which electronic communication aids will most help single-operator business owners increase their business volume. The results show that the type of work or service the business operator provides directly affects the system(s) they should acquire.

Generally, business operators who remain at their place of business most of the time need only simple telephone services and equipment such as call waiting and an answering machine, which cost no more than $200 to buy or $25 a month to rent. Those who operate predominantly away from their place of business and *depend* on continual customer contact require a more advanced system such as a cellular telephone, for a purchase cost of about $900 to $1500 plus an operating cost of $60 to $100 a month. Business owners who frequently send and receive documents to and from other businesses--regardless of the type of service they provide--probably also need a facsimile machine costing $900 to $1800.

Only a few business owners--generally those providing consulting or similar services--need an office copier and a personal or portable computer, each costing about $1500 to $2200.

i

Figure 6-8. (*Continued*)

CONTENTS

APPENDIXES

Figure 6-8. *(Continued)*

EVALUATING ELECTRONIC COMMUNICATION AIDS
FOR SINGLE-OPERATOR BUSINESS OWNERS

INTRODUCTION

In a letter dated September 24, 1992, Evelyn K. Wollasey, Executive Secretary of the Association of Single-Operator Business Owners (ASOBO), asked Macro Engineering Inc to evaluate the electronic office and communication aids currently available and identify which would be most useful for ASOBO members to acquire.

The need for the study evolved from ASOBO members' awareness that having the right electronic office aid--such as a facsimile machine, portable telephone, or electrostatic copier--could significantly increase their business volume. The problem faced by most members, however, is knowing how to evaluate which equipment or service, from the wide range available, is the *correct* choice for their particular line of business. They cannot depend on vendors' opinions, since vendors are predictably most interested in promoting their own products and services.

In our report we will

o   Identify each type of office aid, briefly describe what it does and how it can be useful to a small business owner, and quote a general price range.

o   Divide ASOBO members into groups, based on the kind of service they provide or product they sell.

o   Analyze which electronic office aids will be most useful for the members within each group.

We will be analyzing only general services and products, rather than specific products by brand name, because the technology driving the electronic communication industry is advancing so rapidly, and manufacturers are producing new and upgraded services and products so frequently, that any definitive evaluation we make now would become obsolete in only a few months. Our intention in this report, therefore, is to identify services and products that would be particularly attractive for certain groups of ASOBO members, and the factors members should consider before making a choice.

1

Figure 6-8.   (*Continued*)

THE CURRENT RANGE OF ELECTRONIC SERVICES AND PRODUCTS

The services and products most likely to be of value to ASOBO members fall into three categories:

o   Those that are telephone-based, and therefore are clearly communication aids, ranging from call forwarding through facsimile transmission to portable telephones.

o   Those that are computer-based, including electronic mail and personal and portable computers.

o   Those that are primarily convenience-based, such as office copiers.

The products within each category, and their approximate prices, are described in the remainder of this section and summarized in the illustration on page 3.

TELEPHONE-BASED SERVICES AND PRODUCTS

Telephone Answering Services

Answering services range from those provided at nominal cost by the telephone utilities, to answering machines purchased and installed by users. The most economical in this category are call waiting, call forwarding, and caller identification:

o   **Call Waiting** provides an audible signal to the user if a second person calls while the user is already speaking to an earlier caller (the first caller cannot hear the signal).

o   **Call Forwarding** automatically transfers incoming calls to another number. Before departing to the second location, the user enters (dials) the number where he or she can be reached.

o   **Caller Identification** shows the caller's number on a digital display panel built into the user's telephone, even before the user picks up the handset. It is particularly convenient for screening incoming calls when used in combination with Call Waiting.

Telephone utilities charge from $1.95 to $3.25 a month for each of these services.

An alternative to renting on-line message services is to install an answering machine that informs callers when the owner is not present, and asks them to record a message or leave a telephone number so that the owner can return the call. Most answering machines permit the owner to listen to the recorded messages from a remote telephone. Answering machines cost from $90 to $245 to buy, or $10 to $20 a month to rent, and cost nothing to operate.

2

Figure 6-8. (*Continued*)

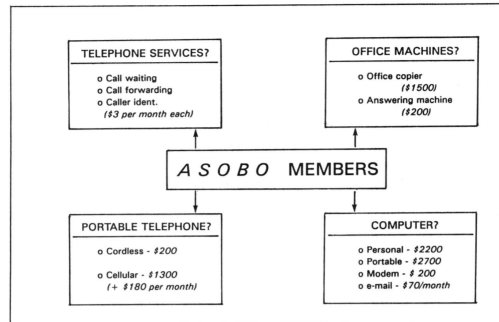

Choices Available to ASOBO Members

Portable Telephones

Portable telephones are for users who move about during the working day and consequently do not have continuous access to a fixed telephone. A portable telephone is similar to but slightly larger than a regular telephone, is battery powered, and has an integral transmitter-receiver and either a concealed or retractable whip antenna. There are two types:

o   The **Cordless Telephone**, which communicates with a transmitter-receiver built into the telephone base. It has a range of about 100 yards and is intended primarily for local use around a business site or residential property. A cordless telephone costs about $200 to buy, and nothing to operate.

o   The **Cellular Telephone** (also known as a **Cellphone**), which communicates with one of several transmitter-receivers located within a city and its outlying areas. Each base transmitter-receiver has a range of up to four miles, depending on the terrain and how much the area is built up, with the area it covers being called a "cell" (from which the name "cellular" telephone has evolved). If a caller drives from one cell to another, his or her calls are transferred automatically from cell to cell. A cellular telephone costs between $900 and $1500 to buy, and about $100 a month to operate, with the rate depending on usage.

3

Figure 6-8. *(Continued)*

Facsimile Machines

The services and products considered so far all convey voice-transmitted messages. Facsimile machines still use telephone lines and equipment to carry messages, but accept the messages and reproduce them in printed form as letters, memos, reports, sketches, and diagrams. When a facsimile machine is connected via a modem to a computer, it can also accept messages electronically and display them on the recipient's computer screen.

The advantage of a fax machine to small business owners has been analyzed by Guy Desrogers of Assiniboine Business Consultants, who writes

> The most significant advantage of facsimile transmission is that a business that owns a fax machine is much less likely to be affected by mail stoppages...(and)...sending a letter across town by fax is significantly cheaper than sending it by courier. It is still cheaper to fax a letter from coast to coast rather than send it by (air courier) services, particularly if the machine is programmed to send messages during the night when off-peak telephone rates apply.[1]

A facsimile machine costs between $900 and $1800 to purchase, and nothing to operate for incoming messages other than the cost of the paper on which messages are printed (about 9 cents a page). For outgoing messages there is no operating cost for local calls, and only long-distance charges for out-of-town calls.

COMPUTER-BASED SERVICES AND PRODUCTS

Electronic Mail (e-mail)

Electronic mail conveys messages entirely electronically and is popular with computer buffs. It requires that both sender and receiver have a computer, plus a modem for linking the computer to the telephone system, and that they subscribe to a computer messaging service such as *iNet* or *Compu-Serve*. The sender keys the message at his or her terminal, reads it on the video screen, and then transmits it through the modem. The receiver reads the message instantaneously on his or her video screen and replies in the same way. Consequently, e-mail is extremely fast.

Assuming that the user already owns a computer, the remaining cost is the purchase of a modem for approximately $200, and subscription to the computer messaging service, which can vary from $12 to $30 a month. There is also a monthly usage fee, which depends on the amount of system log-on time the user accumulates and can range from as little as $5 to as much as $60 or even $100 a month.

Personal Computers

Personal computers (PCs) are stand-alone units comprising a keyboard, data processor, memory, screen, and printer. They can be used for making calculations, performing statistical analyses, keeping records, creating charts and graphs, and writing

4

Figure 6-8. *(Continued)*

letters and reports (i.e. word processing). This report, for example, is being typed directly into a personal computer. The cost for a complete system can range from $1600 to $6000. For most ASOBO members, a $2000-$2200 computer system should suffice.

Portable Computers

Portable computers are similar to PCs but are more compact and normally do not have a printer. They range from "luggable" units weighing 12 to 17 pounds that can fit into an aircraft overhead luggage rack, to "laptop" units weighing 5 to 10 pounds that can be slipped into a briefcase. Many of the larger portables have a memory and operating capability similar to that of a PC; laptops, however, generally have a much smaller memory and a limited capability.

A portable computer is especially suitable for businesspeople who travel a lot, particularly if the portable has a built-in modem and so can be used to transmit data and messages to, and receive them from, an office-based PC or mainframe computer. The cost of a portable computer can range from about $800 for a simple laptop to $7000 for a luggable with top-of-the-line capabilities.

A CONVENIENCE-BASED PRODUCT: THE OFFICE COPIER

We tend to think of an office copier as a large high-speed floor model that produces high-quality images and costs or rents for a high price. For small business owners, however, there is a much more modest range of machines that offer slow-to-moderate speed, moderate-to-high-quality images, and a moderate price of between $1000 and $2000. For the limited quantity of copies that many ASOBO members are likely to make, such a machine should perform admirably.

TYPES OF SINGLE-OPERATOR BUSINESS OWNERS

Single-operator business owners either operate from an office at their place of residence, rent a small office locally, or own or rent a service outlet or store. Generally, they fall into two groups:

o   Those who remain at their place of business all or most of the time, and whose customers come to them for service. Shop owners, technicians who service portable instruments and appliances, and some accountants and independent teachers form this group.

o   Those who take their service to their customers, and so remain at their place of business only a portion of the time. Consultants, surveyors, sales representatives, tradespeople, and technicians who service fixed or large equipment fall into this group.

A few business owners pursue a profession that may place them in either group, or in both groups. These are principally accountants, architects, interior decorators, and some independent teachers and veterinarians.

5

Figure 6-8. *(Continued)*

IDENTIFYING A BUSINESS OWNER'S NEEDS

The four descriptions that follow provide a general "rule of thumb" for determining which group of ASOBO members would most benefit from a particular type of electronic communications service or office aid. Appendix A contains a more definitive breakdown, since it lists 15 representative single-operator business owners and, for each, identifies whether we consider the businessperson would have a strong, moderate, or slight need for a particular service or product, or no need. Appendix B summarizes the approximate purchase, lease, and operating costs for each electronic aid.

CELLULAR TELEPHONES AND TELEPHONE ANSWERING SERVICES

o   Where customers generally bring work to the business owner, as in the case of accountants, shop owners, and music teachers, the business owner normally does not need a cellular telephone but should buy or rent an answering machine and, perhaps, use a call waiting service.

o   Where business owners take their service to the customer, and need to be easily reached by customers, then a cellular telephone would be a useful acquisition, with a call-forwarding service to route calls coming in to their office number to the cellular number. Business owners who do not have a cellular telephone would be wise to purchase an answering machine.

FACSIMILE MACHINES

o   The group in which a business owner belongs has little or no bearing on whether the business will benefit from having a facsimile machine. The criterion is whether the business owner frequently sends documents to clients and other parties, or receives documents from them. If so, the business owner will find a facsimile machine a fast, expedient, and economical means of document transmission.

PERSONAL AND PORTABLE COMPUTERS

o   Similarly, business owners who generate a lot of documentation, either in tabular or text form, regardless of the group in which they "fit," will benefit most from having a personal computer in their office. Accountants, architects, consultants, and writers particularly fall into this group. Writers, and possibly consultants and sales representatives, may also need a portable computer and modem.

OFFICE COPIERS

o   Business owners who make duplicate or multiple copies of multipage documents are most likely to find an in-house office copier a convenient acquisition. Those who only occasionally need to make copies may find it more economical to make copies at a local printing or word-processing facility.

6

Figure 6-8.   *(Continued)*

OVERALL IMPLICATIONS

Clearly, there are too many variables for us to identify a particular communication aid for a particular type of business owner. The general suggestions listed earlier will apply in a majority of cases, but not in all.

A factor each business owner has to evaluate personally is whether he or she can afford a particular communication aid. Most single-operator business owners can afford one or more of the low-cost telephone services such as call waiting, call forwarding, and caller identification, or an answering machine. At the other end of the scale, most single-operator business owners neither need nor can afford a computer or office copier. The only exceptions are professionals and paraprofessionals, such as consultants.

Sometimes, small business owners feel they should purchase a computer to simplify their business accounting. Yet a 1991 study by the Pacific Rim Association of Small Business Owners shows that computerized accounting, when compared with manual accounting or a subcontracted accounting service, generally is not economical for businesses with a staff of less than three and a gross annual income of less than $225,000.[2] Unless a single-operator business owner already owns a computer and is experienced in using it, the time taken to become familiar with and use the accounting software will most likely outweigh the time that would be expended in manual bookkeeping.

For the cellular telephone and the facsimile machine--the two communication aids that comprise the middle, moderately priced group--the decision is more difficult. If, to maintain a constant or increasing business volume, a business owner needs to be constantly available to or at least within easy reach of clients and other businesses, then acquisition of either probably would be justified: a cellular telephone for the business owner who is out of the office much of the time, and a facsimile machine for the business owner who tends to remain in or near the office.

An alternative to outright purchase or long-term lease of a facsimile machine, or of an office copier, is to use the services of a local convenience printer or word-processing house for 6 to 12 months. It is not as convenient as having the facility in-house, but it does offer business owners a low-cost way to determine exactly how many fax messages they would send and receive, and how many copies of documents they would make. From this assessment they can decide whether personal ownership is practicable.

ON-THE-HORIZON DEVELOPMENTS

Today, manufacturers and sellers of telephone-based products are clearly targeting both small and single-operator businesses in their marketing plans. This alone may influence ASOBO members to consider seriously whether either a portable telephone or facsimile messaging should be an essential component of their business strategy. For

Figure 6-8. *(Continued)*

example, Linda Gregg Stulberg, writing in *The Financial Post*,[3] reports that many small business owners now view a portable telephone as an imperative acquisition, and that "...the phone becomes their voice-mail system, their answering system."

Two developments currently being user-tested also may influence ASOBO members' decisions either to purchase an electronic aid now or to delay its acquisition until later:

o   Facsimile machines for use by local businesspeople and passersby are being installed in business centers and shopping malls. They look like automated teller machines and are activated by most major credit cards. *The Financial Post* reports that, in trial installations carried out to date, the cost to the user has been 90 cents per page plus the cost of the telephone call.[4]

o   Cordless telephones that can be operated not only from a home but also through public "Telepoints" installed in, for example, airports, shopping malls, and the lobbies of business buildings, are being tested by at least one manufacturer. *Computing Canada* quotes the manufacturer's prediction that for many business-persons the public cordless telephone will replace the cellular telephone.[5] A cordless telephone used in this way would be significantly cheaper to purchase and operate than a cellular telephone.

8

Figure 6-8. *(Continued)*

CONCLUSIONS

Whether a single-operator business owner should use (i.e. purchase or lease) telephone-based communication aids depends primarily on the service the business-person provides. Most business owners--and particularly those whose customers come to them to purchase a product or service--would benefit from leasing one of the less expensive telephone services such as call forwarding or call waiting, or from purchasing an answering machine.

Business owners who spend much of their time away from their regular place of business would particularly benefit from owning a cellular telephone or, if their area of operation is likely to be serviced with "Telepoints," the less-expensive cordless telephone.

Business owners who frequently communicate documents to other organizations, either locally, nationally, or internationally, would benefit from using a facsimile machine. If their volume is high, the machine probably would be owned outright, or possibly leased. If their volume is low, the business owner would probably find that using a locally available third-party machine is more advantageous. The same holds true for office copiers.

Only writers and professionals such as consultants and architects need either a personal or a portable computer.

RECOMMENDATIONS

Before ASOBO members approach suppliers to purchase or lease electronic office equipment, we recommend they first establish their specific needs by

o   identifying whether customers come to them for service, or they take their serv-
    ice to customers, and

o   determining, for their type of business, which electronic communication aids will
    most help them provide a more efficient service for their customers.

9

Figure 6-8. *(Continued)*

REFERENCES

1. Guy Desrogers, *Evaluation of Facsimile Machines for Use in Small Business Operations*, Report No. 90/1, Assiniboine Business Consultants Inc., Fargo, North Dakota, January 12, 1990.

2. Deidre Parsons, *Calculating the Hidden Costs When Using Business Accounting Software*, Report No. 912, Pacific Rim Association of Small Business Owners, Portland, Oregon, May 10, 1991.

3. Linda Gregg Stulberg, "Portable Phone as Business Tool Key to Selling Cellulars," *The Financial Post*, July 9, 1990, p. 32.

4. Laura Ramsey, "User-Pay Units Make Fax Available to Owner-Operators," *The Financial Post*, May 14, 1990, p. 38.

5. "Public Cordless Phones Debut in Effort to Replace Cellular," *Computing Canada*, June 7, 1990, p. 59.

Figure 6-8. *(Continued)*

APPENDIX A

COMPARISON OF NEEDS: ELECTRONIC OFFICE AIDS FOR SINGLE-OPERATOR BUSINESS OWNERS

Legend:
0 - Strong need
o - Moderate need
. - Slight need

Type of Business Owner	Telephone Eqpt & Services							Computer Eqpt				Office copier
	Call forwarding	Call waiting	Caller identification	Answering machine	Cordless phone	Cellular phone	Facsimile (fax)	Personal computer	Portable computer	Modem	e-mail	
Accountants	o	.	.	0			.	0		o	.	0
Architects	0		.	o		0	o	0		.		o
Consultants	0	.	o	0	.	o	0	0	o	0	o	.
Courier service owners	0	o	o			0						
Interior decorators	0	.		o	0	0						
Land surveyors	0	.	.	o		0	.	.				.
Sales representatives	0	o	o	0		0	.	.	.	.	.	.
Service representatives (large or fixed eqpt)	0	o	.	o		0	o					.
Service suppliers (lawn cutting, property maintenance)	0	.		o	0							
Service shop owners (small, portable equipment service)		0		.	o		o	.				.
Shop owners, retail		0	.	.	o			.				
Teachers, independent (music, language, etc)			.		0	o						
Tradespeople (carpenters, pipe fitters, electricians)	o			0		o						.
Veterinarians	0	.	o	0	o	0		.				.
Writers, freelance	0	.		0	o	.	o	0	o	o	.	o

Figure 6-8. (*Continued*)

APPENDIX B

**APPROXIMATE PURCHASE, LEASE, AND OPERATING COSTS
OF SELECTED ELECTRONIC COMMUNICATION AIDS**

Product or Service	Purchase Cost		Lease Cost (per month)		Operating Cost[+] (per month)	
	Low ($)	High ($)	Low ($)	High ($)	Low ($)	High ($)
Call Waiting) Call Forwarding) Caller Ident.)	-	-	2	3	-	-
Answering Machine	90	245	10	20	-	-
Cordless Telephone	150	250	12	20	-	-
Cellular Telephone	900	1500	60	90	50	150
Facsimile Machine	900	1800	60	120	#	#
Personal Computer	1600	6000	110	350	-	-
Portable Computer	800	7000	55	410	-	-
Modem	170	240	-	-	-	-
e-mail	-	-	12	30	5[*]	100[*]
Office Copier	1000	2000	70	140	-	-

NOTES:

+	Does not include cost of materials
#	Incoming faxes: 9 cents a copy
	Outgoing faxes: local - no cost
	long distance - L.D. charge
*	Depends on usage

Figure 6-8. (Continued)

positioning them immediately after the introduction so that executive readers do not have to leaf through the report to find the terminal summary (the report's outcome). The advantages of the pyramidal approach are immediately evident: Busy readers have only to read the initial pages to learn the main points contained in the report, and the writer can help them along by gradually increasing the technical content of the report, catering to semitechnical executive readers up to the end of the recommendations, and to fully technical readers in the discussion and appendix. Although the natural flow of information that occurs in the traditional arrangement is disrupted, Figure 6-9 shows there is now a reader-oriented flow, with the three compartments of information containing progressively more technical details.

This gradually increasing development of the topic in three separate stages is similar to the newspaper technique described earlier, in which the first one or two paragraphs contain a capsule description of the whole story, the next three or four paragraphs contain a slightly more detailed description, and then the final eight or nine paragraphs repeat the same story, but this time with more names, more peripheral information, and more details. Newspapers also cater to both the busy reader who may not have time to read more than the opening synopsis, and the leisurely reader who wants to read all the available information.

A birds-eye mini report depicting the pyramidal arrangement of report parts is shown in Figure 6-10. Segments of a sample report written using the pyramidal approach are shown in Figure 6-11.

Excerpts from
Formal Report 2
Selecting New Elevators
for the Merrywell Building

Before reading these excerpts, read the client's letter authorizing H. L. Winman and Associates to initiate an engineering investigation:

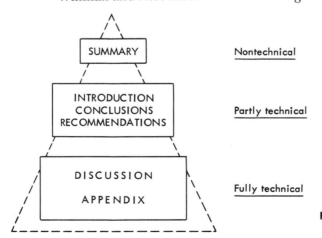

Figure 6-9. The formal report arranged "pyramid style."

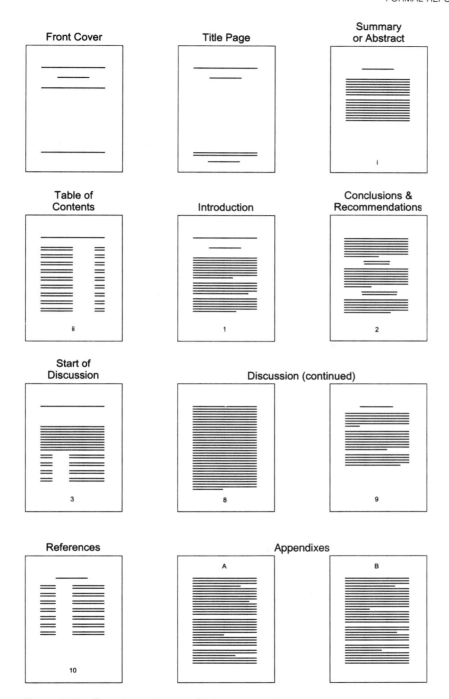

Figure 6-10. Formal report—pyramidal arrangement.

Dear Mr. Bailey:

The elevators in the Merrywell Building are showing their age. Recently we have experienced frequent breakdowns and, even when the elevators are operating properly, it has become increasingly apparent that they do not provide adequate service at the start of work, at noon, and at the end of the working day. I have therefore decided to install a complete range of new elevators, with work starting in mid-August.

Before I proceed further, I would like you to conduct an engineering investigation for me. Specifically, I want you to evaluate the structural condition of my building, assess the elevator requirements of the building's tenants, investigate the types of elevators available, and recommend the best type or combination of elevators that can be purchased and installed within a proposed budget of $500,000.

Please use this letter as your authority to proceed with the investigation. I would appreciate receiving your report by the end of June.

Regards,

David P. Merrywell, President
Merrywell Enterprises Inc

By comparing this letter with the conclusions and recommendations, you can assess how thoroughly Barry Kingsley (the report's author) has answered the client's requests.

COMMENTS ON THE REPORT

The summary is short and direct because it is written primarily for one reader: the President of Merrywell Enterprises Inc. It encourages him to read the report immediately, and to accept its recommendations, by offering the opportunity to save $50,000.

Although the background information contained in the first two paragraphs of the introduction seems to repeat details the client already knows, Barry recognizes he must satisfy the needs of other readers who may not be fully aware of the situation in the Merrywell Building. He then defines the purpose and scope of the investigation by stating the client's terms of reference in paragraph 3 of the introduction. (Note that he has copied them almost verbatim from Mr. Merrywell's letter.)

The conclusions present Barry's answers to Mr. Merrywell's four requests. Their order is different from that in paragraph 3 of the introduction because he has chosen to present the main conclusion first (in this case, the best combination of elevators that can be purchased within the stipulated budget), and to follow it with subsidiary conclusions in descending order of importance. Barry is aware that when using the pyramidal report format he must write conclusions that evolve naturally and logically from the introduction, *because his readers have not yet read the discussion.*

Barry uses the first person plural to open his recommendations because, although he alone is the report's author, he is representing H. L. Winman and Associates' views to the client.

H L WINMAN AND ASSOCIATES

SUMMARY

The elevators in the 71-year-old Merrywell Building are to be replaced. The new elevators must not only improve the present unsatisfactory elevator service, but must do so within a purchase and installation budget of $500,000.

Of the many types and combinations of elevators considered, the most satisfactory proved to be four 8 ft by 7 ft deluxe passenger elevators manufactured by the YoYo Elevator Company, one of which will double as a freight elevator during off-peak traffic times. This combination will provide the fast, efficient service requested by the building's tenants for a total price of $450,000, which will be 10% less than the projected budget.

Figure 6-11. Report 2: Summary page.

SELECTING NEW ELEVATORS FOR THE MERRYWELL BUILDING

INTRODUCTION

When in 1964 Merrywell Enterprises Inc purchased the Wescon property in
Montrose, Ohio, they renamed it "The Merrywell Building" and renovated the
entire exterior and part of the interior. The building's two manually
operated passenger elevators and a freight elevator were left intact, although it was recognized that eventually they would have to be replaced.

Recently the elevators have been showing their age. There have been
frequent breakdowns and passengers have become increasingly dissatisfied
with the inadequate service provided at peak traffic hours.

In a letter dated April 27, 19xx to H. L. Winman and Associates, the
president of Merrywell Enterprises Inc stated his company's intention to
purchase new elevators. He authorized us to evaluate the structural
condition of the building, to assess the elevator requirements of the
building's occupants, to investigate the types of elevators that are
available, and to recommend the best type or combination of elevators that
can be purchased and installed within the proposed budget of $500,000.

CONCLUSIONS

The best combination of elevators that can be installed in the Merrywell
Building will be four deluxe 8 ft by 7 ft passenger models, one of which
will serve as a dual-purpose passenger/freight elevator. This selection
will provide the fast, efficient service desired by the building's tenants,
and will be able to contend with any foreseeable increase in traffic. Its
price at $450,000 will be 10% below the proposed budget.

Installation of special elevators requested by some tenants, such as a full-
size freight elevator and a small but speedy executive elevator, would be
feasible but costly. A freight elevator would restrict passenger-carrying
capability, while an executive elevator would elevate the total price to at
least 20% above the proposed budget.

1

Figure 6-11. Report 2: Introduction and conclusions.

The quality and basic prices of elevators built by the major manufacturers are similar. The YoYo Elevator Company has the most attractive quantity price structure and provides the best maintenance service.

The building is structurally sound, although it will require some minor modifications before the new elevators can be installed.

RECOMMENDATIONS

We recommend that four Model C deluxe 8 ft x 7 ft passenger elevators manufactured by the YoYo Elevator Company be installed in the Merrywell Building. We further recommend that one of these elevators be programmed to provide express passenger service to the top four floors during peak traffic hours, and to serve as a freight elevator at other times.

2

Figure 6-11. Report 2: Conclusions (continued) and recommendations.

EVALUATING BUILDING CONDITION

We have evaluated the condition of the Merrywell Building and find it to be structurally sound. The underpinning done in 1958 by the previous owner was completely successful and there still are no cracks or signs of further settling. Some additional shoring will be required at the head of the elevator drive shaft immediately above the 9th floor, but this will be routine work that the elevator manufacturer would expect to do in an old building.

The existing elevator shaft is only 24 feet wide by 8 feet deep, which is unlikely to be large enough for the new elevators. We have therefore investigated relocating the elevators to a different part of the building, or enlarging the existing shaft. Relocation, though possible, would entail major structural alterations and would be very expensive. Enlarging the elevator shaft could be done economically by removing a staircase that runs up the center of the building immediately east of the shaft. This staircase is used very little and its removal would not conflict with fire regulations. Removal of the staircase will widen the elevator shaft by 11 feet, which will provide sufficient space for the new elevators.

ESTABLISHING TENANTS' NEEDS

To establish the elevator requirements of the building's tenants, we asked a senior executive of each company to answer the questionnaire attached as Appendix A. When we correlated the answers we identified five significant factors that would have to be considered before selecting the new elevators. (There were also several minor suggestions that we did not include in our analysis, either because they were impractical or because they would have been too costly to incorporate.) The five major factors were:

o   Every tenant stated that the new elevators must eliminate the lengthy waits that now occur. We carried out a survey at peak travel times and established that passengers waited for elevators for as much as 70 seconds. Since passengers start becoming impatient after 32 seconds,[1] we estimated that at least three, and probably four, faster passenger elevators would have to be installed to contend with peak-hour traffic.

o   Although all tenants occasionally carry light freight up to their offices, only Rad-Art Graphics and Design Consultants Inc considered that a freight elevator would be essential. However, both agreed that a separate freight elevator would not be necessary if one of the new passenger elevators is large enough to carry their displays. They initially quoted 9 feet as the minimum width they would require, but later conceded that with minor modifications they could reduce the

3

Figure 6-11.   Report 2: Start of discussion.

length of their displays to 7 feet, 6 inches. All tenants agreed that
if a passenger elevator is to double as a freight elevator they would
restrict freight movements to non-peak travel times.

o   The three companies occupying the top four floors of the building
      requested that one elevator be classified as an express elevator
      serving only the ground floor and floors 6, 7, 8, and 9. Because
      these companies represent more than 50% of the building's tenants, we
      considered their request should be entertained.

o   Three companies expressed a preference for deluxe elevators. Rothesay
      Mutual Insurance Company, Design Consultants Inc, and Rad-Art Graphics
      all stated that they had to create an impression of business solidari-
      ty in the eyes of their clients, and feel that deluxe elevators would
      help convey this image.

o   The managements of Rothesay Mutual Insurance Company and Vulcan Oil
      and Fuel Corporation requested that a small key-operated executive
      elevator be included in our selection for the sole use of top
      executives of the building's major tenants. We asked other companies
      to express their views but received only marginal interest. The
      consensus seemed to be that an executive elevator would have only
      limited use and the privilege would too easily be abused. However, we
      retained the idea for further evaluation, even though we recognized
      that an executive elevator would prove costly in relation to passenger
      usage.[2]

We decided that the first two of these factors are requirements that must be
implemented, while the latter three are preferences that should be incorporated if at all feasible. The controlling influence would be the budget
allocation of $500,000 stipulated by the landlord, Merrywell Enterprises
Inc. In decreasing order of importance, the requirements are:

1.   Passenger waiting time must be no longer than 32 seconds.

2.   At least one elevator must be able to accept freight up to 7 feet, 6
      inches long.

3.   An express elevator should serve the top four floors.

4.   The elevators should be deluxe models.

5.   A small private elevator should be provided for company executives.

4

Figure 6-11.   Report 2: Second page of discussion.

The first two pages of the discussion have been included to show how, early in his report, Barry established criteria that will subsequently influence how he selects a combination of elevators that will best meet his client's needs. By carefully identifying the five criteria and describing why each is valid, he shows his readers the direction his report will take. (In later sections of his report—not included in the sample pages—he identifies various combinations of elevators that could be installed and demonstrates which do or do not meet the criteria until he finally reaches an optimum configuration.)

Assignments

PROJECT 6.1: POSITIONING AN ENVIRONMENTAL MONITORING STATION

You are an engineering assistant employed in the local branch of H. L. Winman and Associates. One week ago today, branch manager Vern Rogers invited you into his office and introduced you to Harry Vincent of the Department of the Environment (DOE).

"Mr. Vincent wants us to find a suitable location for positioning an acid rain monitoring station," Vern explained. "His department is installing them on the outskirts of many cities, as part of a nationwide monitoring system."

"They are completely automatic," Harry Vincent broke in. "Each installation consists of a 2-meter diameter shallow concave dish with a hole in its center, a length of tubing leading down from the hole to a digital monitor that analyzes the collected rainwater, and a transmitter that sends the results digitally by telephone line to a control center in Washington."

He and Vern told you that you are to research possible sites, identify three or four that will be suitable, analyze the advantages and disadvantage of each, and recommend which site you feel is most suitable. "Don't just present me with one site," Mr. Vincent added. "Your report should describe all of them so that I can make the final decision."

"What are you looking for in an ideal site?" you asked.

Harry Vincent itemized several requirements:

1. The dish has to be absolutely horizontal, on top of an existing structure. ("We don't plan to build our own structures," he added, "so you will have to negotiate a rental price with the owners of the buildings you examine.") Naturally, the building should have a flat roof.

2. The building does not have to be particularly high, but it should be reasonably free-standing and not have a taller building immediately next to it. ("We don't want grit blowing off the roof of a neighboring building, or smoke from a chimney

depositing soot into the dish. We have to collect *natural* rainfall.") And, of course, the roof you use also should not have a chimney or any exhaust vents.

3. There must be a space inside the building for mounting the monitoring unit. ("You'll have to negotiate a rental price for this, too," Harry added, "and ensure the monitoring unit will be reasonably close to the tubing leading down from the dish. It should also be screened off or behind a locked door, so no one can fiddle with it, and there must be a means for draining the used rainwater after it has been analyzed.")

4. Ideally, there will be telephone lines already leading into the building, and close to the monitoring unit.

5. The monitor must also have an electrical outlet nearby.

"What about temperature?" you asked. "Suppose we find an unheated building: are there limitations for the analyzer?"

"It's okay up to $+42°C$, and down to $+2°C$," Harry Vincent replied. "The unit is sampling rainwater, so both it and the ambient temperature must remain above freezing. If there is any likelihood of the monitor being exposed to freezing temperatures, then it must be kept in an insulated box with internal electric heating."

The following day you started your investigation, and in the ensuing five days you found several locations you thought would be suitable, all on the outskirts of, or just outside, your city, and all with a flat roof. They comprised:

1. A headstone and monument manufacturer, beside a cemetery.
2. A radio transmitting shack, with a transmitting tower beside it.
3. A riding stable with a barn and associated tack room.
4. The projection room/refreshment booth at a drive-in theater.
5. A small "strip" shopping mall.
6. A building in a light-industry industrial park.

For this project you are to research and document *real* locations in or near your city, using three of the six foregoing ideas, plus one of your own, for a total of four possible sites. At each location you will need to make notes so that later you can write an accurate description of

- the building and its distance and bearing from the city center,
- the topography of the land and types of buildings nearby,
- the presence of any nearby industries, their direction and distance from the site, and whether they are likely to emit pollutants,
- whether the proposed building has electrical and telephone connections, and
- the direction of the prevailing wind (so that you can assess whether industrial and other pollutants will be blowing toward the site).

At each site make a note identifying whether you feel the site will be an excellent, good, or only fair location for the environmental monitoring station, and why you have given it a specific rating.

At your "own choice" location, also determine

- exactly where you will locate the dish and the monitoring unit (be sure to jot down exact dimensions),
- whether you will be able to drill through the roof to connect the dish to the monitoring unit, or whether you will need to take the tubing out of a window and up the wall to the roof,
- the length of tubing (in meters) you will need to connect the dish to the monitoring unit, and then to the drain,
- the availability of a drain hole,
- the availability of electricity and telephone, and
- the monthly rental cost to mount the collector dish on the roof and install the monitoring unit in the building.

You need not research information on the other three locations. Instead, you may assume you made the following three sets of notes (see A, B, and C), one for each location, and now you must decide which of the three sets of notes best fits each of the three locations.

Location A:

- Building owner: Wishart Properties (head office: 2020 Main St.); I talked to Marnie Wolski.
- Can we drill through the roof? No. The owner is adamant about this.
- Rental: $125 per month (minimum of 12 months); they want an agreement or lease.
- Where will monitor unit go? In an equipment room at the south side of building. The door has no lock, but a lock can be installed (DOE will have to pay for it).
- Length of tubing required? 210 meters (includes 8-meter distance from monitor to drain hole).
- Both power and telephone are available.

Location B:

- The owner will be happy for us to drill through the roof! (He wants to put up a TV satellite dish and will twin his cable beside DOE's tubing; but he insists there must be no cost to him for doing this and that the hole through the roof *must* be properly sealed and rainproof.)
- He has a storage room at east end of building which is always locked that we can use for the monitoring unit.
- Power and telephone lines are available, but we'll have to bring in our own telephone line (he's suspicious he'll be paying DOE's telephone bills without knowing it if the line goes through his switchboard).
- Building owner: Steve Oliphant.
- Length of tubing required: 180 meters (unfortunately, he wants the satellite dish at the west end of the building, immediately above his office, which means we'll have to run tubing the length of the building—about 90 meters).
- Rental cost: $1020 per year. (Minimum rental: 2 years.)

- There is a drainhole 5 meters from the storage room. I've included this in the tubing length.

Location C:

- Power and telephone are problems here: the owner (Leon Fleischer) insists that DOE has its own meter for electricity and its own telephone lines and is billed separately.
- He can give us roof space and a disused washroom immediately beneath it; washroom has a small window.
- We *cannot* drill through the roof.
- Rent—$99.50 a month (he didn't mention either a lease or a minimum rental period).
- Tubing length required—30 meters (including 2 meters to connect monitoring unit to drain hole).

Note: Should circumstances prevent you from identifying and researching an "own choice" location, you may use the following details and apply them to a fourth location drawn from the list:

- Length of tubing required: 47 meters.
- You can take the tubing through the roof: there's an unused drain pipe you can tap into and travel up, but you cannot drill a new hole.
- Building owner: Tricia Delaney.
- Rental: $90.00 a month (six months minimum lease).
- Drainage is readily available (you can tap back into the drainpipe).
- The owner won't let you bring in new electricity and telephone lines. She will "rent" electricity to you, based on an estimate of your usage. Similarly, she will "rent" telephone lines and long-distance calls to you. No price has been mentioned.

Now you have all the details you require and you are ready to start organizing and writing your report. But one factor still troubles you:

> How long can the tubing reasonably be, from the dish to the monitoring unit, before it starts affecting water flow? The system depends on hydraulic water flow (i.e. the water is not pumped from dish to monitor), and for short distances the drop from the level of the dish to the level of the monitor creates a sufficient "head" to ensure a natural flow. But over a long, almost horizontal length of tubing there may be flow problems.

Now, armed with all your data, write your formal investigation report for Harry Vincent.

PROJECT 6.2: TESTING HIGHWAY MARKING PAINTS

For the past seven years the Highways Department in your state has used "Centrex CL" for marking highway pavement center lines and lanes. Recent advances in paint technology, however, have brought several new products onto the market, which their manufacturers claim are better than Centrex CL. To meet this challenge, Centrex, Inc has developed a new paint ("TL") and has recommended that the Highways Department use it in place of CL.

In a letter dated March 18 of this year, Senior Highways Engineer Morris Hordern commissions the local branch of H. L. Winman and Associates to carry out independent tests of the new paints. Branch Manager Vern Rogers assigns the project to you.

You start your project by obtaining samples of white and yellow highway paint from six manufacturers, transferring the samples into unmarked cans, and then coding the cans like this:

Manufacturer	*Paint Coding*	
	White	*Yellow*
1. Centrex Inc, Hartford, Connecticut Paint type: CL (the "old" paint)	WA	YL
2. Novell Paint Ltd, Moorstown, New Jersey Paint type: 707	WB	YM
3. Hi-Liner Products, Rockford, Illinois Paint type: HILITE	WC	YN
4. Multiple Industries Corporation, Cleveland, Ohio Paint type: MICA	WD	YO
5. Wishart Incorporated, Utica, New York Paint type: ROADMARK 8	WE	YP
6. Provincial Paint Company, Pittsburgh, Pennsylvania Paint type: 81-234	WF	YQ
7. Centrex Inc, Hartford, Connecticut Paint type: TL (the "new" paint)	WG	YR

You then place the coding list into a sealed envelope and lock it away in a safety deposit box at a local bank.

You decide to paint sample stripes on two regularly traveled stretches of highway and to assess the samples in four ways:

Spraying characteristics.
Drying time.
Visibility after three months.
Visibility after six months.

You assess spraying characteristics as Excellent, Very Good, Good, Fair, and Poor. The ratings are:

Very good: WA, WB, WC, WD, WF, YL, YR
Good: WE, WG, YM, YN, YO, YQ
Fair: YP

You assess drying time in minutes:

WA:16   WC:18   WE:14   WG:19   YM:26   YO:14   YQ:12
WB:33   WD:11   WF:13   YL:13   YN:18   YP:10   YR:15

After three months you assess visibility by day and by night. You use five persons (one is yourself) to rate the stripes independently and to place the stripes' visibility on a scale of 1 to 10. You then average the five assessments (night readings are taken with headlights at high beam).

Paint Code	Concrete Pavement Day	Concrete Pavement Night	Asphalt Pavement Day	Asphalt Pavement Night	Paint Code	Concrete Pavement Day	Concrete Pavement Night	Asphalt Pavement Day	Asphalt Pavement Night
WA	8	8	8	9	YL	8	9	9	9
WB	7	8	7	7	YM	7	7	7	9
WC	8	9	10	9	YN	7	9	7	8
WD	9	9	7	8	YO	8	8	7	8
WE	7	6	7	8	YP	7	8	6	8
WF	8	9	9	10	YQ	5	6	4	6
WG	7	7	7	8	YR	9	10	9	9

After another three months the same five persons again assess stripe visibility, with these results:

Paint Code	Concrete Pavement Day	Concrete Pavement Night	Asphalt Pavement Day	Asphalt Pavement Night	Paint Code	Concrete Pavement Day	Concrete Pavement Night	Asphalt Pavement Day	Asphalt Pavement Night
WA	6	6	5	7	YL	6	7	7	8
.WB	4	5	5	5	YM	6	7	6	8
WC	8	8	9	9	YN	6	7	6	7
WD	6	7	6	8	YO	6	7	6	8
WE	6	5	6	7	YP	3	4	4	5
WF	5	7	5	6	YQ	2	3	3	4
WG	6	7	6	7	YR	8	9	8	9

You consolidate all your results into two tables, one for white paint, one for yellow paint, and then:

- Reject any unacceptable paints (see the guidelines in steps 4, 5, and 7 of the list that follows).
- Rank acceptable paints in order of suitability.
- Identify the best paint(s) to use for highway marking.
- Retrieve the paint coding list from the bank deposit box.
- Write your report.

Some factors you use to conduct your study and to write your report are:

1. Senior Highways Engineer Morris Hordern's office address is 416 Inkster Building, 2035 Perimeter Road of your city.
2. The paint stripes were painted on two stretches of highway:
   2.1 Highway 101 (concrete surface), 1½ miles north of the intersection with highway 216.
   2.2 Highway 216 (asphalt surface), ½ mile west of the intersection with highway 101.
3. You are unable to borrow the regular highway paint stripe applicator from the Highways Department. Instead, you mount a regular applicator on a small garden tractor. The paint stripes are applied at night, between midnight and 6:00 a.m.
4. Paint Manufacturers' Association specification PMA-28H states that spraying characteristics for fast-drying highway paint should be at least "Good," and preferably "Very good." To achieve "Very good," the paint must flow smoothly and evenly without forming globules or dripping from the nozzle.
5. You refer to specification ASTM D-711 to establish the maximum acceptable paint drying time, which is 20 minutes.
6. Guidelines you give to the persons assessing paint visibility are:

	Distance Visible	
Rating	Day	Night
10	500 m	200 m
8	400 m	160 m
6	300 m	120 m
4	200 m	80 m
2	100 m	40 m

You then average the five assessments.
7. You establish minimum acceptable visibility levels for the paints to be:
   After three months' traffic wear: 7
   After six months' traffic wear: 6

Note: Calculate real dates for each stage of the study and quote them in your report.

Your report should not only present the results of your tests, but also analyze them, draw conclusions, and make a recommendation.

PROJECT 6.3: STORMWATER DISPOSAL, CAYMAN FLATS

The Fairview Development Company wants to develop an area of virgin land known as Cayman Flats on the southern perimeter of Montrose, Ohio. The City of Montrose displays interest in the proposal and asks for a formal presentation of the company's development plans.

The land is flat, and Fairview Development Company soon realizes that it has a stormwater drainage problem to overcome before it can complete its presentation. It calls in H. L. Winman and Associates to resolve the problem (see Figure 6-12). The project is assigned to you.

You start your investigation by examining Cayman Flats. It is generally flat, low-lying, and frequently waterlogged. The maximum variation in height is 9 feet; its area is 1594 acres. To the east is a railway line (Northern Railways) into Montrose center, to the north a residential area, and to the west and south lie arable land 70% cultivated (see map, Figure 6-13).

You calculate the maximum stormwater runoff for the land in its present condition. (Stormwater is rainwater that must be drained from the land quickly to prevent flooding of low-lying areas and basements. Maximum stormwater runoff is the largest amount of water likely to occur; it is calculated on past records of maximum precipitation accumulated from the heaviest rainstorms.) Using the Rational Formula, you calculate that a 21 foot diameter culvert would be required to handle the heaviest peaks.

Since such a large culvert would offer construction problems, use a lot of property, represent an eyesore, and be a source of danger to small children, you

---

FAIRVIEW DEVELOPMENT COMPANY
212 Bligh Street
Montrose, Ohio
45287

Martin Dawes, President
H. L. Winman and Associates
475 Reston Avenue
Cleveland, Ohio 44104

Dear Mr. Dawes:

We are preparing a feasibility study for the City of Montrose, in which we are proposing to develop the Cayman Flats area to the south of the city as a new residential district. This low, flat land offers drainage problems because of its distance from the Wabagoon River. The storm sewers of the intervening developed areas cannot be used since they have insufficient capacity to handle the additional stormwater runoff that Cayman Flats will generate.

We are asking you to conduct an engineering investigation into this stormwater disposal problem and to recommend an economical method that we can include in our presentation to the City of Montrose.

Yours sincerely,

Frederick C. Magnusson
President

FCM/jms

Figure 6-12. Letter authorizing Cayman Flats project.

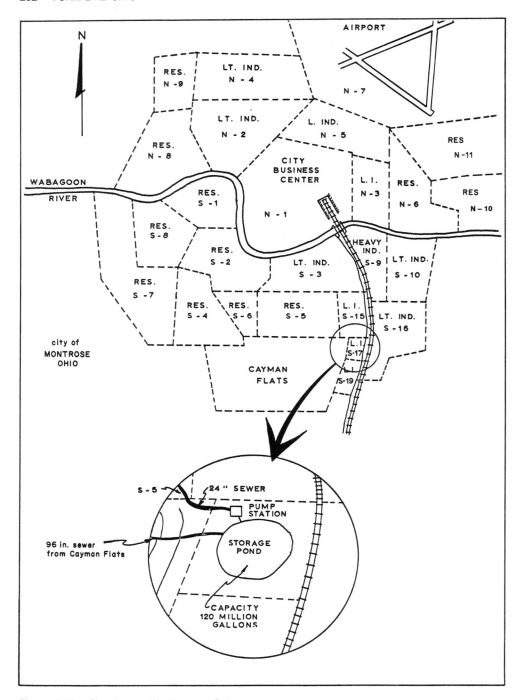

Figure 6-13.   Sketch map of Montrose, Ohio.

look for other methods. The most obvious is to construct storm sewers throughout Cayman Flats before development starts. Because the runoff would then be channeled, the stormwater would be "staged" (that is, reach the outfall, or disposal sewer system, as a series of peaks). You estimate that the largest peak could be handled by a 96 inch storm sewer.

Your next step is to examine the territory between Cayman Flats and the Wabagoon River. The whole area has been developed; to build a sewer of the required size directly to the river would be phenomenally expensive, and you would likely have difficulty in getting municipal approval. The distance is too great to build it around the southwest perimeter of the city. But there is a 40 foot (12 m) wide belt of open land on the west side of the railway line. It is owned by Northern Railways, who agree to lease it to the city for $1200 per year and will guarantee a 20 year lease.

Some other factors you discover are:

1. Storm sewers of the residential areas north of Cayman Flats are of only 18 inch diameter in zones S-4 and S-6, and 24 inch diameter in zone S-5. Those in S-4 feed into the 30 inch sewers in S-2; those in S-6 run directly down to the river, following the boundary line between S-2 and S-3; those from S-5 feed into the 36 inch sewers in S-3.
2. The municipalities view with alarm your inquiries to find an easy route to the river for a 96 inch storm sewer through zones S-3, S-5, and S-6. They fear (rightly) that you could not excavate deeply enough to avoid interfering with the existing services. (Too deep an excavation would foil the natural flow into Wabagoon River.)
3. Cayman Flats slopes very slightly downhill from the southwest corner to the northeast corner. Total drop is 9 feet 3 inches.
4. Although zones S-15, S-17, and S-19 are all classed as "light industry," S-17 has never been developed. It is a small zone of approximately 95 acres, lower in elevation than the surrounding zones, and hence rather swampy. Undoubtedly the amount of fill needed to build it up has hindered its development.

At this point an idea occurs to you. If zone S-17 is already acting as a collection point for some of the stormwater from the surrounding zones (principally Cayman Flats), why not excavate it even further and use it as a quick runoff storage pond? If it were big enough, you could hold all the stormwater runoff from Cayman Flats and then pump it at a controlled rate into a smaller diameter outfall sewer from S-17, along the railway line, to the river. It would mean building a pump station (it would have to be automatic), and fencing the storage pond, but it would be feasible. You make some quick calculations:

> Size of storage pond required: 120,000,000 gallons.
> Size of outfall sewer to river: 24 inch diameter.

To evaluate the ideas so far, you work up some cost estimates:

1. Cost of installing storm sewer system throughout Cayman Flats. $987,000
2. Cost of 96 inch diameter storm sewer along railway line from Cayman Flats to Wabagoon River. $1,272,000

3. Cost of 24 inch diameter storm sewer over same route.	$827,000
4. Cost of purchasing zone S-17, excavating the storage pond and fencing it, and building an automatic pumphouse and pump system.	$621,500
5. Pumphouse annual maintenance and operating costs.	$ 6,500

At this stage you feel your problem has been resolved and you have a reasonable proposal to offer Fairview Development Company. Your report is almost finished before another obvious and logical idea occurs to you: If you are going to *control* the flow of water into the outfall storm sewer, why not control it even further and pump the stored stormwater into the *existing* 24 inch storm sewers of zone S-5? You would have to wait until the stormwaters from S-5 had been fed into the river, but this would be no problem because you have planned an oversize storage pond that could hold the water from the heaviest storms recorded during the past 30 years. All you would have to do is obtain municipal approval and build a 24 inch sewer from the pumphouse beside the pond to one of the storm sewers in Zone S-5 (see Figure 6-13).

You approach the City of Montrose with your unusual idea. The City Engineer hesitates briefly and then gives approval in principal. You then calculate one more cost:

6. Cost of building a 24 inch storm sewer from the pumphouse in zone S-17 to southeast tip of zone S-5 and connecting it to the storm sewer in zone S-5.	$162,000

When you are preparing your report for Fairview Development Company, you should bear in mind how they intend to use it. They will probably attach it to their proposal to the City of Montrose as evidence that they have researched and resolved the stormwater disposal problem. Hence you must write with the knowledge that although you are addressing it to Fairview Development Company, the ultimate readers are likely to be the City Councilors of the City of Montrose.

PROJECT 6.4: PINPOINTING A NOISE PROBLEM

This morning Sydney Volland (your department manager) comes to you and says: "I've had a letter from the area manager of Mirabel Realty. Her staff are complaining it's too noisy in the office—that they get headaches from it and go home tired."

Sydney hands you the letter, dated yesterday:

Dear Sydney

As I mentioned when we met last week, my staff are complaining that for the past three months the noise level in our office has been too high. They claim that it is affecting their work and causing fatigue.

Will you please look into the problem for me to determine whether their complaints are justified. If they are, will you suggest what can be done to remedy the problem, recommend the most suitable method, and include a cost estimate.

Sincerely

Trudy G. Parsenon
Area Manager

"I also talked to her briefly on the phone," Sydney continues. "She figures the staff may be exaggerating a bit, and that the noise isn't nearly as bad as they make out. Just the same, she's worried: staff turnover has been much higher recently, and it's costing her a fortune to train the replacements she has to hire."

Peter asks you to take on the project and assigns it project number C62.

Part 1. At 4:00 p.m. on the same day you visit Mirabel Realty (the office is in Room 210, on the second floor of the Fermore Building at 381 Conway Avenue of your city) and talk to Trudy Parsenon. You notice a background hum, which you consider to be caused by motors in the electric typewriters, computers, and printers. You are still there when the office staff quits at 4:30. After they go, you notice you can still hear the hum, but at a lower level.

You walk around the office with Trudy, who plagues you with questions: "What do you think?" she asks. "Seems like the same noise level you get in any business office, don't you think?"

You suspect she is hoping for a good report from you, which she can use to prove to her staff that their complaints are imaginary.

"I can't tell without taking readings," you hedge. "Noise is a pretty tricky thing. What some people think is too noisy, others hardly notice."

But you do notice that the hum gets significantly louder near the east wall of the office. Then suddenly it stops; or rather, it dies away. The time is 4:45.

"What's on the other side of this wall?" you ask.

"Oh, that's Superior Giftware," Trudy replies. "They distribute cheap imports—that sort of thing."

"Have you talked to them about the hum?"

"Yeah! I asked Saul Ferguson about it—he's the manager next door. Pretty hostile, he was."

"And what time do they quit work?" you ask.

"Right now," Trudy replies. "You can always tell, because they switch their machines off."

You arrange with Trudy that you will take sound-level measurements one week from today. You want to find out how much of the noise in the insurance office is generated by normal office activity and how much by the people next door.

You consider that a visit to Superior Giftware is essential, since you suspect that the machines Trudy mentioned may be the problem. You want to know the sound levels on both sides of the wall between the two companies, to assess the extent of soundproofing you may want to recommend.

Write to Saul Ferguson, Manager of Superior Giftware, to ask permission to carry out sound-level measurements in his offices one week from today. His business is in room 208, 381 Conway Avenue.

Part 2. It is now one week later. You take a Nabuchi Model 1300 sound-level meter with you and return to 381 Conway Avenue. You plan to measure sound levels at various locations in the Mirabel Realty office under four conditions:

- When both businesses are empty.
- When only Superior Giftware is working (4:30–4:45).
- When only Mirabel Realty is working (8:00–8:15).
- When both businesses are working.

You also plan to take readings in Superior Giftware's office.

As Mr. Ferguson has not replied to your letter, yesterday afternoon you telephoned him to ask if you could come in today to take the measurements. He said he was "terribly busy" and that it's "damned inconvenient," but he somewhat reluctantly agreed. (Yet it was worthwhile being persistent. Almost right away you notice a packaging and sealing machine only 7 feet from the wall separating the two business offices.)

Table 6-3 Average Sound Levels, Second Floor, 381 Conway Avenue, Montrose, Ohio

Mirabel Realty	Both Offices Working (dB)	Only Superior Working (dB)	Only Mirabel Working (dB)	No One* Working (dB)
A	74	73	48	27
B	71	69	51	27
C	66	65	50	27
D	64	61	52	26
E	63	59	51	28
F	59	53	49	26
G	54	51	48	28
Superior Giftware				
H	86	—	—	25
I	83	—	—	26

*Mostly air-conditioner noise.
Note: Measurements made with Nabuchi Model 1300 Sound-Level Meter set to "A" scale.

You record the measurements you take (see Table 6-3) and compare them to the general ratings for office noise, which you obtain from City of Montrose standard SL2020, dated January 20, 1991. The recommended sound levels for an urban office are:

Quiet office: 30–40 dB
Average office: 40–55 dB
Noisy office: 55–75 dB

You note that the sound level in Mirabel Realty's office increases as you move toward the dividing wall between the two offices (see Figure 6-14).

You also notice there seem to be two components of noise in Mirabel Realty's offices, some being transmitted through the air and some being transmitted through the structure (from Superior Giftware's machines, through the floor). A hand placed on the walls or floor feels the vibration. Floors in both offices are tiled.

Before leaving you tell Trudy Parsenon that there seems to be a noise problem, but it can be corrected. You warn her, however, that it may prove expensive. She says she will have a job justifying the costs to her head office.

On returning to your office you write a progress report for Sydney Volland. (To do this, you may want to use some of the information in Part 3).

Part 3.   You consider possible ways to reduce the sound levels in Mirabel Realty's office:

1. You could erect a false wall, insulated internally with Corrugon, from floor to ceiling on Mirabel Realty's side of the wall between the two companies.
2. Or 7/16 inch thick black cork panels could be glued on the Mirabel Realty side of the wall.
3. Or you could install carpet throughout Mirabel Realty's office.
4. Or Superior Giftware's packaging machine could be mounted on Vib-O-Rug (insulating rubber that eliminates transmission of vibration from machine to building structure).

You recognize that remedies 1 and 2 are alternatives (they both deal with sound transmitted through the air). Remedies 3 and 4 also are alternatives (they both dampen vibrations and sound carried through the structure). Remedy 3 also quite effectively dampens internal office noise.

You consider the approximate costs:

Remedy 1—$ 8,700
Remedy 2—$ 1,600
Remedy 3—$14,900
Remedy 4—$ 950

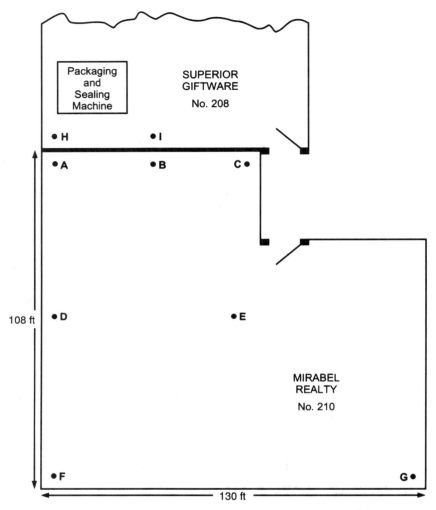

● Points at which sound levels recorded; each point is 4 ft from nearest wall.

Figure 6-14. Plan of Mirabel Realty's office.

You consider possible problems each remedy may present:

1. Corrugon is in short supply; delivery time would be a minimum of 3 months.
2. To some people, cork has an offensive smell; this can be partly corrected by treating the cork with polymethynol.
3. The carpet must be dense and have a good quality rubber underlay (included in the approximate cost).
4. Depends on cooperation of Superior Giftware's manager.

You calculate probable noise reductions for each method:

Remedy 1: 6 to 10 dB
Remedy 2: 4 to 7 dB
Remedy 3: 8 to 12 dB
Remedy 4: 3 to 5 dB

(These anticipated reductions apply only to the Mirabel Realty office, when both businesses are working.)

You consider which alternatives to recommend to Mirabel Realty and then write an investigation report describing your findings and suggested corrective measures. You also write a brief cover letter to Trudy Parsenon summarizing your findings. Prepare the letter for Sydney Volland's signature.

Note: You prepare a *formal* report because Trudy has mentioned she might have difficulty convincing her head office executives that they must authorize the cost of the remedial measures. And, because Trudy knows little about noise and its effects, you decide to include some explanatory information. (You would also be wise to research and document such information at a library to establish positive evidence for the statements you make in your report.)

PROJECT 6.5: MOVING A BUS STOP

H. L. Winman and Associates has been carrying out a transportation study for Montrose (Ohio) Community College, and shortly will be issuing a report recommending changes in traffic patterns and improvements in transit services. These, it is hoped, will ease the traffic bottlenecks that occur in the mornings and evenings at major intersections near the college. Project coordinator is Garth Ruttner.

In response to a telephone call from Darryl Kane, the college's Technical Services Administrator, Garth assigned you to a separate project at the college. He told you that Mr. Kane will give you all the details and that your time is to be charged against Montrose Community College Project No. M71-C.

When you visit Darryl Kane the following morning, he tells you that he has received a series of noisy telephone calls from an indignant cafe owner, Luigi Picolino, who operates a restaurant at 227 Wallace Avenue, just south of Main Street. Yesterday, Mr. Picolino visited the college.

"Indignant" hardly described Mr. Picolino adequately. He was bristling with anger. He had been to the Transit Office; he had been to the Montrose Streets and Traffic Department; he had even called at the home of his City Councilor; and to all he woefully related his tale. They all did the same thing: told him they cannot help him and that he should go and see someone else. Angry at being shuffled from person to person, none of whom showed any real interest in his problem, Mr. Picolino finally turned to the college, because it is college students who are the source of his annoyance.

With some difficulty, Mr. Kane calmed him down and elicited a collection of details from which he pieced together the cause of Mr. Picolino's indignation. Apparently, college students traveling from east and west Main Street disembark from their buses at Wallace Avenue and wait for the college bus at a special stop on Wallace Avenue immediately in front of Picolino's Pantry. On cold or wet mornings the students crowd into the Pantry to keep warm while they wait. Luigi Picolino's complaint is: (1) they don't buy anything (other than an occasional cup of coffee); and (2) they discourage other customers from coming in. The cafe becomes so crowded that passersby think it is full and turn elsewhere to seek their breakfast. This is the time of day when Mr. Picolino used to do a lot of business, but since the bus stop was put in five months ago his early morning business has dropped to virtually nil.

Mr. Kane soothed Mr. Picolino's ruffled surface by promising that the problem would be resolved shortly, and then placed the telephone call to Garth Ruttner that resulted in your visiting the college this morning. He asked you to evaluate Mr. Picolino's complaint and, if it is valid, to investigate ways to resolve it. He also asked for a report of your findings that he can present to city and college officials.

In your investigation you take the following approach, asking yourself pertinent questions as you go along:

1. Is Mr. Picolino's complaint warranted?
   Yes, it seems to be. You check on two consecutive mornings:

Time	No. of Students in Cafe on Wednesday	No. of Students in Cafe on Thursday
07:00	0	0
07:10	5	0
07:20	8	7
07:30	13	10
07:40	15	18
07:50	27	36
08:00	42	39
08:10	23	31
08:20	6	5
08:30	1	0

   a. Mr. Picolino informs you that these are quieter mornings than usual. You check up and find that this week 35% of the college's student population are taking exams in the afternoon and are unlikely to be attending classes in the morning.
   b. There are seats for 36 customers in Picolino's Pantry.

2. How frequent are the buses?
   A bus loads passengers at the bus stop at 07:15, 07:32, 07:45, 07:55, 08:05, 08:15, and 08:28.

3. What is the bus route in this area?
   The bus travels north along Wallace Avenue (a one-way street), and then turns left to travel west along Main Street.
4. Could the bus stop be moved?
   There are four possible directions:
   a. South: Yes. But it would have to be at least 120 yards south; there is a crosswalk immediately south of the present stop.
   b. North: No. No room between the existing stop and Main Street, and Wallace Avenue north of Main Street is not on the bus route.
   c. East: No. Not on bus route.
   d. West: Yes. By Caernarvon Hotel.
5. What factors affect the location of the bus stop?
   a. Connecting bus routes drop passengers at points A and B on Main Street (refer to Figure 6-15).
   b. Montrose Rapid Transit System insists that the connecting stop must be within 100 yards of the intersection of Main Street and Wallace Avenue.

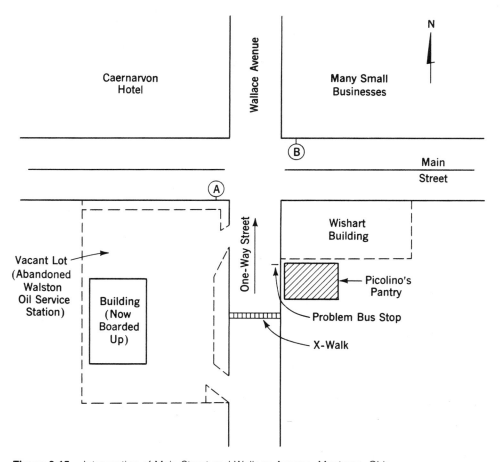

Figure 6-15. Intersection of Main Street and Wallace Avenue, Montrose, Ohio.

    c. There is a taxi stand outside the Caernarvon Hotel, but the Montrose Streets and Traffic Department is adamant: It will not take part of the taxi stand to create a bus stop in that block.

    d. The bus stop could be moved directly across the street, to the west side of Wallace Avenue, where there are no stopping restrictions. But this poses two problems: Bus doors would open onto the traffic side of the bus rather than onto the sidewalk, and it is still too close to Picolino's Pantry.

6. You note that the lot at the southwest corner of the Main and Wallace intersection is vacant. Previously a Walston Oil service station, it is now abandoned and has been for 10 months; the building is boarded up. You investigate ownership. The land was purchased by the City of Montrose six months ago. The city plans to build a car park on the lot, but not for five years, when it plans to acquire adjoining lots.

7. An idea begins to form: Why not build a bus shelter on the vacant lot and route the bus through the lot? (It could use the old entry and exit ramps built for the service station.)

8. You discuss your idea with the Traffic Engineer at the Montrose Streets and Traffic Department (D. V. Botting), who says his Department would not object providing the shelter conforms to City of Montrose, Transit Division, Building Construction Specification BCS 232. You also check with Ken Whiteside of the Montrose Rapid Transit System, who says his department will go along with it. (He also says Transit will pay the cost of building a shelter on the lot, providing the cost is no greater than $2400.) Then you check with the City of Montrose (Planning Division) and get approval to use the land for five years. The city will also authorize expenditure of $500 a year for general upkeep of the lot. You ask all three parties to state their agreement with your plan in writing, and you receive letters from them indicating their willingness to participate.

9. You write to three contractors asking them to give you a price quotation, telling them that the shelter must be able to hold at least 40 people, that the price quotation must include an adequate electric heating system (minimum temperature +50°F), seats, and lighting, and that the work must be completed within three weeks of receipt of an order. You also inform them that the work must conform to City of Montrose Bylaw 61A.

10. The lowest price quotation you receive is submitted by Donovan Construction Company, 2821 Girton Boulevard, Montrose, whose price of $3145 is still $745 above the maximum stipulated by Montrose Transit. (Other quotations were $3367 and $3488.) You visit Gavin M. Donovan, owner-manager of Donovan Construction Company, to see if he can bring his price down to within Montrose Transit's limit.

"Impossible" he says. "No way! Nobody could do it."

Then, as an afterthought, he adds: "But tell you what: We could do as good a job by renovating the service station."

He digs out some files from his desk. "See here," he says, "we ran an estimate on renovating the existing building and installing seats. Electric heat and light are already in it. The price would be $1965 and I'll throw in some landscaping."

Since this resolves the problem, you return to your office and prepare a report for Darryl Kane. You decide to prepare a formal investigation report, since he will want to submit copies to all parties involved and to use it as the document for implementing conversion of the Walston Oil building and moving the bus stop.

CHAPTER 7
Other Technical Documents

This chapter describes how to write a user's manual, provides detailed guidelines for writing a technical instruction, offers suggestions for writing a scientific paper, and describes how to convert your knowledge of a process, equipment, or new technique into an interesting magazine article or technical paper.

USER'S MANUAL

In our technological era, with its increasingly complex range of hardware and software, there is a growing need for manufacturers to write clear technical manuals to accompany what they sell. Most manufacturers issue a user's manual with each of their products, which contains (1) a brief description of the product, (2) instructions on how to use it, and (3) suggestions for remedying problems that may occur. For qualified repair specialists they may also produce a set of maintenance instructions containing detailed service and repair procedures. Both publications perform the same task for different readers: user manuals assume the reader has only slight technical knowledge, while maintenance instructions assume the reader is a technical expert.

The suggestions that follow apply to any basic user's manual. Whether you are writing a manual to accompany heavy construction equipment, a delicate instrument, or a new version of a software program, you must organize your description so that it follows a coherent pattern.

WRITING PLAN

The writing plan for most user manuals has four compartments, as shown in Figure 7-1. The two top compartments describe the product, while the two lower compartments tell the reader how to use it.

The following paragraphs demonstrate how these four compartments are used. The product is assumed to be an overhead projector found in a classroom, conference room, or lecture hall.

DESCRIBING THE PRODUCT

The **summary statement** briefly describes the product and its main purpose:

> An overhead projector projects 8 in. × 10 in. film transparencies containing drawings and illustrations onto a screen, which normally is above and behind the speaker. The projector stands in front of or beside the speaker, so that the speaker can use the projector and view or point to parts of the illustration without facing away from the audience.

The **product description** identifies each part and describes its components. For example, a user manual accompanying a word-processing program sold to owners of personal computers would describe the contents and application of its five 3.5-inch diskettes. For equipment that has several discrete components or parts, the product description may be subdivided into two sections: main parts and detailed description:

- The **main parts** section simply lists the main components:

  > The projector consists of three main parts: a base containing the light source, a ground glass plate for holding the transparency, and a lens assembly for projecting the light onto a screen.

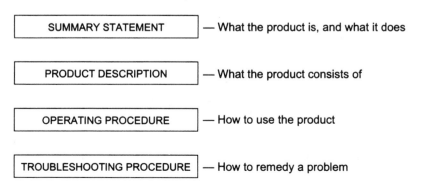

Figure 7-1. Writing plan for a user's manual.

If there are more than three parts, they may be presented in subparagraph or "point" form.

- The **detailed description** provides a lot more information about each part, and in particular draws attention to items the user will operate or use. The parts must be described *in the same sequence* that they have been presented in the main parts section:

> The light source is housed in a 12-inch square metal projection compartment and consists of a high-intensity bulb supported above a concave, highly polished mirror. The assembly includes a spare bulb, a cooling fan, and an interlock which disconnects the power when the cover plate that forms the upper surface of the compartment is opened.
>
> The cover plate is a metal frame attached by hinges to the back of the projection compartment, with an 11-inch square ground glass plate set into it. Suspended immediately above the glass plate is a lens-and-mirror assembly, which takes the projected light, magnifies it, and projects it onto a screen. The assembly is supported by an arm extending from a vertical post attached to the projection compartment. This arm is moved up and down the post to control focus, and the lens assembly is tilted to adjust the height of the projected image on the screen.

There must be a logical pattern to the detailed description, which may be presented in either a spatial or a sequential arrangement. (Spatial means "arranged in space.") In a spatial arrangement, which has been used to describe the overhead projector, the sequence depends on the shape of the equipment or product:

Vertical Subjects — Tall, narrow subjects lend themselves to a vertical description, each item being described in order from top to bottom, or bottom to top. Vertically arranged control panels and electronic equipment racks fall into this category, as does the overhead projector.

Horizontal Subjects — Long, relatively flat subjects demand a horizontal description, with each part being described in order from left to right, or right to left.

Circular Subjects — Round subjects, such as a clock, a pilot's altitude indicator, or a circular slide rule, suit a circular arrangement, with the description of markings starting at a specific point (for example, "12" on a clock) and continuing clockwise until the entire cycle has been described. In effect, this is the same as the horizontal description, if you consider that the circle can be broken at one point and unwound into a straight line. Alternatively, if the subject has a series of items that are arranged more or less concentrically (such as the scales of a circular slide rule), they can be described in sequence starting at the center and working outward, or from the outermost item inward.

A sequential arrangement introduces the parts or items in the sequence in which the reader will use them. If the overhead projector had been described sequentially, its detailed description would have been like this:

> The transparency to be viewed is placed on a ground glass plate set into the hinged cover on top of the overhead projector's 12-inch-square metal base. From inside the base a light source, which sits above a highly polished concave reflector, is controlled by a switch on the front of the projection compartment. Focus and height of the image on the screen are controlled by adjusting the lens assembly,

which is suspended 12 to 15 inches immediately above the glass plate. This assembly is held in position by an arm extending horizontally from a vertical post attached to the back of the projector's base. Focus is adjusted by rotating a knurled knob on the upright post, which raises or lowers the lens assembly. Image height is adjusted by tilting the lens assembly.

USING THE PRODUCT

The **operating procedure** describes in chronological sequence the steps for setting up and using the product:

To ready the projector for use:
1. Place a transparency flat on the ground glass plate, with its image facing up and the top of the transparency farthest away from you (so that you can view the transparency naturally when you are facing the audience and the projector is immediately in front of you).
2. Depress the ON/OFF switch to "ON" (it is on the front of the projector; that is, on the side nearest to you), and check that the projector lamp comes on.
3. Examine the image on the screen and, if it is not in focus, rotate the knurled knob on the vertical post back and forth until the image is in focus.
4. If the image is not centered on the screen, swivel the whole projector left or right to move the image horizontally, and tilt the lens assembly up or down to move the image vertically.

Note that each step is short, has a number, and uses verbs in the imperative mood. (For more information on writing instructional steps, see "Give Your Reader Confidence" on pages 219 and 220 of this chapter.)

The **troubleshooting procedure** tells the reader what to do when, having followed the operating procedure correctly, the equipment does not work. It also has short, numbered steps and verbs in the imperative mood:

If the projector lamp fails to come on when the ON/OFF switch is depressed, follow this procedure:
1. Check that the projector cable is plugged into a wall outlet.
2. Check that the frame holding the cover plate is pressed firmly down onto its housing, so that the safety interlock is engaged.
3. Lift up the front of the frame holding the cover plate, look into the projection compartment, and check that the bulb is serviceable. If it is not, swing the spare bulb into position or insert a new bulb, and then lower the frame gently but firmly onto the housing.
4. Check that the wall outlet is powered up (i.e. that the circuit breaker has not tripped).

If none of these four steps causes the lamp to come on, call a service technician.

TECHNICAL INSTRUCTION

When H. L. Winman special projects engineer Andy Rittman wants a job done, he issues instructions in clear, concise terms: "Take your crew over to the east end of the bridge and lay down control points 3, 4, and 7," he may say to the survey crew chief. If he fails to make himself clear, the crew chief has only to walk back across the bridge to ask questions. But Fred Stokes, chief engineer at Macro Engineering Inc, seldom gives spoken instructions to his electrical crews. Most of the time they work at remote sites and follow printed instructions, with no opportunity to walk across a project site to clarify an ambiguous order.

A technical instruction tells somebody to do something. It may be a simple one-sentence statement that defines what has to be done but leaves the time and the method to the reader. Or it may be a step-by-step procedure that describes exactly what has to be done and tells when and how. It is the latter type of technical instruction that will be described here.

Before attempting to write an instruction, you must first define your readers, or at least establish their level of technical knowledge and familiarity with your subject. Only then can you decide the depth of detail you must provide. If they are familiar with a piece of equipment, you may assume that the simple statement "Open the cover plate" will not pose a problem. But if the equipment is new to them, you may have to broaden the statement to help them first identify and open the cover plate:

> Find the hinged cover plate at the bottom rear of the cabinet. Open it by inserting a Robertson No. 2 screwdriver into the narrow slot just above the hinge and then rotating the screwdriver half a turn counterclockwise.

START WITH A PLAN

A clearly written instruction contains four main compartments, as shown in Figure 7-2. These compartments contain the following information:

- A **summary statement** outlines briefly what has to be done:
  > The 28 Vancourt Model AL-8 overhead projectors in rooms A4 to A32 are to be bolted to their projection tables . . .
- The **purpose** explains why the work is necessary:
  > . . . to reduce the current high damage rate caused by projectors being accidentally knocked onto the floor.

  A technician who understands *why* a job is necessary will much more readily follow an instruction.
- A short paragraph or list describes the **tools and materials** that technicians will need to perform the task (they can use this as a checklist to ensure they have accumulated everything they need before they start work):

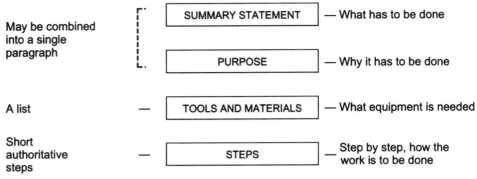

Figure 7-2. Writing plan for an instruction.

To carry out the modification you will require:
- Modification kit OHP4, comprising
  1 template, OHP4-1
  4 bolts, flat head, 2 in. long, ⅛ in. dia
  4 washers, 1 in. dia, with ⁵⁄₃₂ in. dia central hole.
  - A ¼ in. drill with a ³⁄₁₆ in. drill bit
  - A Phillips No. 2 screwdriver
  - A slot-head No. 3 screwdriver
  - A sharp pencil
- The **steps** that readers must follow take them through the whole process:
  Proceed as follows:
  1. Disconnect the projector's power cord from the wall socket, then take the projector to a table and turn it on its side.
  2. Using a slot-head No. 3 screwdriver, unscrew the four bolts that hold the feet onto the base of the projector. Remove but retain the bolts and feet for future use.
  3. Place template OHP4-1 onto the projection table and position it where the projector is to stand. Using a sharp pencil, mark the table through each of the four holes in the template.
  4. Drill four ³⁄₁₆ in. dia holes through the table top, at the places marked in step 3.
  5. Place the overhead projector, right side up and with the lens assembly facing the screen, so that the four screwholes identified in step 2 coincide with the four holes in the table top.
  6. From beneath the table, place a washer under each hole and insert a 2 in. flat head bolt up through the washer and hole until it engages the corresponding hole in the projector base. Tighten the four bolts in place, using a Phillips No. 2 screwdriver.

The writing plan embodying these four compartments, when combined with the following suggestions, will consistently ensure that any instructions you write will be clear, direct, and convincing.

GIVE YOUR READER CONFIDENCE

A well-written technical instruction automatically instills confidence in its readers. They feel they have the ability to do the work even though it may be new to them and highly complex. Consider these examples:

Vague   Before the trap is set, it is a good idea to place a small piece of cheese on the bait pan. If it is too small it may fall off and if it is too big it might not fit under the serrated edge, so make sure you get the right size.

Clear and Concise   Cut a 17-inch length of 10-gauge wire and strip 1 inch of insulation from each end. Solder one end of the wire to terminal 7 and the other end to pin 49.

The first excerpt is much too ambiguous. It only suggests what should be done, it hints where it should instruct (almost inviting readers to nip their fingers), and in 31 explanatory words it fails to define the size of a "small" piece of cheese. The second excerpt is assertive and keeps strictly to the point. Such clear and authoritative writing immediately convinces its readers of the accuracy and validity of the steps they have to perform.

The best way to be authoritative is to write in the imperative mood. This means beginning each step with a strong verb, so that your instructions are commands:

Ignite the mixture. . . .   *Connect* the green wire. . . .
Mount the transit on its tripod. . . .   *Excavate* 3 feet down. . . .
Apply the voltage to. . . .   *Measure* the current at . . .
Cut a 2 inch strip of. . . .   *Count* the number of blips. . . .

The imperative mood in the clear, concise excerpt quoted earlier keeps the instruction taut and definite. Notice how the verbs "cut," "strip," and "solder" make readers feel they have no alternative but to follow the instructions. The vague excerpt would have been equally effective (and much shorter) if it had been written in the imperative mood and if the vague verb "place" had been replaced by an image-conveying verb-adverb combination such as "wedge firmly:"

Before setting the trap, wedge a ⅜ inch cube of cheese firmly under the serrated edge of the bait pan.

The following two statements clearly show the difference between an instruction written in the imperative mood and one that is not:

Disengage the gear, then start the engine.
(definite; uses strong verbs)

> The gear should be disengaged before starting the engine.
> (*indefinite; uses weaker verbs*)

The first statement is strong because it tells readers to do something. The second is weak because it neither instructs them to do anything nor insists that anything need be done ("should" implies it is only *preferable* that the gear be disengaged before the engine is started).

In the imperative mood the first word in a sentence most often will be a strong verb; sometimes, however, the verb may be preceded by an introductory or conditioning clause:

> Before connecting the meter to the power source, *set* all the switches to "zero."

The imperative mood is maintained here because the main verb starts the statement's primary clause (in this case the clause that describes the action to be taken).

If you want to check whether a sentence you have written is in the imperative mood, ask yourself whether it *tells* the reader to *do* something. If it does, then you have written an *instruction*.

AVOID AMBIGUITY

There is no room for ambiguity in technical instructions. You have to assume that the person following your instructions cannot ask questions, and so you must never write anything that could be interpreted more than one way. This statement is open to misinterpretation:

> Align the trace so that it is inclined approximately 30° to the horizontal.

Each technician will align the trace with a different degree of accuracy, depending on his or her interpretation of "approximately." How accurate does "approximately" require the technician to be? Within 5°? Within 2°? Within ½°? Maybe even 10° either side of 30° is acceptable, but the reader does not know this and feels doubtful. Worse still, the reader's confidence in the technical validity of the whole instruction is undermined. Replace such vague references with clearly stated tolerances:

> Align the trace so that it is inclined 30° (±5°) to the horizontal.

More subtle, but equally open to misinterpretation, is this statement:

> Adjust the capstan handle until the rotating head is close to the base.

Here the offending word is "close," which needs to be replaced by a specific distance:

> Adjust the capstan handle until the distance between the rotating head and the base is 2.5 mm.

Similarly, replace vague references such as "*relatively* high," "*near* the top," and "*an adequate* supply" with clearly stated measurements, tolerances, and quantities.

Avoid weak words such as "should," "could," "would," "might," and "may," because they weaken the authority of an instruction and reduce the reader's confidence in the writer. For example:

> Set the meter to the +300 V range. The needle should indicate 120 V (±2 V).

In this example, "should" *implies* that it would be nice if the needle indicated within 2 volts of 120 V, but it is not essential! No doubt the writer meant it *must* be within the specified voltage range, but has failed to say so. Nor has the reader been told to note the reading. The writer has forgotten the cardinal rule of instruction writing: *Tell* the reader to *do* something. To be authoritative, the instruction needs very few changes:

> Set the meter to the +300 V range, then check that its needle indicates 120 V (±2 V).

Notice how the steps in the sample instruction in Figure 7-3 are clear, concise, and definite. You need not be a specialist in the subject to recognize that they would be easy to follow.

WRITE BITE-SIZE STEPS

Technicians working on complex equipment in cramped conditions need easy-to-follow instructions. You can help them by writing short paragraphs, each containing only one main step. If a step is complicated and its paragraph grows unwieldy, divide it into a major step and a series of substeps, numbering the paragraphs and subparagraphs:

> 3. List the documentary evidence in block J of Form 658. Check that blocks A to G have been completed correctly, then sign the form and distribute copies as follows:
>   3.1 Attach the documentary evidence to Copies 1 and 2 and mail them to the Chief Recording Clerk, Room 217, Civic Center, Montrose, Ohio.
>   3.2 Mail Copy 3 to the Computer Data Center, using one of the special pre-addressed envelopes.
>   3.3 File Copy 4 in the "Hold—Pending Receipt" file.

Heathkit®

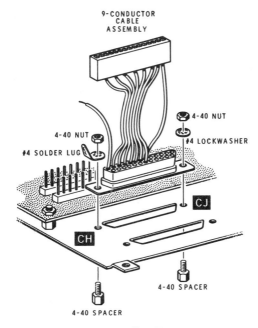

9-CONDUCTOR CABLE ASSEMBLY

4-40 NUT

4-40 NUT

#4 SOLDER LUG

#4 LOCKWASHER

CJ

CH

4-40 SPACER

4-40 SPACER

Detail 4-7B

( ) Refer to Detail 4-7B and mount the 9-conductor cable assembly at CH and CJ. Use a 4-40 spacer, a #4 solder lug, and a 4-40 nut at CH; use a 4-40 spacer, a #6 lockwasher, and a 4-40 nut at CJ.

( ) Position socket S401 at the other end of the assembly with the slotted side up, as shown and install it on plug P401.

( ) Connect the black wire coming from the assembly to solder lug CH (S-1).

( ) Similarly mount the 7-conductor cable assembly at CK and CL with a #4 solder lug at CL. Then install socket S402 on plug P402, again, with the slotted side up.

( ) Connect the black wire coming from the assembly to solder lug CL (S-1).

( ) Refer to Detail 4-7C and position the 6-conductor cable assembly and the connector bracket as shown. Slide the connector bracket onto the connector and mount it at CN with a 6-32 × 1/4" pan head screw, #6 lockwasher and a 6-32 nut.

( ) Position socket S403 at the other end of the assembly with the slotted side as shown and install it on plug P403.

Refer to Pictorial 4-8 (Illustration Booklet, Page 13) for the following steps.

( ) Position the back panel as shown and insert the two end and center tabs into the slots in the back panel of the chassis.

( ) While holding the panel in place, install socket S404 on plug P404. Be sure to match up the lip on the socket and plug when you install it.

( ) Similarly install socket S405 on plug P405.

( ) With the panel still engaged in the chassis, rotate it to a vertical position and fasten it at CP with a #6 × 1/4" sheet metal screw.

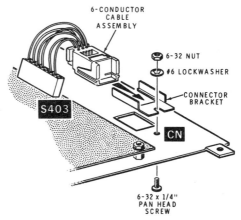

6-CONDUCTOR CABLE ASSEMBLY

6-32 NUT

#6 LOCKWASHER

CONNECTOR BRACKET

S403

CN

6-32 x 1/4" PAN HEAD SCREW

Detail 4-7C

Figure 7-3. Excerpt from an instruction manual. As each step is completed, the user inserts a check mark beside the appropriate paragraph. (Courtesy of the Heath Company, Benton Harbor, Michigan)

3.4 When Copy 2 is returned by the Chief Recording Clerk, attach it to Copy 4 and file them both in the "Action Complete" file.

INSERT FAIL-SAFE PRECAUTIONS

Insert precautionary comments into instructions whenever you need to warn readers of dangerous conditions or of damage that may occur if they do not exercise care. There are three precautionary notices you can use:

Warnings: To alert readers of an element of personal danger (such as the presence of unprotected high voltage terminals).

Cautions: To tell readers when care is needed to prevent equipment damage.

Notes: To make general comments (for example, "On some older models the valve is at the rear of the unit").

Draw attention to a precautionary comment by placing it in the middle of the text, indenting it on both sides, and preceding it with the single word WARNING, CAUTION, or NOTE. Draw a box around the words WARNING and CAUTION to give them extra prominence. For example:

> ┌─────────────┐
> │ WARNING   │
> └─────────────┘

Disconnect the power source before removing the cover plate.

Ensure that every precautionary comment *precedes* the step to which it refers. This will prevent an absorbed reader who concentrates on only one step at a time from acting before reading the warning. Never assume that mechanical devices, such as indentation and the box drawn around the precautionary word, are enough to catch a reader's attention.

Use warnings and cautions sparingly. A single warning will catch a reader's attention. Too many will cause a reader to treat them all as comments rather than as important protective devices.

INSIST ON AN OPERATIONAL CHECK

The final test for any technical instruction is the reader's ease in following it. Since you cannot always peer over a reader's shoulder to correct mistakes, you should find out whether users are likely to run into difficulty *before* you send an instruction out. To obtain an objective check, give the draft instruction to someone of roughly equal competence to the people who eventually will be using it, and observe how well that person performs the task.

Note every time the user hesitates or has difficulty, and then, when he or she has completed the task, ask if any parts need clarification. Rewrite ambiguities, and then recheck your instruction with another person. Repeat these steps until you are confident your readers will be able to follow your instruction easily.

SCIENTIFIC PAPER

Earlier chapters described how technical reports should be planned, organized, written, and presented by engineers and engineering technicians working for industry, business, and government. Within this context one additional report remains to be described: the research report prepared by scientists and technologists working in industrial and university laboratories. Research reports are most often prepared and published as *scientific papers* which, although their parts are similar to those of an investigation report, differ in style, organization, and emphasis.

A scientific paper either identifies and attempts to resolve a scientific problem, or it tests (validates) a scientific theory. It does so by describing the four main stages of the research:

1. Identifying the problem or theory.
2. Setting up and performing the tests.
3. Tabling the test results (the findings).
4. Analyzing and interpreting the findings.

These four stages represent the major divisions of a scientific paper, with each stage preceded by a descriptive heading: **Introduction, Materials and Methods, Results**, and **Discussion**. These stages are similar to those used for the laboratory report in Chapter 4 (p. 95) and the investigation report in Chapter 5 (Figure 5-1, p. 113). There are, however, differences in a scientific paper's appearance and writing style.

APPEARANCE

A scientific paper straddles the borderline between semiformal and formal presentation. Normally the title is centered about two to three inches from the top of the first page. (These dimensions apply to a keystroked or typewritten paper, as in Figure 7-4; for a typeset paper less open space would be used.) The author's name and the name of the company or organization the author works for can also be centered at the top of the page, about one inch below the title. Alternatively, the author's name and affiliation can be placed at the bottom left of the first page, or at the end of the paper.

The abstract (summary) appears next, and starts about 1¼ inches below the title or the author's name. For a keystroked or typewritten paper the abstract should be indented about one inch from both side margins. For a typeset paper, the abstract normally is set in bolder type and may also be indented slightly on both sides.

ACID RAIN TESTING IN THE CITY OF FELDSPAR, OHIO, 19XX

Corrine L. Danzig and Mark M. Weaver
University of Feldspar Research Laboratory

ABSTRACT

Acid rain is a growing concern in the United States, with its effects becoming increasingly noticeable south of Lakes Erie and Ontario. To determine what increases have occurred in the City of Feldspar, Ohio, over the past nine years, acidity levels were measured and compared to measurements recorded in 19xx, when the average pH level was YYY. The current tests showed that the average pH now is ZZZ, but that the acidity levels are not uniform. The greatest toxicity was found to be at the University of Feldspar, in the southeastern area of the city, with the lowest toxicity in the northwestern area of the city. The change was attributed primarily to the increase in fossil-fuel-burning heavy industries in the Poplar Heights industrial park, which is 3 miles northwest (generally upwind) of the principally affected areas.

INTRODUCTION

Over the past 20 years, observers in the states immediately south of Lakes Erie and Ontario have reported an increased incidence of crop spoilage and tree defoliation, which has been attributed to acid rain created primarily by fossil-fuel-burning industries and automobile emissions. Tests were taken initially in 19xx to assess the pH levels of the precipitation falling in and around the City of Felsdpar, Ohio.[1] These were compared with

Figure 7-4. The first page of a scientific paper.

The body of a scientific paper starts about ¾ inch beneath the abstract and looks much like the body of a semiformal report. Normally, a keystroked or typewritten scientific paper is double-spaced throughout, including the abstract (i.e. there is a blank line between each line of type), while a typeset report is single-spaced (i.e. there are no blank lines, except between paragraphs and before and after headings). The first line of every paragraph is indented five columns or typewriter spaces.

WRITING STYLE

The rules for brevity, clarity, and directness suggested in Chapters 1, 3, and 11 for technical letters and reports apply equally to scientific papers. But there is one exception: Where I have recommended that you consistently write in the active voice, in some sections of a scientific paper—particularly the Materials and Methods section—convention requires that you use the passive voice. For example, in a technical report I have advised you to write

I placed the sample in the chamber . . .

or

The technician placed the sample in the chamber. . . .

But in a scientific paper you will more commonly write:

The sample was placed in the chamber. . . .

In other words, the identity of the "doer" is consistently concealed in a scientific paper, and use of the first person ("I" or "we") is carefully avoided. This penchant for the passive voice is traditional and remains unaffected by the change to the active voice for technical reports and papers.

ORGANIZATION

A scientific paper has six main parts:

1. **Abstract** (summarizes the paper, emphasizing the results).
2. **Introduction** (provides background details and outlines the problem or theory tested).
3. **Materials and Methods** (describes how the tests were performed).
4. **Results** (states the findings).
5. **Discussion** (analyzes and interprets the findings).
6. **References** (lists the documents consulted).

Sometimes you may receive significant assistance or guidance from other people during your research and may want to acknowledge their help. Such acknowledgments normally are placed after the discussion but before the references.

If you are accustomed to writing laboratory reports or investigation reports, you have probably noticed that the "Conclusions" heading has been omitted from this list. Where the conclusions in a laboratory or investigation report serve as a *separate* terminal summary (a summing up of the results and their analysis), in a scientific paper they are embedded in both the introduction and discussion and only rarely are preceded by separate headings.

Abstract.   The rules for writing an abstract are almost identical to those for writing the summary of an investigation report. In an abstract you (1) outline the problem and the purpose of your investigation; (2) mention very briefly how you conducted the investigation or tests; (3) describe your main findings; and (4) summarize the conclusions you have drawn. All this must be done in as few words as possible; ideally, your abstract will be about 125 words long and never more than 250 words. A typical abstract is shown in Figure 7-4.

From the abstract, readers must be able to decide whether the information you provide in the remainder of the paper is of particular interest to them and whether they should read further. Because a scientific paper is addressed to readers who generally are familiar with your technical or scientific discipline, you may use technical terminology in the abstract. (This is the major difference between a report summary and a scientific abstract.) The abstract should be written last, when the remainder of the paper has been written, so that you can *abstract* the brief details you need from what you have already written.

Introduction.   In the introduction you prepare readers so that they will readily understand the technical details in the remainder of the paper. The introduction contains four main pieces of information:

1. A definition of the problem and the specific purpose of the investigation or tests you conducted. (This should be a much more detailed definition than appeared in the abstract.)
2. Presentation of background information that will enable the reader to fully understand and evaluate the results. Often it will include—or sometimes may consist entirely of—a review of the previous scientific papers, journal articles, books, and reports on the same subject. (This is known as a literature survey.) Rather than simply list the pertinent documents, you are expected to summarize the main findings of each and their relevance to your investigation. These documents should be cross-referenced to your list of references at the back of the report. Here is an example:

    Previous measurements of acid rain in Montrose were recorded in 19xx by Gershwain (3), who reported an average acidity level of xxxx, and. . . .

3. A short description of your approach to the investigation and why you chose that particular method.
4. A concluding statement that outlines your main findings.

Notice that the introduction of a scientific paper *includes a brief summary of the results*, which is much less common in a semiformal investigation report and rare in a laboratory report. Yet here there is a parallel with the alternative method of presenting a formal report, in which the writer presents the results three times (see Figure 6-9). In a scientific paper the results also are presented three times: very briefly in the abstract and introduction, and fully in the results section.

Materials and Methods. This section has to be thoroughly prepared and presented because, from what you write here, readers must be able to replicate (perform) an identical investigation or series of tests. The materials list must include *all* equipment used and specimens or samples tested, which may range from a whole-body nuclear radiation counter to a tiny microorganism. For ease of reference they should be listed in representative groups, such as:

Equipment	Plants
Instruments	Animals
Chemicals	Birds or fishes
Specimens	Humans

Precede this section with the subheading "Materials." If the list is long, you should consider using a subordinate heading before each group of materials, instruments, or subjects.

Precede the specific methods with the subheading "Method." Describe the tests chronologically in paragraph and subparagraph form, using a main paragraph to introduce a test or part of a test and short, numbered subparagraphs to describe the specific steps you took. Avoid writing in the imperative mood, so that you do not inadvertently start writing an instruction. For example:

Write:	The test unit was connected to the X-Y terminals of the recorder.
Not:	Connect the test unit to the X-Y terminals of the recorder.

Results. The results often may be the shortest section of your paper. If your methods section has described clearly how the investigation or tests were conducted, the results section has only to state the result:

Acid rain is above average in the southeastern part of the city, below average in the northwestern part, and average in the southwestern and northeastern parts. Tables 1 through 8 show the measurements recorded at the eight metering stations.

(Tables and charts depicting your findings often will be a major part of the results section.)

Never comment on the results, because analysis and interpretation belong *only* in the discussion.

Discussion.   Your readers now expect you to analyze and interpret your findings. You will be expected to discuss:

- The results you obtained, compared to the results obtained by previous researchers.
- Any significant correlation, or lack of correlation, between parts of your own findings.
- Factors that may have caused the differences.
- Any trends that seem to be evident or to be developing (ideally, by referring to graphs and charts you have presented in the results).
- The conclusions you draw from your analysis and interpretation.

The conclusions should show how you have responded to the problem stated at the start of your introduction, and identify clearly what your investigation, tests, and analyses show. As such, they should form a fitting close to the narrative portion of your research paper.

References.   The final section of your paper is a list of the documents you have referred to earlier or from which you have extracted information. Chapter 6 provides general instructions for preparing a list of references, which is the preferred method for most technical reports, or a bibliography. You can use Chapter 6 as a guideline, but you should check first with the editor of the journal in which your scientific paper is likely to be published to determine: (1) whether an endnote or bibliographical listing is preferred, and (2) the exact format the journal uses for listing authors' names, book and journal titles, publisher details, and so on.

TECHNICAL PAPERS AND ARTICLES

WHY WRITE FOR PUBLICATION?

The likelihood that one day you may be asked to write a technical paper for publication, or even want to do so, may seem so remote to you now that you might be justified in skipping this section. Yet this is something you should think about, for getting one's name into print is one of the fastest ways to obtain recognition. Suddenly you become an expert in your field and are of more value to your employer, who is happy because the company's name appears in print beneath yours. You become of more value to prospective employers, who rate authors of technical papers more highly than equally qualified persons who have not published. And you have positive proof of your competence, and sometimes a few extra dollars from the publisher.

Mickey Wendell has an interesting topic to write about: As a senior technician in H. L. Winman and Associates' Materials Testing Laboratory, he has been testing concretes with various additives to find a grout that can be installed in frozen soil during the Alaskan winter. One mixture that he analyzed but discarded contained a new product known as Aluminum KL. As a byproduct of his tests he has discovered that mixing Aluminum KL with cement in the right proportions results in a concrete with very high salt resistance. He reasons that such concrete could prove invaluable to builders of concrete pavements in snow-affected areas of the United States and Canada, where salt mixtures are applied in winter to melt the snow.

Mickey has been thinking about publishing this particular aspect of his findings and has jotted down a few headings as a preliminary outline. Here are the four steps he must take before his ideas appear in print.

STEP 1: SOLICIT COMPANY APPROVAL

Most companies encourage their employees to write for publication, and some even offer incentives such as cash awards to those who do get into print. However, they expect prospective authors to ask for permission before they submit their manuscripts.

To obtain permission, Mickey must write a brief memorandum outlining his ideas to John Wood, his department head. He should ask for approval to submit a paper, explain what he wants to write about and why he thinks the information should be published, and outline where he intends to send it. His proposal is shown in Figure 7-5. John Wood will discuss the matter at management level, and then signify the company's approval or denial in writing. Mickey knows he must have *written* consent to publish his findings.

STEP 2: CONSIDER THE MARKET

Mickey must decide very early where he will try to place his paper. If he prefers to present his findings as a technical paper before a society meeting, as suggested by John Wood (see his comment in Figure 7-5), he will be writing for a limited audience with specialized interests. If he decides to publish in the journal of a technical society, he will be writing for a larger audience, but still within a limited field. If he plans to publish in a technical magazine, he will be appealing to a wide readership with a broad range of technical knowledge. His approach must therefore differ, depending on the type of publication and level of reader.

A guiding factor may be Mickey's writing capability. A paper to be published by a technical society requires high-quality writing. The editor of a society journal normally does not do much prepublication editing, other than making minor changes to suit the format and style of the society's publications. A technical magazine article, however, will be edited—sometimes quite fiercely—by a professional editor who knows the exact style that readers expect. Such an editor prefers authors to approximate that style and expects them to organize their work well and to write coherently; but he or she is always ready to prune or graft, and sometimes even completely rewrite portions of a manuscript. Hence, the pressure on the authors of magazine articles is not so great.

H L WINMAN AND ASSOCIATES

INTER-OFFICE MEMORANDUM

To: John Wood ....

....

Date: June 24, 19xx

From: Mickey Wendell ....

Subject: Approval for Proposed ....

Technical Article

May I have company approval to write an article on concrete
additives for publication in a technical journal? Specifically, I want to describe our experiments with Aluminum KL
and the salt-corrosion resistance it imparted to the concrete
samples we tested for the Alaska transmission tower project.
I believe that our findings will be of interest to many
municipal engineers in the northern United States and Canada,
who for years have been trying to combat pavement erosion
caused by the application of salt during snow removal.

I was thinking of submitting the article to the editor of
"Municipal Engineering," but I'm open to suggestions if you
can think of a more suitable magazine.

M.W.

Approval granted. Let me see an outline and the first draft before you submit them.

I suggest you also consider writing a technical paper for presentation at the Combined Conference on Concrete to be held in Chicago next March.

John Wood
July 6, 19xx

Figure 7-5. Soliciting company approval to publish an article or paper.

A secondary consideration may be the state of Mickey's wallet: if it is thin and he needs a new set of tires for his Honda, he may choose to write a magazine article. Normally, there is no pay for writers of technical papers, other than recognition by one's peers.

Perhaps the most important factor is for Mickey to be able to identify a potential audience for his information. Readers of society journals and technical magazines may be the same people, but they expect different information coverage in a technical paper than they do in a magazine article.

Technical Paper.   Readers of society journals are looking for facts. They neither expect nor want explanations of basic theories, and they can accept a strongly technical vocabulary. A technical paper can be very specific. It can describe a minute aspect of a large project without seeming incomplete, or it can outline in bold terms the findings of a major experiment. No topic is too large or too small, too specialized or too complete, to be published as a technical paper.

Technical Article.   Most readers of magazine articles are looking for information that will keep them up to date on new developments. Some will have definite interest in a specific topic and would welcome a lot of technical details. Others will be looking mainly for general information, with no more than just the highlights of a new idea. Magazine authors must therefore appeal to a maximum number of readers. Their articles should be of general interest; their style can be brief and informal; their vocabulary must be understandable; and they should sketch in background details for readers whose technical knowledge is only marginal.

STEP 3: WRITE AN ABSTRACT AND OUTLINE

Many editors prefer to read either a summary of a proposed paper or an abstract and outline before the author submits the complete manuscript. They may want to suggest a change in emphasis to suit editorial policy, or even decline to print an interesting paper because someone else is working on a similar topic.

This type of summary is much longer than the summary at the head of a technical report; the abstract, however, usually is quite short. The summary contains a condensed version of the full paper in about 500 to 1000 words. An abstract contains only very brief highlights and the main conclusion (rather like the summary of a report), since it is supported by a comprehensive topic outline.

Some authors write the complete first draft of the paper before attempting to write a summary or abstract, and then leave the revising and final polishing until after the paper has been accepted by an editor. Others prepare a fairly comprehensive outline, often using the freewheeling approach suggested in Chapter 1, and leave the writing until after acceptance. Both methods leave room for the authors to incorporate changes before the final manuscript is written.

Since the summary or abstract and outline have to "sell" an editor on the newsworthiness of his topic, Mickey Wendell must make sure that the material he submits is complete and informative. In addition he must indicate clearly:

- Why the topic will be of interest to readers.
- How deeply the topic will be covered.
- How the article or paper will be organized.
- How long it will be (in words).
- His capability to write it.

Mickey can cover the first four items in a single paragraph. The fifth he will have to prove in two ways. He can prove his technical capability by mentioning his involvement in the topic and experience in similar projects. He can demonstrate his ability to write well by submitting a clear, well-written summary or abstract.

If Mickey later decides to prepare his paper for presentation before the Combined Conference on Concrete, as John Wood has suggested, he will have to prepare a summary in response to a "call for papers" letter sent out by the society and submit it to the papers selection committee. His summary has to convince the committee that the subject is original, topical, and interesting, and that he has the technical capability to prepare it, if his paper is to be accepted.

STEP 4: WRITE THE ARTICLE OR PAPER

A good technical paper is written in an interesting narrative style that combines storytelling with factual reporting. Articles published in general interest magazines tend to be written like feature newspaper stories, whereas technical papers more nearly resemble formal reports. If the article deals with a factual or established topic, the writing is likely to be crisp, definite, and authoritative. If it deals with development of a new idea or concept, the narrative will generally be more persuasive, since the writer is trying to convince the reader of the logic of his or her argument.

The parts of an article or paper are similar to those for a report. Mickey's article, for example, contains four main parts:

Summary A synopsis that tells very briefly what the article is about. It should summarize the three major sections that follow. Like the summary of a report, it should catch and hold the reader's interest.

Introduction Circumstances that led up to the event, discovery, or concept that Mickey will describe. It should contain all the facts readers need to understand the discussion that follows.

Discussion How Mickey went about the project, what he found out, and what inferences he drew from his findings. The topic can be described chronologically (for a series of events that led up to a result), by subject (for descriptive analyses of experiments, processes, equipments, or methods), or by concept (for the development of an idea from concept to fruition). The methods are very similar to those used for writing the discussion of the formal report (see Chapter 6).

Conclusion   A summing-up, in which Mickey draws conclusions from and discusses the implications of his major findings. Although he will not normally make recommendations, he may suggest what he feels needs to be done in the future, or outline work that he or others have already started if there is a subsequent stage to the project.

Illustrations are a useful way to convey ideas quickly, to draw attention to an article, and to break up heavy blocks of type. They should be instantly clear and usefully *supplement* the narrative. They should never be inserted simply to save writing time; neither should they convey exactly the same message as the written words. For examples of effective illustrating, turn to any major publication in your technical field and study how its authors have used charts, graphs, sketches, and photographs as part of the story. For further suggestions on how to prepare illustrative material, see Chapter 8.

Mickey should not be surprised when the editor who handles his manuscript makes some changes. An editor who feels the material is too long, too detailed, or wrongly emphasized for the journal's readers, will revise it to bring it up to the expected standard. Mickey may feel that the alterations have ruined his carefully chosen phrases, but readers will not even be aware that changes have been made. They will simply recognize a well-written paper, for which Mickey rather than the editor will reap the compliments.

Assignments

PROJECT 7.1: WRITING AN AUTOMATED TELLER MANUAL

Recently you were having a conversation with your bank manager about automated teller machines.

"We *encourage* people to use them," she said, "but some of our customers seem afraid to do so." She added that this was particularly true of older people, who hesitate to deposit their pension checks in one of the machines.

"Maybe you need clearer instructions," you suggested, "in a little handbook. Some people who are not accustomed to computers may feel uncomfortable reading instructions on a video screen."

The manager liked your idea and asked you to write a short user's manual (you were to be paid for it), just for making deposits (checks and cash) and for withdrawing cash.

"Limit it to those two activities," she suggested, "to keep the booklet simple. I'll have a few hundred printed up for my customers, and we'll see how they react."

Write the user's manual for the bank manager. Assume it is for an automated teller machine you are familiar with.

Note: There can be no troubleshooting procedure other than to tell the user to call the bank's 24-hour central office, using the direct-line telephone beside the machine.

PROJECT 7.2: USER MANUAL FOR A MEETING TIMER

Mechanical engineer Darwin Haraptiniuk places a drawing on your desk (see Figure 7-6) and says: "You're a good writer. Write me a user's manual to go along with this."

You ask what "this" is.

"It's a meeting timer I designed several years ago. Now I'm having 10 prototypes built, to test in various companies locally. If their response is positive, I'm going to redesign it using microprocessor technology and go into mass production. I want the user manual to ensure that whoever tests the timer uses it properly."

"How does it work?" you ask.

"Like a gasoline pump. You work out the *average* hourly rate for the kind of people attending a particular meeting and set it in one window, and the number of people sitting around the table goes in the other window."

"How?"

"By rotating the knobs on the left. Then you just depress the buttons, one at the start of the meeting, and one at the end."

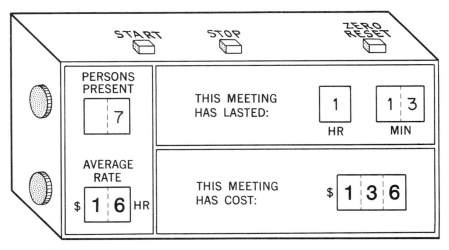

Figure 7-6.   Darwin Haraptiniuk's meeting timer.

"But what's it for?" you ask.

"To help speed up business meetings," Darwin says. "You put it on the table so everyone can see it. When they can *see* how long a meeting has lasted and what it's cost so far, I bet they'll concentrate on getting things done quickly."

"All right," you say. "I'll get right on to it."

"One thing more," Darwin interjects. "You'd better warn users not to change the number of people attending the meeting or the average rate while the machine is running. That tends to damage the mechanism."

"What do you do if you have a problem, then?"

"Switch the machine off, turn everything back to zero, and start all over again."

"And if that doesn't work?"

"Call me. My number is 774-1685."

"Okay," you say. And then you ask: "What am I to call your machine?"

"Ah! Good point. What about the 'Darwin Meeting Timer Model No. 1'?"

"Fine!"

Now write the user manual.

PROJECT 7.3: RESEARCHING A NEW MANUFACTURING MATERIAL OR PROCESS

You are to research information on a topic allied to your technology and then prepare it for both written and oral presentation. The topic may be a new manufacturing material, method, or process. The written and spoken presentations must:

1. Introduce the topic.
2. State why it is worth evaluating.
3. Describe the material, method, or process.
4. Discuss its uniqueness and usefulness.
5. Show how it can be applied in your particular field.

To obtain data for your topic, you will have to research current literature and probably talk to industrial users, manufacturers, and suppliers. Typical examples of topics are: a new oil that can be used at very low temperatures; a method for supporting the deck of a bridge during concrete-pouring by building up a base on compacted fill; a new paint for use on concrete surfaces; and a new materials-handling system.

You may assume that both your readers and your audience are technicians to whom the topic will be entirely new.

PROJECT 7.4: INSTRUCTIONS FOR A BLUEPRINTER

Your company has an outdated blueprinter in its drawing office, which the company keeps as a standby blueprinting machine in case one of the newer machines breaks down. (The resale or trade-in value of the old machine is so low it has not been worth trying to dispose of it.) Figure 7-7 shows the blueprinter viewed from the side. Here are some other facts you know about the blueprinter:

- The exposure roller is made of glass and has a high intensity light inside it.
- The light burns off the yellow coating on the blueprint paper (except where the image of the tracing prevents light from reaching it).
- The developing roller forces the paper through ammonia fumes (produced by heating liquid ammonia).
- The blueprinter has a chimney to exhaust the fumes (because they are toxic).
- Fumes "develop" the blueprint (the yellow areas turn blue).
- The number set on the speed selector is the number of feet of paper the blueprinter produces in a minute.

A tattered card containing a descriptive procedure hangs on a hook beside the blueprinter. It says:

OPERATING PROCEDURE

The procedure for making blueprints using the Fallows Model 207B Blueprinter is as follows:

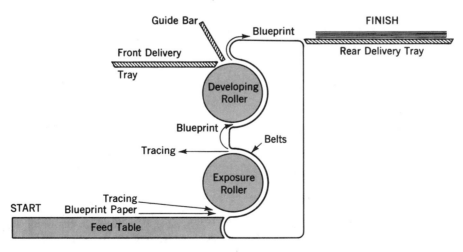

BLUEPRINTER (VIEWED FROM SIDE)

Figure 7-7. The Fallows Model 207B Blueprinter (side view).

The blueprinter is relatively simple to operate. Once the power is on and the machine is operating, the steps involved in making blueprints are easy to follow. In the first place, for each print a sheet of blueprint paper is placed on the feed table, face side up. (The face side is the yellow side.) The original tracing, right way up, is then placed onto the blueprint paper. Before pushing the two sheets of paper into the machine, the speed selector is set to the correct speed, which is 15 for medium speed blueprint paper. The two sheets are then pushed toward the machine until they are grasped between the belts and the exposure roller. When the two sheets emerge they are separated and the blueprint paper is folded back and fed up and over a developing roller. The original tracing comes right out and is removed in readiness for making another blueprint. This printing procedure is repeated until all the copies you need are made. At the top of the machine there is a guide bar that directs the prints to one of two delivery trays. The front delivery tray is at the front of the machine immediately in front of the operator; the back delivery tray is used when you want to stack a succession of prints without removing them. A person who has, say, 20 prints to make, would select "rear delivery." When finished printing, the speed selector is set back to 5.

As newcomers to the department have not found the procedure helpful, rewrite it as an instruction. Include a warning about the danger of inhaling ammonia fumes.

Note: The type of paper described here is more correctly known as "whiteprint" or "blueline" paper. In true blueprint paper the image appears white against a dark blue background.

PROJECT 7.5: TESTING CABLE CONNECTORS

Write an instruction to all installation supervisors at sites 1 through 17 (see Project 4.7 of Chapter 4) telling them to inspect all cable connectors on site. Here is some additional information you may need:

1. Background to this project is contained in Project 4.7 of Chapter 4.
2. The instruction is to be written as an interoffice memorandum.
3. Don Gibbon, electrical engineering coordinator at H. L. Winman and Associates, will sign the memorandum.
4. Tell the site installation supervisors that they are to report the number of GLA connectors they find to Don Gibbon.
5. Tell them to replace all connectors marked GLA with connectors marked MVK.
6. All connectors on site are to be checked. (It would be best to check those in stock first, then use checked connectors to replace those in use that are found to be faulty.)
7. Inform them that faulty GLA connectors are to be sent to the contractor with a note that they are to be held for analysis under project 92A7.
8. The site supervisors are to complete their tests within seven days.

PROJECT 7.6: INSTALLING A MINI-MINDER

You are employed in the construction department of Midstate Telephone System, and you have been asked to write an installation instruction sheet to be shipped with the "Mini-Minder" (an electronic intrusion detection device developed by MTS's engineering department). The Mini-Minder is a small detection unit which

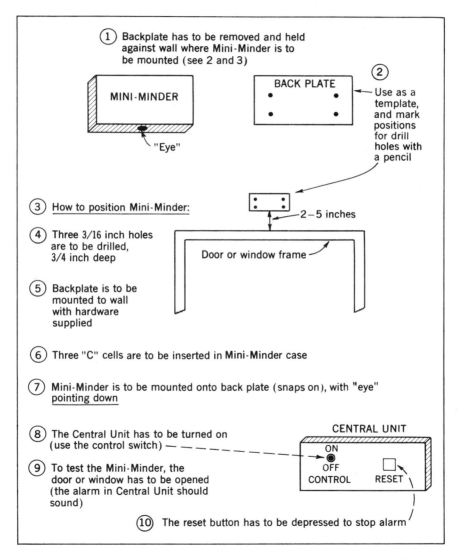

Figure 7-8. Installing and testing a Mini-Minder.

is mounted above windows, doors, or any other entries, where it automatically detects movement and transmits an alarm signal to a central unit. The central unit is concealed within the building and, when activated, both sounds an audible alarm and sends a message to police headquarters.

The Mini-Minder is shown in Figure 7-8. It is fixed to the wall above the door or window by removing the backplate and screwing the plate to the wall. The

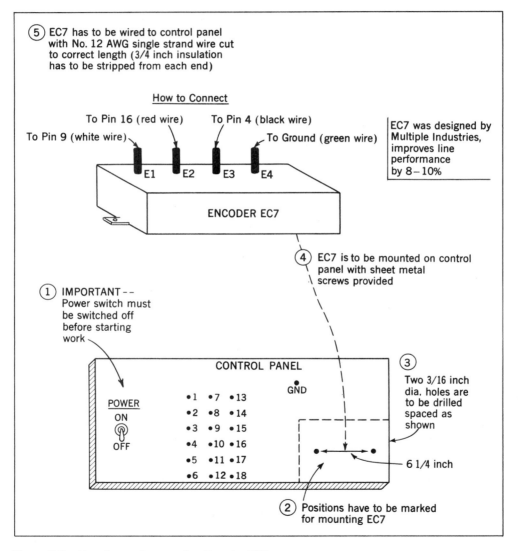

Figure 7-9. Mounting and connecting Encoder EC7.

unit is then snapped onto the backplate (removing the unit from the backplate also sounds the alarm). The unit is battery operated.

All materials and hardware are supplied with the Mini-Minder but the installer will need an electric drill with a ³⁄₁₆ inch masonry drill bit, a Robertson No. 2 screwdriver, and a sharp pencil to do the job. The sequence in which the installation should occur is shown by the circled numbers in Figure 7-8.

PROJECT 7.7: INSTALLING ENCODER EC7

You work for Midstate Telephone System and you have been asked to write instructions for installing an EC7 encoder at all MTS microwave transmission sites. The instructions are to accompany the encoder, which is the box illustrated in Figure 7-9. The encoder removes unwanted signals and improves transmission performance by 8% to 10%.

In your instructions, tell the site technicians that they must: find a suitable location for the encoder at the bottom right of the control panel (see Figure 7-9); drill two holes for mounting the encoder; and connect the encoder with four wires (the connections are shown on the diagram). All materials (such as mounting hardware and wire) are supplied with each encoder but the technician will need a soldering iron, some Ersin 60/40 resin core solder, a drill with a ³⁄₁₆-inch bit, and a Robertson No. 2 screwdriver to carry out the work. The sequence in which the installation is to be carried out is shown by the circled numbers on the diagram.

When the installation is complete you should remind the technician to turn on the power to the control panel.

CHAPTER 8
Illustrating Technical Documents

If you open any well-known technical magazine, you will immediately notice that illustrations are an integral part of most articles. Some are photographs that display a new product, a process, or the result of some action; others are line drawings that illustrate a new concept; some demonstrate how a test or an experiment was tackled; while still others are charts and graphs that show progress of a project or depict technical data in an easy-to-visualize form.

Illustrations serve an equally useful purpose in technical reports, where their primary role is to help readers understand the topic. Interesting illustrations attract readers' eyes and encourage them to read a report. They can also break up dull-looking pages of narrative that lack eye appeal.

This chapter discusses the types of illustrations seen most often in technical reports, indicates the overlaps that exist between illustration types, suggests occasions when they can be used most beneficially, and provides guidelines for preparing them.

PRIMARY GUIDELINES

The criterion for any illustration is that it should help explain the narrative; the narrative should *never* have to explain an illustration. Hence, an illustration must be simple enough for readers to understand quickly and easily. When I flip through the pages of *Scientific American* and stop at a particular page, usually I look first at the drawings or photographs. If they immediately tell a story and catch my interest (and in *Scientific American* they regularly do), I am encouraged to read the article.

To help readers readily understand the illustrations you insert into your reports, follow these seven guidelines:

1. Before selecting or designing an illustration, first consider the specific audience for whom you are writing and what you want your readers to learn from each illustration.
2. Keep every illustration simple and uncluttered.
3. Let each illustration depict only one main point.
4. Position each illustration as near as possible to the narrative it supports (see "Positioning the Illustrations," at the end of this chapter).
5. Label each illustration clearly with a figure or table number and a title (with the figure number and title centered *beneath* a graph or chart, and the table number and title centered *above* a table).
6. Add a caption (that is, comments or remarks) beneath a figure title, to draw attention to significant aspects of the illustration.
7. Refer to every illustration at least once in the report narrative.

GRAPHS

Graphs are a simple means for showing a change in one function in relation to a change in another. A function used frequently in such comparisons is time. The other function may be temperature, erosion, wear, speed, strength, or any of many factors that vary as time passes.

Engineering technician John Greene wants to determine how long it takes a newly painted manometer case to cool down after it comes out of the drying oven. He goes to the paint shop armed with a stopwatch and a special thermometer. When the next manometer case comes out of the oven, he starts taking readings at half-minute intervals and records the results as in Table 8-1.

Table 8-1   Cooling Rate, Manometer Case MM-7

TIME ELAPSED (min:sec)	TEMPERATURE (deg C)	TIME ELAPSED (min:sec)	TEMPERATURE (deg C)
:30	152.9	5:30	43.9
1:00	123.4	6:00	41.1
1:30	106.7	6:30	38.9
2:00	91.2	7:00	37.2
2:30	77.8	7:30	35.6
3:00	69.5	8:00	34.5
3:30	61.7	8:30	33.4
4:00	55.6	9:00	32.8
4:30	51.5	9:30	31.7
5:00	47.3	10:00	31.1

Ambient temperature 22.8°C   Oven temperature 180°C

These readings are part of a study he is undertaking on the cooling rate of different components manufactured by Macro Engineering Inc. The information will also be used by the production department to establish how long manometer cases must cool before assemblers can start working on them with bare hands (the maximum bare-hand temperature has been established by management/union negotiation to be 38°C).

A quick inspection of this table shows that the temperature of the case drops continuously, is within 8.3 degrees of the ambient temperature after 10 minutes (*ambient* means "surrounding environment"), and is down to the bare-hand temperature after seven minutes. A closer examination identifies that the temperature drops rapidly at first, then progressively more slowly as time passes.

Single Curve. John Greene can make the data he has recorded in Table 8-1 much more readily understood if he converts it into the graph in Figure 8-1. Now it is immediately evident that the temperature drops rapidly at first, then slows down until the rate of change is almost negligible. (There is no point in John measuring or showing any further drops in temperature unless he wants to demonstrate how long it takes the component to cool right down to the ambient temperature, which is probably 30 minutes or more.)

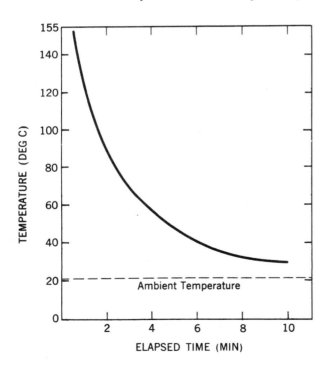

COOLING RATE – MANOMETER CASE MODEL NO. MM-7   **Figure 8-1.** Graph with a single curve.

Multiple Curves. In Table 8-2 John compares the temperature readings he has recorded for the cooling manometer case with measurements he has taken under similar conditions for a cover plate and a panel board. This time, however, he simplifies the table slightly by showing the temperature at one-minute intervals.

What can we assess from this table? The most obvious conclusion is that in 10 minutes the cover plate has cooled down less than the manometer case, and even less than the panel board. We can also see that the initial rate at which the components cooled varied considerably: the panel board, very quickly; the manometer case, fairly quickly; the cover plate, seemingly quite slowly. But it is difficult to assess whether there were any *changes in the rate of cooling* as time progressed.

This data again can be shown more effectively in a graph, which John Greene has plotted in Figure 8-2. The rapid initial drop in temperature is evident from the initial steepness of the three curves, with each curve flattening out to a slower rate of cooling after two to four minutes. The difference in cooling rates for the three components is much more obvious from the curves than in the table. (The curved lines in Figures 8-1 and 8-2 are commonly referred to as curves, even though in some cases they may be straight lines or a series of short straight lines joining points plotted on the graph.)

Eventually John will have to present this data in a report. If he plans to present only a general description of temperature trends, his narrative can be accompanied by graphs like these. But if he also wants to discuss exact temperatures at specific times for each material, then the narrative and graphs will have to be supported by figures similar to those in Table 8-2.

Constructing a graph usually offers no problems to technical people because they recognize a graph as a logical means for conveying statistical data. But if they are to construct a graph that both tells a story *and* emphasizes the right information, they must know the tools they will be working with.

Table 8-2 Cooling Rates for Three Components

TIME ELAPSED (MINUTES)	COVER PLATE	PANEL BOARD	MANOMETER CASE
		TEMPERATURE (DEG C)	
0:30	154.5	145.1	152.9
1	136.7	97.3	123.4
2	112.3	67.8	91.2
3	95.1	51.7	69.5
4	82.3	42.3	55.6
5	71.7	36.1	47.3
6	63.9	32.2	41.1
7	55.6	29.5	37.2
8	49.5	27.2	34.5
9	43.4	26.1	32.8
10	38.9	25.0	31.1

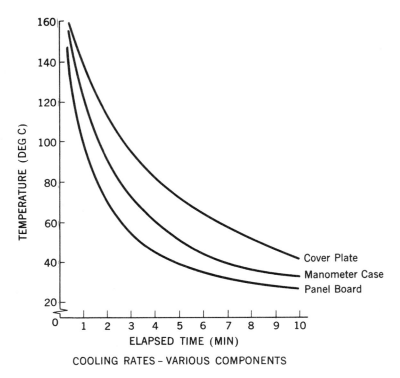

COOLING RATES – VARIOUS COMPONENTS

Figure 8-2. Graph with multiple curves.

Scales.   The two functions to be compared in the graph are entered on two scales: a horizontal scale along the bottom and a vertical scale along the left side. (On large graphs the vertical scale is sometimes repeated on the right side to simplify interpretation.) The scales meet at the bottom left corner, which normally—but not always—is designated as the zero point for both.

The two functions are commonly known as the dependent and independent variables, so named because a change in the dependent variable *depends* on a change in the independent variable. For example, if I want to show how the fuel consumption of my automobile increases with speed, I can enter speed as the independent variable along the bottom scale, and fuel consumption as the dependent variable along the left side, as in Figure 8-3. (Fuel consumption *depends* on speed; speed does not depend on fuel consumption.) The same applies to John Greene's temperature measurement graphs: Temperature is the dependent variable because it depends on the *time that has elapsed* since the components came out of the oven (the independent variable).

When you construct a graph, the first step is to identify which function should form the horizontal scale and which the vertical scale. Table 8-3 lists some typical situations which show that the same function can be an independent

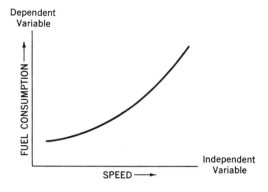

Figure 8-3. The dependent variable depends on the independent variable.

variable in one situation and dependent in another. Selection of the independent variable depends on which function can be more readily identified as influencing the other function in the comparison.

The second factor to consider is scale interval. Poorly selected scale intervals, particularly scale intervals that are not balanced between the two variables, can defeat the purpose of a graph by distorting the story it conveys.

Table 8-3 Identifying Dependent and Independent Variables

GRAPH ILLUSTRATES	DEPENDENT VARIABLE (VERTICAL SCALE)	INDEPENDENT VARIABLE (HORIZONTAL SCALE)
1. The effect that frequency has on the gain of a transducer	Gain	Frequency
2. How much a motor's speed affects the noise it produces	Noise	Speed
3. How attendance at a ball game varies with temperature	Attendance	Temperature
4. The changes in temperature brought about by changes in pressure	Temperature	Pressure
5. How much an increase in payload reduces an aircraft's range by limiting the amount of fuel it can carry	Aircraft range (or fuel load)	Payload
6. How much increasing the fuel load of an aircraft to achieve greater range reduces its effective payload	Payload	Fuel load (or aircraft range)

Note: A function can be either dependent or independent, depending on its role in the comparison (see temperature in examples 3 and 4, and both functions in examples 5 and 6).

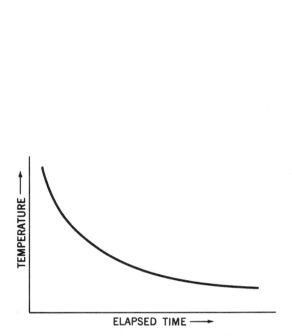

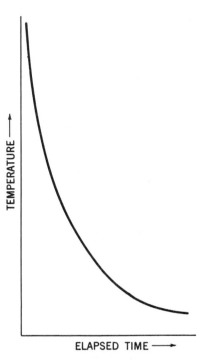

Figure 8-4(a). Effect of compressed vertical scale.

Figure 8-4(b). Effect of compressed horizontal scale.

Suppose John Greene had made the vertical scale interval of his time vs. temperature graph in Figure 8-1 much more compact, but had retained the same spacing for the horizontal scale. The result is shown in Figure 8-4(a). Now the rapid initial decrease in temperature is no longer evident; indeed, the impression conveyed by the curve is that temperature dropped only moderately at first, remained almost constant for the last three minutes, and will never drop to the ambient temperature. The reverse occurs in Figure 8-4(b), which shows the effect of compressing the horizontal scale: Now the curve seems to say that temperature plummets downward and it will be only a minute or two until the ambient temperature is reached. Neither curve creates the correct impression, although technically the graphs are accurate.

      Normally both scales start at zero, which would be the case when the curve is comfortably balanced in the graph area. If it is crowded against the top or right side, then a zero starting point is unrealistic. In Figure 8-1 the curve occupies the top 75% of the graph area. Since no points will ever be plotted below the ambient temperature (which will hover around 23°C), the bottom portion of the vertical scale is unnecessary. This can be corrected by starting the vertical scale at a higher

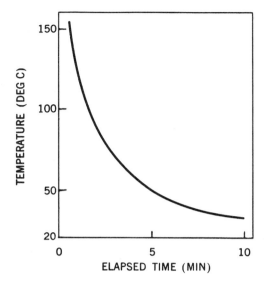

Figure 8-5. A correctly centered curve.

value (say 20°, as in Figure 8-5), or by breaking the scale to indicate that some scale values have been omitted (Figures 8-2 and 8-6).

Multiple-curve graphs should have no more than four curves, otherwise they will be difficult to interpret, particularly if the curves cross one another. You can help a reader identify the most important curve by making it heavier than the others (Figure 8-6) and can differentiate among curves that cross by using different symbols for each (Figure 8-7):

Most important curve	— (bold line)
Next most important curve	— (light line)
Third curve	--- (dashes)
Least important curve	··· (dots)

Avoid using colored lines because the average copier reproduces all the lines in only one color (usually black).

Simplicity.   Simplicity is important in graph construction. If a graph illustrates only trends or comparisons, and the reader is not expected to extract specific data from it, then you may omit the grid, as shown in Figure 8-5. But if the reader will want to extrapolate quantities, you should include a grid, as in Figures 8-6 and 8-7. Note that Figure 8-1 has an *implied* grid that only suggests the grid pattern for the occasional reader who may want to draw in a grid. Note also that you may omit the top and right-hand borders on graphs without grids, as in Figure 8-3.

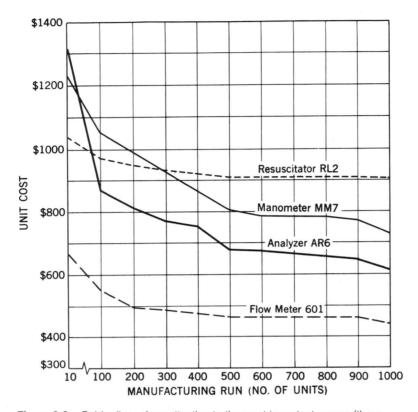

Figure 8-6. Bolder lines draw attention to the most important curves (those showing maximum benefit from quantity manufacturing). The grid permits readers to draw reasonably accurate data from the graph. (Courtesy Marco Engineering Inc, Phoenix, Arizona)

Omit plot points and ensure that all captions are *horizontal*. (The only one that may be entered vertically is the caption for the vertical scale function.) Insert captions for the curves at the end of the curve whenever possible (Figure 8-2) or, alternatively, above or below the curve (Figures 8-6 and 8-7). Never write a caption along the slope of the curve.

CHARTS

Most charts show trends or compare only general quantities. They include bar charts, histograms, surface charts, and pie charts.

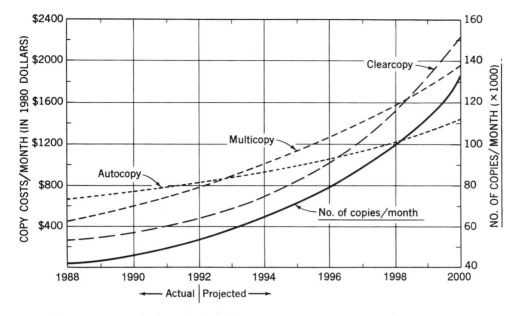

Figure 8-7   Different symbols distinguish between curves showing current and
projected copying costs for three copiers. Note the two vertical
scales, which permit three functions to be shown on one graph.
(Courtesy H. L. Winman and Associates, Cleveland, Ohio)

BAR CHARTS

You can use bar charts to compare functions that do not necessarily vary
continuously. In the graph in Figure 8-1, John Greene plotted a curve to show
how temperature decreased continuously with time. He could do this because
both functions were varying continuously (time was passing and temperature was
decreasing). For the production department, however, he has to prepare a report
on how long it takes various components coming from the oven to cool to a safe
temperature for bare-hand work. He prepares a bar chart to depict this because
he knows the report will be read by both management and union representatives,
and some of the readers may need easy-to-interpret data. He also has only one
continuous variable to plot: elapsed time. The other variable is noncontinuous
because it represents the various components he has tested. In this case elapsed
time is the dependent variable, and the components are the independent variable.
The bar chart John constructs is shown in Figure 8-8.

Scales for a bar chart can be made up of such diverse functions as time, age
groups, heat resistance, employment categories, percentages of population, types
of soil, and quantities (of products manufactured, components sold, software

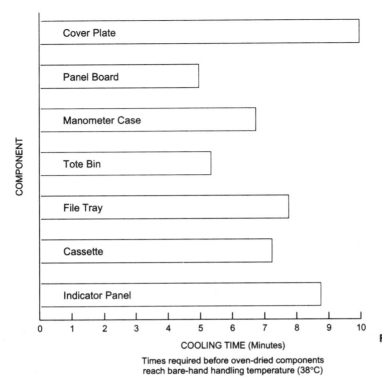

Figure 8-8. Horizontal bar chart with one continuous variable (cooling time).

COMPONENT

COOLING TIME (Minutes)

Times required before oven-dried components reach bare-hand handling temperature (38°C)

programs used, and so on). Charts can be arranged with either vertical or horizontal bars depending on the type of information they portray; when time is one of the variables, usually it is plotted along the horizontal axis, as in Figure 8-8. The bars normally are separated by spaces the same width as each of the bars.

In a complex bar chart, the bars may be shaded to indicate comparisons within each factor being considered. The vertical bar chart in Figure 8-9 uses two shades to describe two factors on the one chart. Individual bars can also be shaded to show proportional content, as has been done in Figure 8-10, in which case a legend must be inserted beside or below the graph to show readers what each bar represents. Alternatively, each segment may be labelled, as in Figure 8-11.

Horizontal bar charts can be used in an unconventional way by arranging the bars on either side of a zero point. This can be done, for example, to compare negative and positive quantities, satisfactory and defective products, or passed and failed students. The chart in Figure 8-12 divides products returned for repair into two groups: those that are covered by warranty, and those that are not. Each bar represents 100% of the total number of items repaired in a particular product age group and is positioned about the zero line depending on the percentage of warranty and nonwarranty repairs.

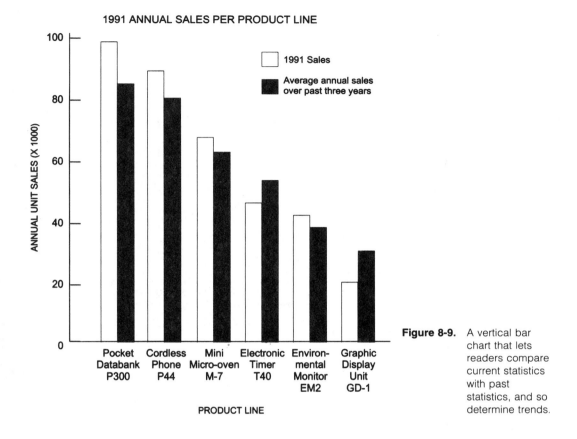

1991 ANNUAL SALES PER PRODUCT LINE

Figure 8-9. A vertical bar chart that lets readers compare current statistics with past statistics, and so determine trends.

HISTOGRAMS

A histogram looks like a bar chart, but functionally it is similar to a graph because it deals with two continuous variables (functions that can be shown on a scale to be increasing or decreasing). It is usually plotted like a bar chart because it does not have enough data on which to plot a continuous curve (see Figure 8-13). The chief visible difference between a histogram and a bar chart is that there are no spaces between the bars of a histogram.

SURFACE CHARTS

A surface chart (Figure 8-14) is like a graph, in that it has two continuous variables that form the scales against which the curves are plotted. But, unlike a graph, individual curves cannot be read directly from the scales.

The uppermost curve on a surface chart shows the *total* of the data being presented. This curve is achieved as follows:

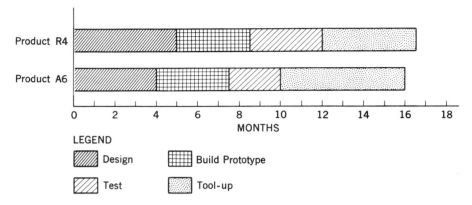

Figure 8-10. The bars in this chart show development times for proposed new products. The legend is included with the chart.

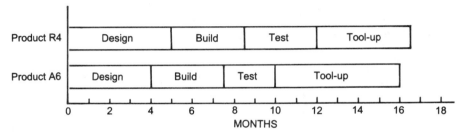

Figure 8-11. A segmented bar chart with internal labelling.

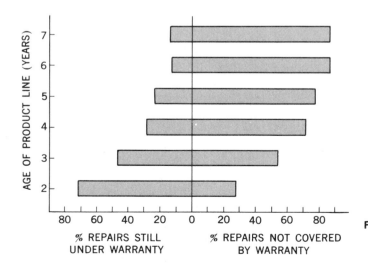

Figure 8-12. A bar chart constructed on both sides of a zero point.

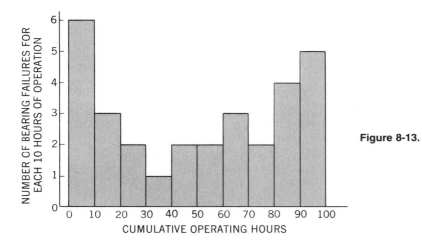

Figure 8-13. This histogram shows the number of bearing failures for every 10 hours of operation. Considerably more data would have been required to construct a curve.

1. The curve containing the most important or largest quantity of data is drawn in first, in the normal way. This is the Thermal curve in Figure 8-14.
2. The next curve is drawn in above the first curve, using the first curve as a base (or "zero") and adding the second set of data to it. For example, the energy resources shown as being variable in 1990 are:

    Thermal Power: 33,000 MW
    Hydro Power: 14,500 MW

In Figure 8-14, the lower curve for 1990 is plotted at 33,000 MW. The 1990 data for the next curve is 14,500 MW, which is added to the first set of data so that the second curve indicates a *total* of 47,500 MW. (If there were a third set of data, it would be added in the same way.)

The area between curves is shaded to indicate that the curves represent the boundaries of a cumulative set of data. Normally, the lowest set of data has the darkest shade, and each set above it is progressively lighter.

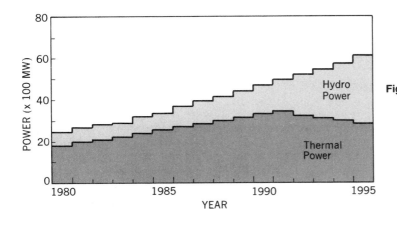

Figure 8-14. Surface chart adds hydro data to thermal data to show actual and projected energy resources of a power utility. (Courtesy Crestview Power Company)

PIE CHARTS

A pie chart is aptly named, because it looks like a whole pie viewed from above with cuts in it ready for people of varying appetites. It is a pictorial device for showing approximate divisions of a whole unit. The pie chart in Figure 8-15 depicts the percentage of electronic equipment manufactured under various product categories.

If a pie chart has a lot of tiny wedges which would be difficult to draw and hard to read, you may combine some of them into a larger single wedge and give it a general heading, such as "miscellaneous expenses," "other services," or "minor effects." All the wedges must add up to a whole unit, such as 100%, $1.00, or 1 (unity).

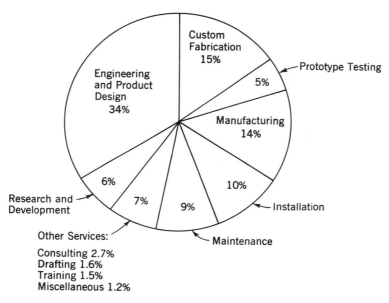

Figure 8-15. Pie chart shows Macro Engineering Inc's products and services. "Slices" add up to 100%.

DIAGRAMS

Diagrams include any illustration that helps the reader understand the narrative yet does not fall within the category of graph, chart, or table. They can range from a schematic drawing of a complex circuit to a simple plan of an intersection. But there is an important restriction: If included in the narrative part of a report, a diagram must be clear enough to read easily. This means that complex drawings should be placed in an appendix and treated as supporting data.

Diagrams should be simple, easy to follow, and contribute to the story. They can comprise organization charts (see Chapter 2), flow diagrams (Figures 2-4, 8-16, and 9-1), site plans, and sketches.

PHOTOGRAPHS

A photograph can do much to help a reader visualize shape, appearance, complexity, or size. For example, the photographs of Harvey Winman, Tina Mactiere, and Wayne Robertson add depth to the narrative description of the two engineering companies in Chapter 2. The criterion when selecting any photograph is that it must be clear and contain no extraneous information that might distract the reader's attention.

Photographs, however, create difficulties during the printing process. Most photocopiers do not reproduce them well, and preparing them for offset printing, which permits them to be rendered clearly, is time-consuming and can

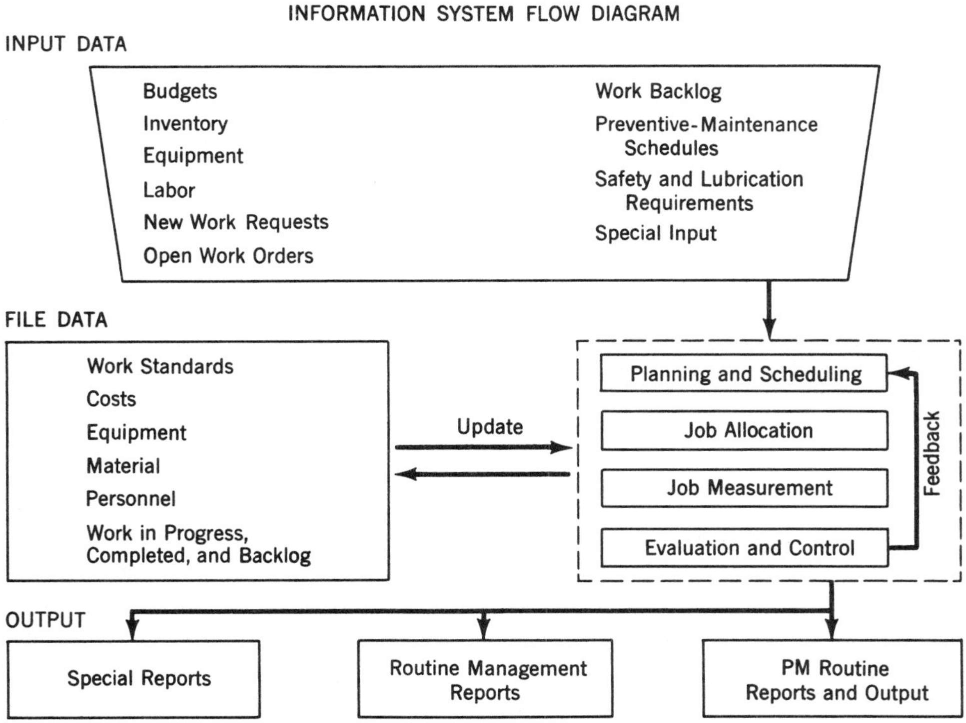

Figure 8-16. Flow diagram of a Maintenance Information System. (Courtesy Mechanical Engineering.)

be expensive. Consequently, before deciding to use photographs, check whether your printer can reproduce them clearly and economically. If the printer cannot handle them easily, you may have to glue individual copies of the photographs into your report (an awkward process), place them in a sleeve or envelope at the back of the report (a cumbersome arrangement), have copies printed professionally (an expensive proposition), or replace the photographs with good sketches.

TABLES

A table may be a collection of technical data, as in Tables 8-1 and 8-2 on pages 243 and 245, or a series of short narrative statements, as in Tables 8-3 and 11-1 on pages 247 and 328. Whether you should insert a table into the report narrative or place it in an attachment or appendix depends on three factors:

1. If the table is short (i.e. less than half a page) and readers need to refer to it as they read the report, then include it as part of the report narrative (i.e. in the discussion), preferably on the same page as the text that refers to it.
2. If readers will be able to understand the discussion without referring to the table as they read the report, but may want to consult the table later, then place the table in an attachment or appendix.
3. If readers will need to refer to a table but the data you have will occupy a full page or more, then
   • summarize the table's key points in a short table to be inserted into the report narrative, and
   • place the full table in an attachment or appendix.

There are five additional guidelines you can use to create tables, particularly if they contain columns of numerical data:

• Keep a table simple by limiting it only to data the readers will *really* need, and create as few columns as possible.
• Decide whether the table is to be "open" (i.e. without lines separating the columns, as in Tables 8-1 and 8-2) or "closed" (with lines separating the columns, as in Table 8-3).
• Insert units of measurement, such as decibels, volts, kilograms, or seconds, at the head of each column rather than after each column entry (see the "min:sec" and "deg C" entries at the top of the columns in Table 8-1).
• Center a table number and an appropriate title above the table.
• Draw readers' attention to a table be referring to it in the report narrative and commenting on a specific inference to be drawn from the table. For example:

   The voltage fluctuations were recorded at 10-minute intervals and entered in column 3 of Table 7, which shows that fluctuations were most marked between 8:15 and 11:20 a.m.

POSITIONING THE ILLUSTRATIONS

Whenever possible, place each illustration on the same page as or facing the narrative it supports. A reader who has to keep flipping pages back and forth between narrative and illustrations will soon tire, and your reasons for including the illustrations will be defeated.

When reports are printed on only one side of the paper, full-page illustrations can be difficult to position. The only feasible way to place them conveniently near the narrative is to print them on the back of the preceding page, facing the words they support. But this in turn may pose a printing problem. A more logical solution is to limit the size of illustrations so that they can be placed beside, above, or below the words, and then to ensure that you create spaces for them at the appropriate places as you keystroke the report (or inform the person who is typing the report what the figures' and tables' dimensions are and exactly where they are to be placed).

When an illustration is too large to fit on a normal page, or *is* going to be referred to frequently, consider printing it on a fold-out sheet and inserting it at the back of the report (see Figure 8-17). If the illustration is printed only on the extension panels of the foldout, the page can be left opened out for continual reference while the report is being read. This technique is particularly suitable for circuit diagrams and flow charts.

Position horizontal full-page illustrations sideways on a page but so that they are read from the right (see Figure 8-18). This holds true whether they are placed on a left or right page.

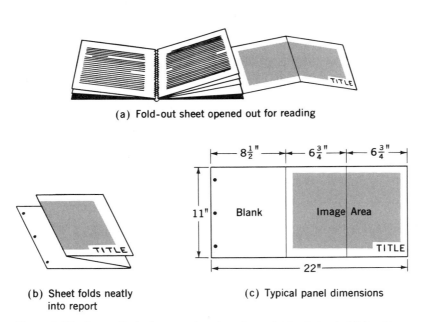

(a) Fold-out sheet opened out for reading

(b) Sheet folds neatly into report

(c) Typical panel dimensions

Figure 8-17.   Large illustrations can be placed on a fold-out sheet at the rear of a report.

ILLUSTRATING A TALK

An illustration for a talk must be utterly simple. Its message must be so clear that the audience can grasp it in seconds. An illustration that forces an audience to read and puzzle over a curve detracts from a talk rather than complements it.

If you want to convert any illustration to a large visual aid or slide, you will have to observe the following limitations:

- Make each illustration *tell only one story*. Avoid the temptation to save preparation time by inserting too much data on one chart. Be prepared to make two or three simple charts in place of a single complex one.
- Use bold letters large enough to be read easily by the back row of your audience.
- Use very few words, and separate them with plenty of white space.
- Give the illustration a short title.
- Insert only the essential points on a graph. Let the curves tell the story, rather than bury them in construction detail.
- Accentuate key figures and curves with a bold or colored pen (but remember that from a distance some colors look very similar).
- Avoid clutter—a simple illustration will draw attention to important facts, whereas a busy one will hide them.

If you are preparing to illustrate an in-plant briefing, you may hand-letter your tables, charts, and graphs with a felt pen on large sheets. But if you are preparing to present a technical paper before a conference audience, you must prepare them more professionally. In the first instance, you are expected to produce a standard job at no great expense. In the second, you are expected to convey an image of yourself and the company you represent; in effect, the quality of your illustrations will demonstrate the quality of your company's products or services.

If you prepare your own illustrations, you can use one of the currently available software programs to produce good quality tables, charts, and diagrams on a personal computer. The programs are fast and easy to use, and many of them work well with a dot-matrix printer. (Of course, if you want really crisp-looking

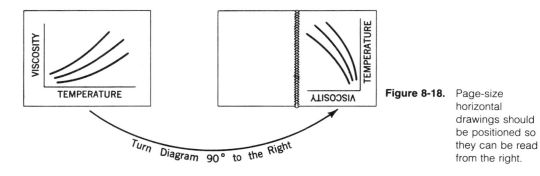

Figure 8-18. Page-size horizontal drawings should be positioned so they can be read from the right.

originals, a laser jet printer will produce superior hard copy for you.) The printouts can then be converted into transparencies or 35 mm slides.

WORKING WITH AN ILLUSTRATOR

Although you may prepare your own illustrations, in a large technical organization you may work with a company draftsperson or illustrator who will prepare your drawings, graphs, and charts according to your requirements. Good communication between you and the illustrator is essential if the drawings you want are to appear in the form you visualized when you wrote your report.

An illustrator needs to have much more than a bare, roughed-out sketch to work from. You will have to describe your project and its outcome in detail, so that the illustrator will know:

- The background to and purpose of the report or oral presentation.
- Who the readers will be, what their technical knowledge is, and how they will use the information contained in the report or presentation.
- What each illustration is to portray, and what particular aspects are to be emphasized.
- What size each illustration is to be (vertical and horizontal dimensions) and, for a talk, the size of the expected audience and their likely distance from the screen.
- What printing method is to be used (for a report).
- How much the illustrations will be reduced in size when they are printed or converted into slides or transparencies. (A drawing that is to be reduced must have lines that are not too fine.)
- When you need the illustrations.

You can help an illustrator even more by providing a sketch of the proposed illustration and a copy of the words the illustration is to support. Better still, talk to the illustrator *before* you write your report or make your speaking notes, describe what illustrations you plan to use, and ask for suggestions for their preparation.

Assignments

PROJECT 8.1: WHO BUYS "PLANIT"?

You work for a very successful software company which publishes a monthly user magazine. The company's most successful software has been Planit, a program for organizing a user's business operations.

Part 1. The editor of the user magazine asks you to provide an illustration showing the breakdown of buyers by user groups for last year, when there were 14,236 sales, and suggests you keep the illustration simple. The buyers were:

Hospitals	1588
Small businesses, generally	1011
Public utilities	2165
Consultants, generally	233
Manufacturers	3176
Architects	217
Writers/editors	116
Land surveyors	245
Sales representatives	866
City/town planners	1155
Engineers	1732
Miscellaneous	433
Radio & television stations	1299

Part 2. When you give your illustration to the editor, the response is: "I like that. But I'd also like another one comparing last year's buyers with those of four years ago, which was the first year we marketed Planit."

The sales of Planit four years ago were:

Hospitals	174
Small businesses, generally	1393
Public utilities	1132
Consultants, generally	174
Manufacturers	2351
Sales representatives	958
City/town planners	1306
Engineers	784
Miscellaneous	261
Radio & television stations	174

PROJECT 8.2: COMPARING ELECTRICITY COSTS

Your company markets heat pumps and wants to demonstrate to electricity users that a heat pump can significantly reduce their electricity bills. Prepare an illustration based on the following actual monthly bills for four different dwellings last year. The residences are identical five-room bungalows built at the same time and in the same block on Margusson Avenue. The only differences are that No. 216 and 234 do not have air conditioning (marked "No AC" in the table), while No. 227 and 248 do (marked "+AC"), and No. 234 and 248 each have a heat pump.

Monthly Electricity Bills

| | ACTUAL BILLS | | AMOUNT SAVED | |
| | No Heat Pump | | With Heat Pump | |
	216, No AC ($)	227, +AC ($)	234, No AC ($)	248, +AC ($)
January	48.07	49.23	12.06	11.59
February	45.15	46.01	11.68	11.86
March	43.20	42.86	10.90	11.13
April	41.11	42.20	4.15	3.80
May	38.37	42.15	0.36	3.48
June	35.20	48.16	—	6.10
July	33.06	57.19	—	18.26
August	32.87	56.80	—	17.44
September	34.11	47.10	2.64	7.21
October	38.62	39.20	7.85	8.60
November	41.67	41.10	9.62	10.51
December	43.20	42.89	11.58	11.77

PROJECT 8.3: ASSESSING RELAY LIFE

Eight years ago the Interstate Telephone Company bought and installed 90,000 relays, to be used in a long-range testing program that would assess failure rates. The relays purchased were:

Type	Quantity
Nestor 221	40,000
Vancourt 1200	20,000
Macro R40	20,000
Camrose Series 8	10,000

As relays failed they were replaced and the failures were recorded and totalled for each year:

Number of Failed Relays

YEAR	NESTOR 221	VANCOURT 1200	MACRO R40	CAMROSE 8
1	1,123	901	1,180	105
2	1,080	805	690	124
3	1,007	762	321	131
4	1,076	813	279	306
5	1,140	878	322	402
6	1,656	910	415	545
7	2,210	956	478	609
8	3,303	1,012	584	891
Totals	12,595	7,037	4,269	3,113

Prepare a graph or chart depicting these failure rates and predicting failures for the next five years.

PROJECT 8.4: MEASURING RADAR RANGE

During the past six months you have been conducting tests to assess the range of a new radar system known as the Search R20. Your tests were taken with a radar set on the ground at various elevations and an aircraft approaching at various altitudes. Two sets of tests were taken, each with the radar set positioned at four different elevations but equipped with two different magnetrons:

Altitude of Radar Set (ft)	MAXIMUM RANGE IN MILES MEASURED WITH AIRCRAFT AT:				
	5,000 ft (mi)	10,000 ft (mi)	20,000 ft (mi)	40,000 ft (mi)	60,000 ft (mi)
MAGNETRON QA					
Sea Level	48	86	112	134	161
1200	73	105	136	151	179
2400	87	138	166	182	189
4800	101	159	193	211	220
MAGNETRON QB					
Sea Level	54	101	127	156	188
1200	79	118	160	175	201
2400	96	151	183	198	213
4800	111	170	214	229	244

Prepare graphs or charts to demonstrate:

1. The effect that transmitter height has on the range measured.
2. The differences between magnetrons QA and KB.

PROJECT 8.5: ILLUSTRATIONS RELATED TO OTHER PROJECTS

Prepare a graph, chart, table, or diagram to accompany the report you would write for the following projects in previous chapters.

1. Project 4.8.   Prepare a drawing illustrating the positions of Sylvan Lake and the fishing resort, the gravel road (you will have to imagine its exact shape and how it winds through the bush), the landing strip, and the points along the road where repairs need to be carried out. Remember also to identify the source of gravel.

2. Project 5.2. You feel your proposal/report on portable computers for the WEMS project needs livening up. Design an illustration to be inserted into the report. Explain why you chose that particular type of illustration.

3. Project 6.1. Prepare a site map of your city (or part of the city) showing the four potential locations you have examined for the environmental monitoring station.

4. Project 6.2. Prepare a comparison table to accompany the report evaluating highway paints.

5. Project 6.4. Prepare a graph or chart to show how sound levels differ under the four sets of conditions at Mirabel Realty:

- When only Mirabel Realty is working.
- When both Mirabel Realty and Superior Giftware are working.
- When only Superior Giftware is working.
- When neither office is working.

(Prepare more than one illustration if necessary.)

Include a "normal" office working sound level. Remember that your illustration(s) should *compare* the sound levels, not simply show the levels recorded for the four sets of conditions.

Explain why you have chosen a particular type of graph or chart. Also explain how your graph or chart helps convey the idea to the reader that, when both offices are working or only Superior Giftware is working, the sound level *increases* as you approach the north wall.

CHAPTER 9

Technically—Speak!

This chapter covers two facets of public speaking, both concerned with the oral presentation of technical information. The first is the oral report, sometimes called the technical briefing, delivered to a client or to colleagues. The second is the technical paper presented before a meeting of scientific or engineering-oriented persons. Both depend on public speaking techniques for their effectiveness, although neither requires vast experience or knowledge in this field. Also included in this chapter are suggestions on how to present information at and contribute properly to the meetings you attend.

THE TECHNICAL BRIEFING

Your department head approaches your desk, a letter in hand, and says:

> "Mr. Winman has had a letter from the RAFAC Corporation. They're sending in some representatives next Tuesday. I'd like you to give them a rundown on the project you're working on."

Every day visitors are being shown around industrial organizations, and every day engineers and technicians are being called upon to stand up and say a few words about their work. On paper, this sounds straightforward enough, but to those who have to make the oral presentation it can be a traumatic experience. Much of their nervousness can be reduced (it can seldom be entirely eliminated, as any experienced speaker will tell you) if they are given some hints on public speaking. The best training, of course, is practical experience, which can be gained only by standing up and doing the job.

ESTABLISH THE CIRCUMSTANCES

Your first step after being told that you have to deliver an oral report is to establish the circumstances affecting your presentation. Go to your department head—or the person who is arranging the event—and ask four questions. The first two deal with your audience:

1. **Who Will I Be Speaking to?**
   If you are to focus your presentation properly and use appropriate terminology for the people you will be addressing, you need to know whether they will be engineers and technologists knowledgeable in your area of expertise, technical managers with only a general appreciation of the subject, or laypersons with no or very little technical knowledge.

2. **What Will They Know Already?**
   If you are to avoid boring you listeners by repeating information they already know, or confusing them by omitting essential background details, you need to find out how much they know about your subject or will have been told before you address them.

The second two questions deal with the briefing itself:

3. **How Long Do You Want Me to Talk?**
   Find out if you are to describe the project in detail or simply touch on the highlights. The answer will directly influence how deeply you cover the topic.

4. **Where Is the Presentation to Take Place?**
   Identify whether you are to make your presentation in your company's conference room or training room, or at a client's or some other premises. Within your own company, you can easily identify what audiovisual facilities are available and where your audience will be seated in relation to you. If you will be speaking at another location, you should ask for a description of the facilities or, even better, a chance to view them in advance.

Only when these factors have been firmly established can you start making notes. Jot down the topics you intend to discuss, and arrange them in an interesting, logical order. Try to avoid letting your familiarity with the subject blind you to characteristics that are unimportant to you but might be interesting to an unenlightened listener.

FIND A PATTERN

The best technical briefings follow an identifiable pattern, just as written formal reports do. You can establish a pattern for your briefing by mentally placing yourself in your listeners' shoes and asking three questions they are likely to have in mind:

1. What Are You Trying to Do?   Use the answer to this question to build an *introduction*, as you would for a formal report. Offer your listeners some background information, which may comprise:

- How your company became involved in the project (with, perhaps, a comment on your own involvement, to add a personal touch).
- Exactly what you are attempting to do (in more formal terms, your objectives).
- The extent or depth of the project.

2. What Have You Done So Far?   This would be the *discussion* section of the formal report. The answers to this question should cover:

- How you set about tackling the project.
- What you have accomplished to date (work done, objectives achieved, results obtained, and so on).
- Preliminary conclusions you have reached as a result of the work done (if the work is complete, these will be the final conclusions).

3. What Remains to Be Done? (or What Do You Plan to Do Next?)   This question is relevant only if the project is still in progress, in which case it is equivalent to the *future plans* section of a written progress report. Answers to this question should cover:

- The scope of future work.
- Results you hope to achieve.
- A time schedule for reaching specific targets and final completion.

If the project is complete, this question is not relevant and is replaced by an alternative question: **What Are the Results of Your Project?** The answer would then be combined with the final answer to question 2 and would be equivalent to the *conclusions* section of a written report.

Now you have a pattern for the main part of your presentation. But you still have to give it a beginning and an ending. Start with a quick synopsis of the project in easy-to-understand terms—the equivalent of a report summary. End by repeating the project's highlights—the key points you made—and its main outcome (this becomes a terminal summary), and then invite your listeners to ask questions. The complete pattern is shown as a flow diagram in Figure 9-1.

PREPARE TO SPEAK

Make brief speaking notes on prompt cards no smaller than 6 × 4 inches. Write in large, bold letters that you can see at a glance, using a series of brief headings to develop the information in sufficient detail. A specimen prompt card is shown in Figure 9-2.

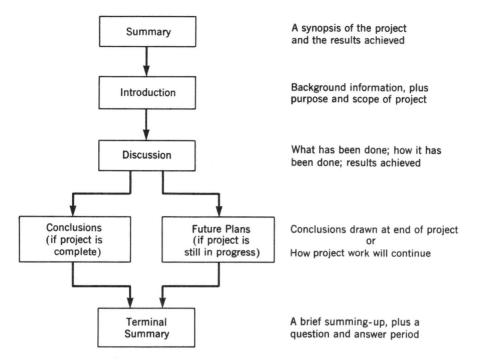

Summary	A synopsis of the project and the results achieved
Introduction	Background information, plus purpose and scope of project
Discussion	What has been done; how it has been done; results achieved
Conclusions (if project is complete) / Future Plans (if project is still in progress)	Conclusions drawn at end of project or How project work will continue
Terminal Summary	A brief summing-up, plus a question and answer period

Figure 9-1. Flow diagram for a technical briefing.

The amount of information you provide in your notes will depend on the complexity of the subject, on your familiarity with it, and on your previous speaking experience. As a general rule, the notes should not be so detailed that you cannot extract pertinent points at a glance. Neither should they be so skimpy that you have to rely too much on your memory, which may cause you to stumble haltingly through your presentation.

Don't overlook the practical aspects of the briefing. If you have equipment to demonstrate, consider its layout in relation to a logically organized description. Try to arrange the briefing so that you will move progressively from one side of the display area to the other, instead of jumping back and forth in a disorganized way. If the display is large and easy to see, let it remain unobtrusively at the back of the area. If it is small, consider moving it forward and talking from beside or behind it.

Prepare visual aids if they will help you give a clearer, more readily understood briefing. They may range from a series of steps listed as headings on a flip chart to a working model that demonstrates a complex process. Strive for simplicity; let each visual aid support your commentary rather than make the commentary explain an overly complex aid. Suggestions for preparing visual aids such as graphs, charts, and diagrams are contained in Chapter 8.

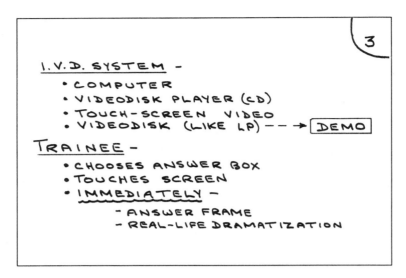

Figure 9-2. Prompt card for an oral report.

If you plan to use visual aids, practice working with them, first on their own and then as part of the whole presentation. This gives you a chance to check whether you have keyed them in at the correct places, and whether the entries in your notes are sufficiently clear to permit you to adjust from speech to a visual aid and then back again without losing continuity.

Practice your briefing. Run through it several times, working entirely from your prompt cards, until you can do so without undue hesitation or stumbling over awkward words. If the cards are too hard to follow, or contain too much detail, amend them. Then ask a colleague to sit through your demonstration and give critical comments.

After each practice reading, modify your notes to include more information or even delete unnecessary words. As you grow familiar with the notes, you will find that your confidence increases and certain sentences and phrases spring readily to mind at the sight of a single word or topic heading. This will help you maintain oral continuity.

Time yourself each time you rehearse your presentation. Aim to speak for slightly less time than allowed; for example, plan to speak for 17 or 18 minutes for a talk scheduled to last 20 minutes. This will give you time to include some previously unanticipated remarks, should you want to do so at the last minute.

NOW MAKE YOUR PRESENTATION

Coping with Nervousness. There are very few people who are not at least a little nervous when the time comes to stand up and speak before an audience. Some nervous tension is perfectly normal and can even help a speaker give a better performance. You will find that, once you start speaking, your nervousness will gradually decrease. A lot depends on the quality of your speaking notes: if you have done a thorough job preparing them, know they are reliable, and have practiced using them, you will find the familiar phrases and sentences form easily. Then you will begin to relax and so speak with even greater confidence, which in turn will help you relax even more.

Improving Your Platform Manner. Knowing some elementary platform techniques can help improve your performance. There are nine:

- Be businesslike, yet cheerful.
- Let your enthusiasm for your subject *show*. If your audience sees that you really enjoy talking about your subject—it will be apparent from the vigorous manner in which you present your material—they will listen more attentively.
- Speak extemporaneously from notes, never from a fully written speech, or you will lose contact with your audience.
- Look at your audience. Try to speak to individuals in turn, rather than the group as a whole, picking them out in different parts of the room so that every listener will feel he or she is being addressed personally.
- Use humor sparingly, and *only* if it fits naturally into your presentation. Never insert a joke to "warm up" an audience: you want to make sure your audience is laughing *with* you and not *at* you.
- Speak at a moderate rate—120 to 140 words per minute is recommended.
- Speak up. If possible, try speaking without a microphone, since this gives you much greater freedom of movement and tonal flexibility. If the room is large and you have to use one, try to obtain a lavalier (travelling) microphone that clips onto your clothing and has a long cord. Better still, ask for a radio microphone.
- Pause occasionally to study your speaker's notes. Never be afraid to stop speaking for a few moments while consolidating your position and establishing that every major topic has been covered. This also gives you an opportunity to check elapsed time.
- Avoid distracting habits that tend to divert audience attention. Examples are pacing back and forth or balancing precariously on the edge of the platform (the audience will be far more interested in seeing whether you fall off than in following your paper). Also avoid nervous afflictions, such as jingling keys or coins in your pocket (put them in a back pocket, out of reach), playing with objects on the speaker's table (remove them before you start speaking), or cracking your knuckles.

Reaching Out to Your Audience. Although good pre-platform preparation and knowledge of platform techniques can give you confidence, they are not sufficient in themselves to break down the initial barrier between speaker and audience. Successful speakers develop a well-rounded personality which they use continuously and unconsciously to establish a sound speaker-audience relationship. When you speak to an audience only once and then only briefly, probably the most important attributes to develop are enthusiasm and sincerity: enthusiasm about your topic, and sincerity in wanting to help your audience learn about it.

The time and effort you invest in preparing for a briefing will depend on your confidence as a speaker and your familiarity with the subject. The more confident you are, the less time you will need. No one will expect you to give a fully professional briefing at your first attempt, but your listeners will appreciate your efforts when they can see that you have prepared your talk carefully and are presenting it interestingly.

THE TECHNICAL PAPER

Chapter 7 discussed the steps Mickey Wendell would have to take to publish a magazine article or a technical paper. (He is a senior technician in H. L. Winman and Associates' Materials Testing Laboratory, and he has discovered that an additive called Aluminum KL mixed with cement in the right proportions produces a concrete with high salt resistance.) This chapter assumes that the papers committee of the Combined Conference on Concrete liked Mickey's abstract and summary, and the chairperson of the committee has notified him that his paper has been selected for presentation at the forthcoming Chicago conference. Mickey has four months to prepare for it.

Presenting a paper before a society meeting is much more demanding than delivering the same information at a technical briefing. The occasion is more formal, the audience usually is much larger, and the speaker is working in unfamiliar surroundings. Yet the guidelines for preparing and presenting a technical briefing, discussed earlier in this chapter, still apply.

Unfortunately, many experienced engineers and scientists duck their responsibility to the audience when faced with such a situation and simply read their papers verbatim. This can result in a dull, monotonous delivery that would turn even a superior technical paper into a dreary, uninteresting recital.

The key is to start preparing early, to make good speaker's notes, and to practice speaking from them—much as Mickey Wendell would for a technical briefing. First, however, Mickey must write the publication version of his paper, because he will need it to prepare his speaking notes. He must do this almost immediately, as soon as he hears that his paper has been accepted for presentation at the conference.

The spoken version of a technical paper does not have to cover every point encompassed by the written version. In the 15 to 20 minutes allotted to speakers at many society meetings, there is time to present only the highlights—to trigger interest in the listeners so that they will want to read the published version. Mickey has to consider how he is to stimulate and hold that interest.

In effect, Mickey needs to prepare three versions of his speaking notes. The first will consist mainly of brief topic headings derived from the written version of his paper. He should jot these headings onto a sheet of paper and then study them with four questions in mind:

1. Which points will prove of most interest to the audience?
2. Which are the most important points?

3. How many can I discuss in the limited time available?
4. In what order should I present them?

When Mickey was writing his paper, he was preparing information for a reader. Now he is preparing the same information for a listener, and the rules that guided him before may not apply. The logical and orderly arrangement of material prepared for publication is not necessarily that which an audience will find either interesting or easy to digest.

Mickey may assume that the audience at a society meeting is technically knowledgeable, has some background information in the subject area, and is interested in the topic. Most of his audience will likely be civil engineers and technologists, with a sprinkling of sales, construction, and management people. He must keep this in mind as he examines his list of headings, identifies which points he intends to talk about, and arranges them in the order he feels will most suit his listeners. He will use this version for his practice sessions. (Mickey should never take the shortcut of simply entering the headings in the margin of the typed copy of his paper. The sequence may not only seem illogical to Mickey's audience but, if he is very nervous, he may be tempted to start reading his paper. And once he starts reading, he will find it is extremely difficult to return to extemporaneous speech.)

Mickey's next task is to practice speaking from the notes, just as he would for a technical briefing but more often. It will not be enough simply to scan or read the notes and assume that he is becoming familiar with them. He must speak from them aloud, as though he is presenting his paper to an audience.

When he has practiced enough so that he feels his preliminary notes are satisfactory, Mickey can prepare his final speaking notes. These should be typed with a large typeface or hand-lettered (in ink) in clearly legible capital letters.

Mickey should plan to have these final notes ready at least three days before leaving for the conference. To a certain extent the headings in the first set of notes have helped to trigger familiar phrases and sentences. Now he has to familiarize himself with new pages. During these last practice sessions, he should attempt a full dress rehearsal by presenting the paper before some of his colleagues, among whom there may be someone qualified to comment on his platform techniques. If this is not possible, he should at least try speaking the paper alone, standing at a rostrum or desk to simulate actual conditions. This "dry run" will also give him the opportunity to check his speaking time.

TAKING PART IN MEETINGS

We all have occasion to attend meetings. In industry you may be asked to sit on a committee set up for a multitude of reasons, from resolving technical problems that are tying up production to organizing the company's annual picnic. The effectiveness of such meetings is controlled entirely by those taking part. Meetings attended by persons *aware of their role* as participants can move quickly and achieve good results; those attended by individuals who seize the opportunity to air personal complaints can be deadly dull and cripple action. Unfortunately, cumbersome, long-winded meetings are much more common than short, efficient ones.

Meetings can be either structured or unstructured, depending on their purpose. A structured meeting follows a predetermined pattern: Its chairperson prepares an agenda that defines the purpose and objectives of the meeting and the topics to be covered. The meeting then proceeds logically to each point. An unstructured meeting uses a conceptual approach to derive new ideas. Only its purpose is defined, since its participants are expected to introduce suggestions and comments which may generate new concepts (this approach is sometimes known as "brainstorming"). I will discuss the structured meeting here, because you are much more likely to encounter it in industry.

A meeting is composed of a chairperson and two or more meeting participants, one of whom often is appointed to be secretary for that particular meeting. (The secretary makes notes of what transpires during the meeting and, after the meeting, writes the "minutes," or meeting record.) Each person's role is discussed here.

THE CHAIRPERSON'S ROLE

Good chairpersons are difficult to find. A good chairperson controls the direction of a meeting with a firm hand, yet leaves ample room for the participants to feel they are making the major contribution. The chairperson must be a good organizer, an effective administrator, and a diplomat (to smooth ruffled feathers if opinions differ too widely). Much of the success of a meeting will result from the chairperson's preparation before the meeting starts and ability to maintain control as it proceeds.

Prepare an Agenda.   Approximately two days before the meeting, the chairperson should prepare an agenda of topics to be discussed and circulate it to all committee members. The agenda should identify:

- The date, time, place and purpose of the meeting.
- The topics that will be discussed (numbered, and in the sequence they will be addressed), divided into two groups:
    Action items
    Discussion items
  This arrangement ensures that the committee deals with the items most needing attention early in the meeting.
- The person who is delegated to record the meeting's minutes.

A typical agenda is shown in Figure 9-3.

Run the Meeting.   The chairperson's first responsibility is to start the meeting on time: a person with a reputation for being slow in getting meetings

MACRO ENGINEERING INC

To: Members, Electronic Facsimile Research Committee

The monthly meeting of the Electronic Facsimile Research Committee will be held in conference room B at 3 p.m. on Friday September 18, 19xx. The agenda will be:

Action Items:

1.   Accelerated completion of DPS-2A ultra-narrowbeam system installation. *(R. Taylor)*

2.   Purchase of three portable computers. *(W. Frayne)*

3.   Proposals: Papers for 19xx International Electronics Conference. *(D. Thomashewski)*

Discussion Items:

4.   Test program report: high speed feed HS-4. *(C. Bundt)*

5.   Proposal for development of CD-ROM data storage unit. *(R. Mohammed)*

6.   Plans for annual Research Division Banquet. *(C. Tripp; J. Kosty)*

7.   Other business. (Please send topics to me by 10 a.m. Thursday, September 17.)

This month's meeting secretary: J. Kosty.

D. Thomashewski
Daniel K. Thomashewski
Chairperson, EFR Committee
September 14, 19xx

Figure 9-3. An agenda for a meeting.

started will encourage latecomers. The second responsibility is to keep the meeting as short as possible without seeming to "railroad" decisions. The third responsibility is to maintain adequate control.

The meeting should be run roughly according to the rules of parliamentary procedure. (Since most in-plant meetings are relatively informal, full parlia-

mentary procedure would be too cumbersome.) The chairperson should introduce each topic on the agenda in turn, invite the person specializing in the topic to
present a report, and then open the topic for discussion. The discussion offers the
greatest challenge, for the chairperson must permit a good debate to generate
among the members, yet be able to steer a member who digresses back to the main
topic. Chairpersons must be able to sense when a discussion on a subject has gone
on too long, and be ready to break in and ask for a decision. Similarly, they must
know when strong opinions are likely to block resolution of a knotty problem, and
assign a person or subcommittee to investigate further.

Sum Up.   Before proceeding from one topic to the next, the chairperson
should summarize the outcome of the discussion on the first topic. The outcome
can be a general conclusion, a consensus of members' opinions, a decision, or a
statement of action defining who is to do what, and when. In this way all members
will be aware of the outcome, and the secretary will know what to enter into the
minutes.

The chairperson should also sum up at the end of the meeting, this time
reviewing major issues which were discussed and the main results. The chairperson
should also point the way forward by mentioning any important actions to be taken
and the time, date, and place of the next meeting (presuming there is to be one).

The best way to learn to be a good chairperson is to watch others under-
take the role. Study those who seem to get a lot of business done without appearing
to intrude too much in the decision-making. Learn what you should not do from
those whose meetings seem to wander from topic to topic before a decision is
made, have many "contributors" all speaking at the same time, and last far too long.

THE PARTICIPANT'S ROLE

You can contribute most to a meeting by arriving prepared, stating clearly
your facts, ideas, and opinions when called on, and keeping quiet the remainder
of the time. If you observe these three basic rules, you will do much to speed up
affairs. Let's examine them more closely.

Come Prepared.   If a meeting is scheduled to start at 3:00 p.m., do not
wait until 2:30 to gather the information you need. Arriving with a sheaf of papers
in hand and shuffling through them for the first 15 minutes creates a disturbance
and makes you miss much of what is being said. Start gathering information as
soon as the agenda arrives, sort your information to identify the specific items you
need, and then jot down topic headings and details you may want to quote.

Preparation becomes even more important if you have been researching
data on a particular topic and will be expected to present your findings at the
meeting. Start by dividing your information into two compartments:

1. Facts your listeners *must* have if they are to fully understand the case you are
   making and reach a decision (if, for example, you are requesting their approval to
   take a specific course of action). These become "**need to know**" facts.

2. Details your listeners only *may* be interested in and do not necessarily need to understand your case or reach a decision. These are "**nice to know**" facts.

Your plan will be to present only the **need to know** facts in your prepared presentation, but to have the **nice to know** facts ready in case one of the listeners asks questions about them.

The second step is to examine the **need to know** facts to identify the two or three pieces of information your listeners will *most* need to hear. These become your main message or summary statement.

The three compartments form a pyramid-style speaking plan, as illustrated in Figure 9-4.

Be Brief. In your opening remarks, summarize what your listeners most need to hear from you (this is your main message), and then follow it immediately with facts and details from the **need to know** compartment. If you have small amounts of statistical data to offer, either project a transparency onto a screen or print copies and distribute them when you begin to speak. (If you have a lot of information to distribute, print copies ahead of the meeting and ask the chairperson to distribute them with the agenda, so that everyone can examine your data before coming to the meeting.)

At the end of your presentation, invite questions from the meeting participants. For your answers, draw on the data you have prepared for the **nice to know** compartment, but be ready to analyze your data in depth. If some listeners seem to resist your ideas, try to avoid becoming defensive. Say that you can see their point of view, and then explain why your approach is sound or offers a better alternative than they may be suggesting. In particular, avoid getting into a personal confrontation with one or more of the meeting participants. When the chairperson calls on the members to indicate whether they accept your ideas or will approve your proposal, if they respond negatively be ready to accept their decision *gracefully*.

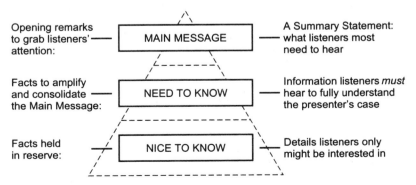

Figure 9-4. Plan for presenting information at a meeting.

Keep Quiet. There are many parts of a meeting when your role is to be only an interested observer. At these times you should keep quiet unless you have a relevant question, an additional piece of evidence, or an educated opinion. Avoid the annoying habit of always having something to add to the information others are presenting (recognize what others already know: you are not an authority on everything). At the same time, do not withhold information if it would be a genuine contribution. Be ready to present an opinion when the chairperson indicates that a topic should be discussed, but only if you have thought it out and are sure of its validity. Recognize, too, that a discussion should be a one-to-one conversation between you and the chairperson, or sometimes between you and the topic specialist. It should never become a free-for-all with each person arguing a point with his or her neighbor.

THE SECRETARY'S ROLE

Sometimes a stenographer is brought in to act as secretary and record the minutes of a meeting, but more often the chairperson appoints one of the participants to take minutes. If you happen to be selected, you should know how to go about it.

Recording minutes does not mean writing down everything that is said. Minutes should be brief (otherwise they will not be read), so there is room only to mention the highlights of each topic discussed. Items that must be recorded are: (1) main conclusions reached; (2) decisions made (with, if necessary, the name[s] of the person[s] who made them, or the results of a vote); (3) what is to be done next and who is to do it; and, sometimes, (4) the exact wording of policy statements derived during the meeting.

The best way to get this information quickly is to write the agenda topics on a lined sheet of paper, spacing them about two inches vertically. In these spaces jot down the highlights in note form, leaving room to write in more information from memory immediately after the meeting.

The completed minutes should be distributed to everyone present, preferably within 24 hours. They should be a permanent record on which the chairperson can base the agenda for the next meeting (if there is to be one) and on which participants can depend for a reminder of what they are supposed to do. I like the format shown in Figure 9-5, which provides an "action" column to draw participants' attention to their particular responsibilities.

Assignments

Speaking situations you are likely to encounter in industry will develop from projects on which you are working. Hence assignments for this chapter are assumed to grow naturally out of the major writing assignments presented in other chapters.

MACRO ENGINEERING INC

ELECTRONIC FACSIMILE RESEARCH COMMITTEE

Minutes of Meeting

Friday, September 18, 19xx, 3:00 p.m.

In Attendance:

C. Bundt	R. Mohammed
W. Feldman	R. Taylor
W. Frayne	D. Thomashewski (Chair)
J. Kosty (Secretary)	C. Tripp

Minutes	*Action*

Action Items:

1. The DPS-2A ultra-narrowbeam system is not yet operational because the S-76 interface needs further modification. R. Taylor expects modifications to be complete September 24, and the system to be operational September 28. — *R. Taylor*

2. The committee approved W. Frayne's proposal to purchase three portable computers, two out of this year's budget, one out of next year's. W. Frayne and R. Mohammed will survey available computers and present a definitive proposal at the October 16 meeting. — *W. Frayne* / *R. Mohammed*

3. Three papers are to be submitted to the program committee for next year's International Electronics Conference. Deadline: October 9. — *C. Bundt* / *J. Tripp* / *W. Feldman*

Discussion Items:

4. C. Bundt reported that phase 1 of the HS-4 test program was completed August 27, but

8. The Research Division banquet will be held in collaboration with the annual Awards Dinner on March 13, 19xx. The banquet committee will establish a joint plan with the Awards Committee. — *C. Tripp* / *J. Kosty*

J. Kosty, Secretary

Figure 9-5. Minutes of a meeting. Note the "Action" column, which draws individuals' attention to their postmeeting responsibilities.

PROJECT 9.1: SPEAKING SITUATIONS EVOLVING FROM OTHER PROJECTS

You have to attend a meeting to present the results of a study or investigation you have carried out, when you will brief managers or a client on your findings and recommendations. In each of the following instances, which are drawn from projects in Chapters 4, 5, and 6, you are to list in point form the information you would convey

1. as your "main message," and
2. as your "need to know" details.

In some projects—mainly in Chapter 4—you will be able to draw on the details provided in the assignment instructions without first doing the report writing project itself. In others—primarily in Chapters 5 and 6—it will help if you have first completed the study and written the report.

Be ready also to present this information orally.

1. Project 4.4, Part 1. On December 11 Vern Rogers asks you to drive over from the Multiple Industries's office where you are installing a Local Area Network to attend the monthly meeting at H. L. Winman and Associates' local branch office. You will be expected to describe progress of the LAN installation project to Vern and the other department heads.

2. Project 4.6. You have just returned from Moorhaven when Ray Korvan from the Department of Communications calls and asks you to attend a D of C project managers' meeting. The managers want to hear whether you have determined the source of the telephone radio interference, so they can decide whether to assign funds to research the project further.

3. Project 4.7. Maintenance crew supervisors from microwave sites 1 through 8 are attending meetings at H. L. Winman and Associates. As you have just returned from microwave site 14, where you have been investigating connector problems, Andy Rittman asks you to come in to one of the meetings and brief the supervisors on your findings.

4. Project 4.8. When you arrive home following the long drive back from Harmonsville (and the chopper flight to and from Sylvan Lake), you find a message from Vern Rogers on your telephone recorder:

"Research engineer David Yanchyn and geologist Frances Cheem, both of Triton Mining Corporation, will be in our office tomorrow morning. Can you come to a meeting at 9 a.m. to tell them what condition the access road is in at Sylvan Lake? Thanks!"

5. Project 5.1.   Robert Delorme telephones you and asks: "Have you finished the Quillicom landfill study?"

You tell him you have, but you have not yet written the report.

"Then I want you to go to the Quillicom Town Council meeting at 7:30 p.m. tonight. The Councilors want to hear what recommendations you will be making."

6. Project 5.2.   You are the technologist assigned to the WEMS project for your area, and you have just completed your report proposing that H. L. Winman and Associates buys five portable computers, one for each of the WEMS technologists across the country.

Andy Rittman telephones you from Cleveland and says: "Head office will be allocating capital purchase funds at the semiannual budget meeting the day after tomorrow. I want you to fly in and come to the meeting so you can brief the department heads on your requirements."

7. Project 5.3.   Gerald Dirksen, your boss and the company owner, tells you that Paulette Machon (vice-president of operations at Baldur Agri-Chemicals, BAC) will be coming to your company's offices tomorrow and bringing BAC's manager of human resources with her. Rather than wait for your report to reach them, they want you to brief them on your findings into BAC's power house problems, and then to discuss them and their implications with you. Mr. Dirksen also will attend the meeting.

8. Project 5.4.   You and Mel Timlick are to attend a meeting with the chief engineer of the Department of Highways, at which you will outline your proposal to conduct a highway intersection lighting study, and hopefully convince the chief to approve your idea.

9. Project 6.1.   The Department of the Environment (DOE) is holding a meeting of six project heads, one of whom is Harry Vincent. Harry asks you to join him at the meeting, because he wants you to present the results of your just-completed study in which you have identified possible locations for an acid rain monitoring station.

10. Project 6.2.   Highways Engineer Morris Hordern asks you to brief his highway engineers on the results of your highway paint study.

11. Project 6.3.   Frederick C. Magnusson, president of Fairview Development Company, has been asked by the City of Montrose, Ohio, to present his proposal for developing the Cayman Flats area of the city to a subcommittee of the City Council. He asks you to attend the meeting with him, since he anticipates there will be questions about drainage of the low-lying area, and he wants you to be there to present your plan as part of his overall proposal.

PROJECT 9.2: INFORMING TECHNICIANS OF A NEW PRODUCT

You have researched information on a new manufacturing material, method, or process, as described in Project 7.3, and have prepared a written description. Now you have to inform other technicians about the product at a lunchtime briefing organized by your department head.

Part 1. On a sheet of paper, write brief notes in point form identifying what you will say under each of the following topic headings:

- Summary statement
- Purpose (of product, material, or process)
- Details (what it does and why it is unique)
- Conclusions

Part 2. Make your presentation.

CHAPTER 10

Communicating with Prospective Employers

As head of the Administration and Personnel Department of H. L. Winman and Associates, Tanys Young is responsible for hiring new staff. Recently she advertised for an engineer to coordinate a new project and received 48 applications from across the country. Since it would have been impractical to interview all the applicants, she narrowed the field down to the nine applicants she felt had the best qualifications, basing her selection on the information contained in the 48 application letters and resumes she had received.

One of the applicants was Eugene Koenig of Indianapolis, Indiana, who believed—quite rightly—that he probably was the most qualified person for the job. But his name was not one of the nine on Tanys's short list, and so he was not even interviewed. Tanys has since met the nine selected applicants and has offered the job to the most promising person. She will never know that Eugene would have been a better person to hire. And Eugene will never know why he was not considered.

What went wrong? The fault was entirely Eugene's, whose letter and resume failed to persuade Tanys that she should talk to him.

In today's highly competitive employment market, job seekers have to tailor *each* resume and application letter they write to capture the interest of a *particular* employer. (To mail copies of an identical resume and similar letter to every employer is a wasted effort.) They must carefully orchestrate the whole employment-seeking process, from preparing their resumes to presenting themselves personally at an interview.

This chapter describes the various stages of the job-seeking process and the careful steps that you, as an applicant, must take if you want every employer you contact to consider you seriously as a potential employee. The process starts with the preparation of a personal data record and then proceeds through preparing a resume, writing an application letter, completing an application form, attending interviews, and accepting or declining a job offer.

THE EMPLOYMENT-SEEKING PROCESS

Figure 10-1 illustrates the five steps a *successful* applicant has to take before being employed. At each step the number of contenders for a particular job is reduced until only one person remains. Tanys Young received letters and resumes from 48 potential employees, but asked only 25 of them—those she felt most nearly met the company's requirements—to complete a company application form. Thus in one step she cut the field in half. Then, from the application forms and resumes, she selected nine persons to interview.

The significant factor here is that Tanys narrowed the field down to only 19% of the original applicants *based solely on their written presentations*. Like Eugene Koenig, you cannot afford to be one of the over 80% who were eliminated from the interview stages because they prepared inadequate written credentials.

The five steps of the job-seeking process are outlined briefly here. At each step you have to present a confident, positive image of yourself if you are to proceed to the next step.

1. Initial Contact. Your first step as a job seeker is to approach a prospective employer and ask to be considered for employment, either by responding to an advertisement or by approaching the employer "cold" (in the

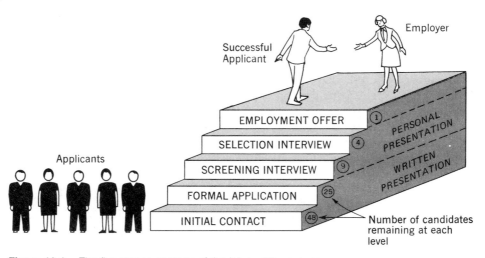

Figure 10-1. The five stages or steps of the job-seeking process.

hope that the employer either has or shortly will have an opening). You may make this initial contact by presenting yourself at the employer's door, by writing a letter, or by telephoning. The personal visit and the letter are better, because they let you *place your resume in the employer's hands.*

2. Formal Application. In step 2 of the job-seeking process, the employer will likely ask you to complete a company application form so that all applicants are documented in the same way.

3. Screening Interview. The first interview you attend helps the employer identify which applicants have the strongest potential. In a large firm such as H. L. Winman and Associates, the screening interview may be conducted by only one person, usually the employment manager or an employment representative.

4. Selection Interview. The most promising candidates are asked to attend a second interview. This time the manager of the department where the successful applicant will work also is present and sometimes is accompanied by technical specialists.

5. Job Offer. The employer makes a formal offer of employment to the successful applicant, often first by telephone and then by letter. The applicant responds, also by telephone and letter, to confirm his or her acceptance.

Not all employment-seeking processes follow exactly along these lines. Sometimes an applicant will obtain a company application form beforehand and submit it in step 1, with his or her letter and resume. At other times there may be only one job interview or, in some cases, there may be three.

DEVELOPING A PERSONAL DATA RECORD

There are three ways you can go about writing a resume: you can rely solely on your memory; you can dust off and update a previous resume; or you can create a new resume from a permanent Personal Data Record (PDR). Using a PDR is best, because it provides a much broader information base for you to draw on.

A PDR becomes particularly useful in future years, when one's ability to recall names, addresses, dates, and specific details of earlier employment diminishes. It can also be invaluable if, when calling initially on a potential employer, you are asked to complete an application form on the premises.

If you do not already have a PDR, prepare one now. You may find it is a pain to start, but is not difficult to update and keep current. There are four topic areas for which you will need to record details (and update them approximately once a year). You can write them on paper or cards, or store them on a computer hard disk or diskette.

EDUCATION

List the schools, colleges, and universities you have attended or are attending. Start at junior high school and record the name of each school, the address and telephone number, the dates you were there, and for high school, your graduation date and your area of specialization. For college and university, particularly note courses taken, special options, and the full name of the degree, diploma, or certificate you were awarded.

WORK EXPERIENCE

For each job you have held in the past—and, if you are currently employed, the job you now hold—list:

- The full name, address and telephone number of the company or organization, and the full name and title of each supervisor you worked for.
- The dates you started and finished employment and, if you held several positions within the company, the name of the position and the date you were appointed to it.
- Your job title, or titles if you held several positions.
- Your specific responsibilities and duties for each position, paying particular attention to the supervisory aspects and responsibilities of any job that you carried out without supervision.
- Any special skills you learned on the job.
- Special commendations you received, or results you achieved.

EXTRACURRICULAR ACTIVITIES

List your activities in organizations that were not necessarily part of the jobs you have held or your education. Such information can often demonstrate your leadership qualities. For example:

- Membership in a club, society, or group, particularly noting your responsibilities as an active participator or committee member. (For example, member of sports committee or secretary of administrative committee.)
- Participation in community activities such as the Big Brother or Sister organizations, 4-H Club, Red Cross Society, Parent-Teacher Association (PTA), or local community club. Particularly describe any executive or administrative positions you have held, with special responsibilities and dates.
- Involvement in a technical society on a local or national level, with particular mention of any conferences you have attended or papers you have presented or published.
- Participation as a sports enthusiast, with special mention of your role as a team leader or coach.

- Involvement in hobby activities such as stock car racing or rebuilding, a computer club, or dog breeding.
- Awards you have received for any activities you have been involved in.

For each activity, include the dates of your involvement and the name, address, and telephone number of a person who can vouch for your participation.

REFEREES

List the names of people you feel are best fitted to speak on your behalf. They fall into two groups: those who can vouch for your *capabilities* (as an employee, student, or committee member), and those who can speak for your *character*. For each person, write down:

- Full name, professional title (such as chief engineer), place of employment, and job position.
- Employer's address and telephone number.
- Home address and telephone number.

(If a referee has changed jobs, list details of both the previous and current employer.)

For each person you worked with in an extracurricular activity, also list:

- The name of the organization you both were involved with, and the referee's position within that organization.
- Whether the person prefers to be called or written to at home or at work.

PREPARING A RESUME

A resume contains key information about yourself, carefully assembled and presented so that prospective employers will be impressed not only by your qualifications but also by your ability to display your wares effectively. (The correct spelling is "résumé," but common usage in the United States has made the accentless "resume" acceptable.)

Technical people tend to be conservative when they write their resumes, yet today's employment environment really demands they be *competitive*. If a resume is to capture an employer's attention, it must display its writer's wares to full advantage.

The three resumes shown here range from fairly conservative to clearly provocative. You will have to decide which you want to use, keeping four factors in mind: which will best represent you as an individual; which will best present your qualifications; which will most suit the position you are applying for; and which will most likely appeal to the particular employer.

For ease of reference here, I will refer to the three styles as the traditional resume, the focused resume, and the functional resume. All three have one important feature in common: they open with a summary statement that (1) describes the applicant's strongest qualifications from the *employer's* point of view, and (2) identifies that the writer is seeking work in a particular field. Ideally, there is a logical connection or development between these two pieces of information, and they are presented in a short paragraph of no more than two or three sentences. For example:

Objective

Following graduation as an engineering technician I spent seven years installing and testing transmission line towers in Minnesota, North Dakota, and Alaska. I now hold a Bachelor of Science degree in Civil Engineering and want to apply my experience and education to researching grouts for tower anchors in permafrost areas.

An assertive statement such as this at the start of a resume draws the employer's attention rapidly to the applicant's primary experience and education and to the employment direction the applicant wants to pursue. If the resume "hits its mark" successfully, the employer automatically reads further into the resume to learn more about the applicant.

To be of most value, the opening statement is focused to suit the needs of a particular employer, or sometimes a group of employers engaged in similar work. The implications for job applicants are far-reaching: now they have to invest much more time, care, and research into resume preparation to ensure their resumes are clearly directed toward a specific audience.

THE TRADITIONAL RESUME

For decades the most widely recognized approach to resume writing has been to divide a job applicant's information into five parts, each preceded by an appropriate heading:

Objective
Education
Experience
Extracurricular Activities
References

The traditional resume is particularly suitable for recent university or college graduates with limited work experience, or for students who shortly expect to graduate. Alison Witney is a biological sciences undergraduate who has held two previous jobs totalling three years of full-time employment. Her resume is shown in Figure 10-2. Comments on the resume, plus guidelines you can use to

①

②

BIOGRAPHICAL DETAILS

ALISON V. WITNEY
210, 1670 Fulham Boulevard
Amiento, Florida 32704
Tel: (305) 474 6318

OBJECTIVE

To work in a position related to Animal Biology or Health Science, where I can use to good advantage both my Diploma in Biological Science and my experience as a veterinary assistant.

EDUCATION AND TRAINING

③
o   Graduate of Morton Stanley High School, Corisand, Nova Scotia, 1989.
o   Will graduate with a Diploma in Biological Science from Amiento Technical College, June 1993

WORK EXPERIENCE

④
1991 to date   **Animal Treatment Center**, Amiento, FL. Veterinary assistant, responsible for reception, grooming, and exercising of animals, assisting veterinarian during

⑤
operations, changing dressings, administering injections and anesthetics, and performing administrative duties such as accounting and ordering of supplies. (One year full time, two years part-time.)

1989 to 1991   **Remick Airlines**, Orlando, FL. Accounts clerk in air freight department; coordinating billings, preparing invoices, following-up lost shipments, assisting clients,
⑥
and writing monthly reports. For nine months assisted in payroll preparation.

1984 to date   **Bar None Riding Stables**, Corisand, FL. Part-time employ-ment teaching the care and handling of horses, and basic riding techniques, to young riders. Assisted in grooming, cleaning, feeding, and saddling-up.

⑦ **ADDITIONAL INFORMATION**

o   Winner of two educational awards: Morton Stanley Science Scholarship (1988) and Amiento Technical College Biology Scholarship (1992).
o   Member of YWCA since 1984, where I now teach swimming and lifesaving.
o   Interests: horseback riding and jumping, swimming, and water skiing.

REFERENCES

The following persons have agreed to act as referees on my behalf:

⑧
Dr. Alex Gavin   Mr. Charles Devereaux
Veterinary Surgeon   Owner-Manager
Animal Treatment Center   Bar None Riding Stables
2230 Wolverine Drive   2881 Westshore Drive
Amiento, FL 32704   Corisand, FL 32726
Tel: (305) 474 1260   Tel: (305) 632 2292

Figure 10-2. A traditional resume or biography of experience.

write a resume of your own in the traditional format, are presented here and are keyed to the circled numbers beside the figure.

1. Job applicants with only limited work experience should try to keep their resumes down to one page. The words "Biographical Details" are an alternative to "Resume."

2. Each line of Alison's name, address, and telephone number is centered about the page centerline to give the top of the resume a balanced appearance.

3. There is no need for Alison to list all the primary and secondary schools she attended: it is enough to state the name of her senior high school and the year she graduated. Then she should list each college or university she attended, plus the type of course enrolled in, the diploma or degree received, and her year of graduation (or expected graduation).

4. Experience is usually presented in reverse order, with an applicant's most recent work experience appearing first and earliest experience last. You should provide more details about recent experience (as Alison has done), and about earlier work that is similar to that of the position you are seeking, than about less-related work. Quote dates as whole years for long periods of employment, but as month and year for short periods; for example: Jun 1991–Feb 1992.

5. For each employer, state the name of the company or organization first, underline it, and then identify the city and state in which it is located. Then describe the position held (give the job title), and what the work involved. Particularly draw attention to the *responsibilities* of the job rather than merely list the duties you performed. Use words that create strong images of your self-reliance, such as:

coordinated	organized
monitored	implemented
presented	supervised
planned	directed

    (Note that Alison uses "responsible for," "administering," "coordinating," and "teaching.")
    If you have held several short part-time jobs, describe them together and draw attention to the most important, like this: "Several after-school jobs, primarily as a stock clerk in a grocery store."

6. The two-column arrangement of dates and work experience is important because it gives a much "lighter" appearance to the page. If the job descriptions were carried to the left—under the dates—the job details would appear as much heavier, less visually appealing blocks of information.

7. Employers are particularly interested in an applicant's activities and interests outside normal work. They want to know if the person is more than a routine employee who arrives at 8 a.m., works until 4:30 p.m., then drives home, eats supper, and presumably watches television all evening. Information on your hobbies, interests, and participation in sports and community activities tells prospective employers that you recognize your role in society, are not too rigid or to narrow, and adapt well to your environment. Employers reason that such an applicant will make an interesting, active employee who will not only contribute much to the company, but also take part in social and sports functions.

8. Try to draw your list of references from a cross section of people you have worked for, been taught by, or served with on committees, and ensure that their relevance

is apparent (their connection to one of your previous jobs or activities must be clear). Before including them in your list, check that all are willing to act as referees.

Both Alison Witney and Dennis Hammond (whose resume appears in Figure 10-3) are well aware of the important role a resume's appearance plays in a prospective employer's readiness to consider an applicant. A carefully arranged and printed resume, typed by a professional typist on a high-quality typewriter, implies that the applicant has high-quality credentials and should be interviewed.

THE FOCUSED RESUME

Job applicants who have more extensive experience to describe do better if they focus an employer's attention on their particular strengths and aims. This means asking themselves what a prospective employer is *most* likely to want to know after reading their opening statement. (Probably it will be: "What have you done that specifically qualifies you to achieve the objective you have presented?") To answer, they must focus on their work experience rather than their education, and particularly on work that is *relevant* to the position they are seeking.

If their experience is sufficiently varied, then they can go one step further and divide the "Work Experience" section of their resume into two parts: (1) work related to the position they are seeking; and (2) work in unrelated areas. They must place all of this information *ahead of* the "Education" section, so that there is a natural flow from their **objective** to their **related experience**. Thus, the parts of a focused resume are:

> **Objective** (or **Aim**)
> **Related Experience**
> **Other Experience**
> **Education**
> **Extracurricular Activities**
> **References**

Dennis Hammond's two-page resume in Figure 10-3 adopts this sequence. The circled numbers beside the resume refer to the following comments.

1. Dennis has sufficient information to warrant preparing a two-page resume, but he should not run over onto a third page. (A third page can be used, however, if an applicant has published papers and articles or has obtained patents for new inventions. These can be listed on a separate sheet, which is identified as an attachment.)
2. Dennis's summary statement clearly shows his thrust toward fiber optics engineering and his desire to obtain employment in that field.
3. The positions described within each "Experience" section should be listed in reverse order, the most recent experience being described first and the earliest

<u>Resume</u>

DENNIS G. HAMMOND, P.E.
310 - 508 Medwin Street
St. Cloud, Minnesota 56301
Tel: (612) 548 1612

OBJECTIVE

After four years supervising the installation and testing of wire and fiber optic telephone communication systems, I returned to college to obtain an M.S. in electronics engineering with a major in fiber optics. I am now seeking employment where I can apply my knowledge and experience in fiber optics engineering.

RELATED WORK EXPERIENCE

June 1990 to
September 1991
and
May 1992 to date

Ebby, Little and Associates, Engineering Consultants, St. Cloud, Minnesota. Supervising engineer, responsible for installation, testing, and analysis of tandem wire and fiber optic telephone communication systems between Brainerd and Little Falls, Minnesota. Currently, carrying out performance tests on installed links.

June 1983 to
August 1987

Southcentral Installation Contractors Inc, Lincoln, Nebraska. For first four years, member of team installing high voltage transmission lines and transformer stations along power grid between Weekaskasing Falls, Nebraska, and Bismarck, North Dakota. After 18 months appointed crew chief in charge of team installing interconnecting and distribution systems to townsites along the route; responsible for: hiring, training, and supervising local labor; ordering and monitoring delivery of parts and materials; arranging and supervising contract work; and preparing progress and job completion reports. From June 1986 to August 1987, assigned as supervisor of team working under contract to Ohio Utilities Corporation, installing and testing fiber optic links between towns up to 28 miles apart.

OTHER WORK EXPERIENCE

January 1977 to
February 1981

United States Air Force. Enlisted serviceman with Construction and Maintenance Directorate. For first two years, member of crew installing basic antenna systems and associated structures. For final two years, site technician responsible for maintenance of transmission lines and antennas at a midwestern USAF base. Attained rank of corporal.

June 1973 to
December 1976

Bowlands Stores Inc., Duluth, Minnesota. Stock clerk in grocery store No. 16. Full time for two summers and June to December 1976; part-time while attending high school.

/2...

Figure 10-3. A focused resume for a job applicant with a varied background.

<u>**Dennis G. Hammond**</u> - page 2

EDUCATION

⑥

o   Master of Science in Electronics Engineering with major in fiber optics, University of
    Minnesota, 1992.
o   Bachelor of Science in Electrical Engineering, University of Montrose, Montrose, Ohio,
    1990.
o   Graduate Electrical Engineering Technician, Walter Halstadt Community College, Reece,
    Minnesota, 1983.
o   Graduate of Winona Collegiate, Duluth, Minnesota, 1976.

ADDITIONAL ACTIVITIES/INFORMATION

o   Member, Institute of Electrical and Electronics Engineers Inc. (IEEE), 1983 to date.
    Secretary, St. Cloud, Minnesota Section, 1990-91.
o   Awards:
    *   Orton R. Smith Scholarship for proficiency in applied mathematics, Walter
        Halstadt Community College, 1982.
    *   Power and Light Scholarship for achievement in communications engineering,
        University of Montrose, 1989.

⑦

o   Technical paper: "Accuracies of Computer Data Transmissions Attainable at High Baud
    Rates Over Fiber Optic Communication Links," in *Communication Technology*, 13:07,
    July 1992. (Paper based on thesis written as part of M.S. program, University of
    Montrose.)
o   Military courses attended while in USAF:
    *   Transmission Line Installation Techniques, 1977.
    *   Supervisory Skills Development, 1979.
    *   First Aid and Safety Methods (various courses), 1978 to 1980.
o   Junior Leader, Duluth, Minnesota, YMCA, 1974 to 1977, teaching swimming and aquatic
    activities to boys age 9 to 15. Awarded Red Cross Bronze Medallion, 1975.

REFERENCES

⑧

The following persons have agreed to provide information regarding my qualifications and
work capabilities:

Martin F. Ebby, P.E.	Philip G. Karlowsky
Project Coordinator	Contracts Manager
Ebby, Little and Associates	Southcentral Installation
360 Rosser Avenue	Contractors Inc.
St. Cloud, Minnesota 56302	1335 Westfair Drive
Tel: (612) 544 1867	Lincoln, Nebraska 68528
Fax: (612) 544 2133	Tel: (402) 632 1450
	Fax: (402) 632 0067

Figure 10-3. *(Continued)*

experience described last. The most recent and most relevant experience should be described in considerably greater depth than early or unrelated experience (compare the descriptions of Dennis's Southcentral Contractors' experience with his Bowlands Stores' experience).

4. As in the traditional resume, each employer's name is listed first, underlined, and followed by the city and state. The person's position or job title is identified next, and then a description of what the job involved. If several positions have been held within the same firm, each is named and its duration stated so that the applicant's progress within the firm is clear.

5. Each position should draw particular attention to the personal responsibilities and supervisory aspects of the job, rather than just list specific duties. Verbs should be chosen carefully so they make the position sound as comprehensive and self-directed as possible. If the paragraph grows too long (and Dennis's paragraph here is rather long), it can be broken into short subparagraphs like these:

   . . . appointed crew chief responsible for
   - installing interconnecting and distribution systems
   - hiring, training, and supervising local labor
   - ordering and monitoring delivery of parts and materials
   - arranging and supervising subcontract work
   - preparing progress and job completion reports.

6. Single-spaced typing should be used as much as possible to keep the resume compact. At the same time there should be a reasonable amount of white space on each side and between major paragraphs to avoid a crowded effect.

7. Education can be listed either in chronological or reverse sequence. If a resume is to be sent out of state, or if the applicant was educated out of state, he or she should identify the city and state of each educational institution attended.

8. Employers are *interested* in a job applicant's accomplishments and extracurricular activities, particularly those describing community involvement and awards or commendations. This part of a resume can be preceded by a heading such as "Additional Information" rather than "Extracurricular Activities."

9. Both of the people Dennis has chosen as referees can be cross-referenced to his previous work experience. Telephone numbers are important, because most employers prefer to talk to rather than receive a letter from a referee.

THE FUNCTIONAL RESUME

Of the three resumes discussed here, the functional resume goes furthest in *marketing* a job applicant's attributes. For some employers its approach may seem too forthright—too blatantly "pushy"; for others—particularly employers seeking someone for a technical sales position—its approach helps demonstrate that the applicant has strong capabilities.

It is the only resume to offer *opinions:* its objective identifies in general terms what the applicant believes he or she can do to improve the quality of the employer's product or service, and then follows immediately with the applicant's key qualifications—the capabilities the applicant believes best demonstrate that he or she is qualified to do what the objective proclaims.

To prove that the applicant's opinions are valid, the third section establishes—with clear facts and figures—what he or she has done for previous employers or organizations. This results in a revised arrangement of the resume's parts:

Objective
Qualifications
Major Achievements
Employment Experience
Education
Awards/Other Activities
References

The intent of this arrangement is to target the resume not just for a particular employer but also for a particular position. It is especially useful under two circumstances: for job applicants who have experience in marketing and want to be employed in technical sales; and for applicants who have a lean educational background but have proven and demonstrable practical experience that can be of value to a specific employer.

The resume in Figure 10-4 shows how Reid Qually uses the functional method to capture the attention of the marketing manager of a company engaged in selling cellular telephone services. The circled numbers beside his resume are keyed to the following comments.

1. Reid has written his objective with a specific employer in mind. He has heard that King-Cell—a relatively new West Coast player in the cellular telecommunications field—is planning to expand and hopes to become a major provider of cellular telephone services across the country. By echoing the company's philosophy, he is almost certain to catch their attention.

2. Reid is aware that, as soon as the personnel manager at King-Cell has read his objective, she is likely to murmur to herself: "You have told me what you want to do. Now tell me *why* you think you can do it." So he immediately offers five reasons, each demonstrating that he can handle the job. Note particularly that

   • each is short, so that the reader assimilates the information quickly,
   • each starts with a strong "action" verb (i.e. "identify," "create," "establish"), which creates a strong, definite, image, and
   • each is an opinion (although not recommended for other types of resumes, opinions can be used here because Reid will follow immediately with *evidence* to support his assertions).

3. Reid's evidence provides *provable* facts, which demonstrate he has already established a solid track record. Reid keeps each piece of evidence short and offers definitive details (i.e. percentages, names, and dates), which adds credibility to his statements.

4. Reid can keep details of his work experience short because he has already identified his major accomplishments. For each employer he provides:

Resume

REID G. QUALLY
7 - 2617 Partridge Avenue
Seattle, Washington, 98105
Tel: (206) 263 4250

OBJECTIVE

(1) To increase market share for a west coast company providing cross-country cellular telephone services and selling cellular telephone systems.

QUALIFICATIONS

I have the particular capability to

(2) o   Identify special-interest client groups and develop innovative marketing strategies for them.

o   Create results-oriented proposals and focus them to meet specific client needs.

o   Follow-through with clients, both before and after a sale.

o   Supervise and coordinate the efforts of small groups.

o   Establish strong inter-personal relations with clients, management, and sales staff.

MAJOR ACHIEVEMENTS

For previous employers and organizations I have

(3) o   Devised an innovative lease/purchase marketing plan for first-time customers, resulting in a 34% increase in lease agreements and a 23% increase in follow-on sales over a 12-month period (for Morton Sales and Leasing, in 1991).

o   Increased sales and leases of facsimile machines by 31%, and answering machines by 26%, over a nine-month period (for Advent Communications Limited, in 1992-1993).

o   Received a company-wide "Salesperson of the Year" award (from Provo Department Stores, in 1987).

o   Advised and coordinated Electronic/Computer Technology students who won a nationwide IEEE "Carillon Communication Award" (for Pacific Rim Community College, 1992).

/2...

Figure 10-4.   A functional resume identifies in detail what an applicant feels he or she can do for a particular employer.

Resume of Reid G. Qually - _page 2_

EMPLOYMENT EXPERIENCE

④ June 1992 to the present Advent Communications Limited, Seattle, Washington.
Assistant Marketing Manager, responsible for coordinating four representatives selling facsimile transmission (fax) and telephone answering equipment to commercial customers.

November 1988 to June 1991 Morton Sales and Leasing, Seattle, Washington.
Sales representative marketing, fax machines and cordless telephones to business accounts and private customers.

July 1985 to October 1988 Provo Department Stores, Store No. 17, Portland, Oregon.
Sales representative in Home Electronics Department. Responsible for over-the-counter sales of stereos, video-cassette recorders, and portable radios.

EDUCATION

⑤ June 1992 Certificate in Commercial and Industrial Sales, Pacific Rim Community College, Seattle, WA (placed 2nd in course with GPA of 3.84).

1987 to 1991 Various courses in theoretical and applied electronics, at Pacific Rim Community College, Extension Division (partial credit toward electronics technician certificate).

June 1985 Graduated from Rosemount High School, Seattle, WA.

AWARDS AND OTHER ACTIVITIES

⑥ October 1991 and November 1992 Coordinator, IEEE "Papers Night," Pacific Rim Community College, at which students of Electronics and Computer Technology presented term projects.

1989 to present Associate Member, Institute of Electrical and Electronics Engineers Inc (IEEE).

1988 to present Member, Pacific Northwest Sales and Advertising Association; currently vice-president.

⑦ REFERENCES

Two persons will provide immediate references; other names are available.

James B. Morton
President, Morton Sales & Leasing
330 Pruden Avenue
Seattle, WA 98107
   Tel: (206) 475 3166
   Fax: (206) 475 2807

Dr. Fergus Radji
(Chairman, Burnaby Section, IEEE)
Pacific West HV Power Consultants
1920 - 784 Thurlow Street
Seattle, WA 98102
   Tel: (206) 488 1066

Figure 10-4. _(Continued)_

- start and finish dates (by month),
- employer's name (underlined),
- employer's location (city and state), and
- his job title and major responsibilities.

To maintain continuity, he lists his employment experience in reverse sequence.

5. Reid has only limited formal education, so he draws attention to his high grade point average (GPA) on returning to school after a long absence.

6. In a functional resume, the "Other Activities" section provides additional information to support statements in the "Qualifications" and "Major Achievements" sections.

7. Reid has asked several people to act as referees but lists only two, partly because they are best able to speak of his qualifications and partly to keep his resume down to two pages.

Reid's use of bullets on page 1, and a two-column format with dates on the left of page 2, provides variety in his resume's overall layout yet continuity within each page. The bulleted items on page 1 can be read easily—just when Reid wants his readers to learn quickly about him—while the facts on page 2 can be examined in more detail.

The functional resume is an effective way to present oneself to a particular employer, but it *must* be done well if it is to create the right impact. Ideally, an applicant should use it only if he or she is confident that the employer will not be "turned off" by its nontraditional approach.

Never be afraid to use a display technique for your resume that will enhance its professional quality and make it stand out among other resumes. (I do not mean you should make your resume "flashy," because an overdone appearance can spark a negative reaction from a reader.) An engineer with technical editing experience recently prepared a two-page resume which he had printed side-by-side on 11 × 17 inch paper, and then folded the sheet so that the resume was inside. On the outside front he printed only his name and the single word "Resume" (see Figure 10-5). On the back he created a table in which he listed the major projects he had worked on and, for each, itemized his degree of involvement. When employment managers placed his resume among other resumes submitted for a particular job opening, its professional experience captured their interest and resulted in the engineer being called in for more interviews than he had anticipated.

WRITING A LETTER OF APPLICATION

Although some resumes may be delivered personally, the majority are mailed with a covering letter. Because potential employers will probably read the letter first, it

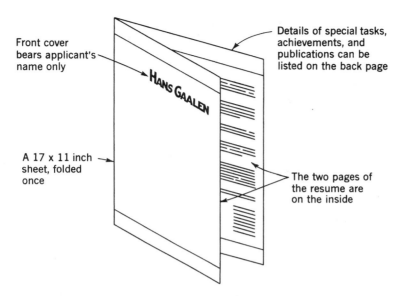

Front cover bears applicant's name only

A 17 x 11 inch sheet, folded once

Details of special tasks, achievements, and publications can be listed on the back page

The two pages of the resume are on the inside

HANS GAALEN

Figure 10-5. An imaginatively prepared resume.

must do much more than simply introduce the resume. The letter needs to state your purpose for writing (that you are applying for a job) and demonstrate that you have some very useful qualifications that the reader should take the time to consider.

An assertive, interesting, and well-planned application letter can prompt employers to place yours among those whose authors they want to interview. On the other hand, a dull, unemphatic letter may cause the same employers to drop your application on a pile of "also rans," because its approach and style seem to imply that you are a dull, unemphatic person.

A letter of application should adopt the pyramid method of writing: it should open with a brief summary that defines the purpose of the letter, continue with strong, positive details to support the opening statement, and close with a brief remark that identifies what action the writer wants the reader to take. These three parts are illustrated in Figure 10-6.

There are two types of application letters: Those written in response to an advertisement for a job that is known to be open, or at the employer's specific invitation, are called "solicited" letters. Those written without an advertisement or invitation, on the chance that the employer might be interested in your background and experience even though no job is known to be open, are referred to as "unsolicited" letters. The overall approach and shape of both letters are similar, but the unsolicited letter generally is more difficult to write.

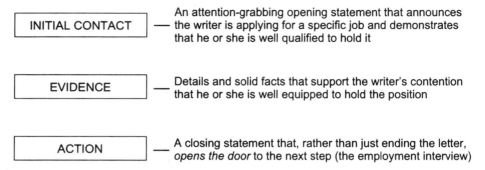

Figure 10-6. Writing plan for a job application letter.

THE SOLICITED APPLICATION LETTER

The main advantage in responding to an advertisement, or applying for a position that you know to be open, is that you can focus your letter on facts that specifically meet the employer's requirements. This has been done by Alison Witney in the letter illustrated in Figure 10-7, which she has written in response to an advertisement in a Florida local newspaper.

The following comments and guidelines are keyed to the circled numbers beside Alison's application letter.

1. For a letter that will have a personal address at the top, you would be wiser to use the modified block style shown here rather than the full block style in which every line starts at the left-hand margin. (See Figure 3-4 of Chapter 3 for information on the modified block letter.) Because this style helps balance a personal letter on the page, it provides a more pleasant initial impression. Each line of the applicant's name, address, and telephone number, and the signature block at the end of the letter, should start at the page centerline.

2. Whenever possible, personalize an application letter by addressing it by name to the personnel manager or the person named in the advertisement. This gives you an edge over applicants who address theirs impersonally to the "Personnel Manager" or "Chief Engineer." If the job advertisement does not give the person's name, invest in a telephone call to the advertiser and ask the receptionist for the person's name and complete title. (You may have to decide whether to send your letter and resume to someone in the personnel department or to a technical manager who is more likely to be aware of the quality of your qualifications and how you could fit into his or her organization.)

3. This is the **initial contact**, in which Alison Witney summarizes the key points about herself that she believes will most interest her reader and states that she is applying for the advertised position. Note particularly that she creates a purposeful image by stating confidently "I am applying. . . ." This is much better than writing "I wish to apply . . . ," "I would like to apply . . . ," or "I am interested in applying . . . ," all of which create weak, wishy-washy images because they only imply interest rather than purposefully apply for a job. An equally confident opening is "Please accept my application for. . . ."

(1)

Alison V. Witney
1670 Fulham Boulevard
Amiento, FL 32704

Tel: (305) 474 6318

March 23, 1993

(2)

Dr. Eugene Cartwright
Animal Science Experimental Institute
Mount Ashburn University
Three Hills, Alabama 35107

Dear Dr Cartwright:

(3)

    I am applying for the position of Research Technician (Animal Sciences) advertised in the March 18, 1993 issue of the Amiento County Herald. I have been involved with animals and their care and treatment for many years, and shortly will receive my Diploma in Biological Science.

(4)

(5)

    My interest in animals dates back to 1979, when I first learned to care for, groom, and ride horses. I now teach horseback riding in my spare time. For the past three years my employer has been Dr. Alex Gavin, veterinary surgeon at the Amiento County Animal Treatment Center, where I assist in the medical treatment of small animals. It was my interest in horses, plus Dr. Gavin's influence, that led to my enrolment in the two-year Biological Sciences course at Amiento Technical College, from which I will graduate in early June. The attached biographical details provide further information on my education, employment background, and work experience.

(6)

(7)

    I will be visiting your research station from April 21 to 23, as part of my college term research project. May I call on you then, while I am at Three Hills?

        Regards,

Alison Witney.

(8)

        Alison V. Witney

enc

Figure 10-7. A solicited letter of application prepared by an undergraduate.

4. The **evidence** section starts here. It should offer facts drawn from the resume and expand on the statements made in the first paragraph. Avoid broad generalizations such as "I have 13 years experience in a metrology laboratory," replacing them with shorter-term descriptions that describe your exact role and responsibilities and stress the supervisory aspects of each position. The name of a person for whom you have worked on a particular project can be usefully inserted here because it adds credibility to the role and responsibilities you are describing.

5. You may indent the first line of each paragraph about five spaces or computer columns, as Alison has done, or start each line flush with the margin, as Dennis Hammond has done in Figure 10-8.

6. The **evidence** section covers the key points an employer is likely to be interested in and draws the reader's attention to the attached resume. If the paragraph grows too long, divide it into two shorter paragraphs (as Dennis Hammond has done).

7. This paragraph is Alison Witney's **action statement**, in which she effectively opens the door to an interview by drawing attention to her imminent visit to the advertiser's premises. She avoids using dull, routine remarks such as "I look forward to hearing from you at your earliest convenience" or "I would appreciate an interview in the near future," both of which tend to close rather than open the door to the next step.

8. Contemporary usage suggests that most business letters should end with a single-word complimentary close such as "Regards," "Sincerely," or "Cordially," rather than the more formal but less meaningful "Yours very truly."

THE UNSOLICITED APPLICATION LETTER

An unsolicited application letter has the same three main parts as a solicited letter and looks very much the same to the reader. To the writer, however, there is a subtle but important difference, in that it cannot be focused to fit the requirements of a particular position an employer needs to fill. This means the job applicant has to take particular care to make the letter sound both positive and directed. Here are some guidelines to help you shape a letter that you are submitting "blind."

- Make a particular point of addressing your letter by name and title to the person who would most likely be interested in you. This may mean selecting a particular department or project head, who will immediately recognize the quality of your qualifications and how you would fit into the organization, rather than applying to the personnel manager. Never address an unsolicited letter to a general title such as "Manager, Human Resources," because, if the company does not use such a title and you have not used a personal name, it will likely be the mail clerk who decides who should receive your letter.

- Try to find out enough information about a firm so that you can visualize the type of work it does and how you and your qualifications would fit the company's needs. This will enable you to focus your letter on factors likely to be of most interest to the employment manager or selected department head.

- Try to make your initial contact positive and interesting even though you are not applying for a particular position, as Dennis Hammond has done in his unsolicited letter in Figure 10-8.

Like Alison Witney, Dennis has used the modified block format for his letter. It is longer than Alison's because he has more information to present, and to do so he has created two **evidence** paragraphs.

Dennis G. Hammond, P.E.
310 - 408 Medwin Street
St. Cloud, MN 56301

Tel: (612) 548 1612

August 1, 1992

Cory D. Richardson, P.E.
Chief Engineer
Minnesota Data Transmission Systems Inc.
440 Barker Tower
1600 Winston Drive S
Minneapolis, MN 55426

Dear Mr. Richardson:

Ⓐ Can you use a project engineer who has specialized in fiber optic transmission systems for the past five years?

Ⓑ My experience evolves from two periods of employment and my area of specialization at the University of Minnesota. For two years I was responsible for installing and testing fiber optic communication links for Ohio Utilities Corporation, then for three years I supervised the installation, testing, and analysis of parallel wire and fiber optic telephone transmission lines for Ebby, Little and Associates of St. Cloud, Minnesota. For my M.S. in Electronics Engineering I majored in fiber optic transmission of information, analyzing signal losses at high baud rates over fiber optic lines up to 12 miles long.

I also hold a B.S. in Electrical Engineering from the University of Montrose, Ohio, and have had six years experience installing high voltage transmission lines and antenna systems. The enclosed resume describes my experience and responsibilities in greater detail.

Ⓒ I would welcome the opportunity to meet you and learn more about your projects in fiber optic communications. I travel frequently between St. Cloud and Minneapolis, and will call you the next time I expect to be in your city.

Sincerely,

Dennis Hammond.

Dennis G. Hammond

enc

Figure 10-8. An unsolicited letter of application prepared by an experienced engineer. A is the initial contact, B is the evidence, and C is the action statement.

COMPLETING A COMPANY APPLICATION FORM

Filling in company application forms can become a boring and repetitive task, yet any carelessness on an applicant's part can draw a negative reaction from readers. Each company or organization usually uses its own specially designed form which, although it asks for generally the same basic information, may vary in detail. Consequently the following suggestions apply primarily to the *approach* you should take rather than suggest what you should write.

- When visiting prospective employers, always carry your personal data file with you so that you can readily search for details such as dates, telephone numbers, and names of supervisors.
- Treat every application form as though it is the *first* one you are completing—write carefully, neatly, and legibly. Never let an untidy application form subconsciously prepare an employer to meet an untidy worker.
- Complete *every* space on the form, entering N/A (not applicable), "Not Known," or a short horizontal line in spaces that do not apply to you or for which you genuinely do not have information. This will prevent an employer from thinking you carelessly (or, worse, intentionally) omitted answering the question.
- Take care that your familiarity with your city and street names does not cause you to abbreviate or omit them. If you write "Mpls" for Minneapolis or "St. Pete" for St. Petersburg, or omit the "St.," "Ave.," or "Crescent" from your street name (because you *know* it is a street, avenue, or crescent), you may create the impression that your approach to work is to take shortcuts whenever possible.
- Use words that describe the responsibility and supervisory aspects of each job you have held rather than list only the duties you performed (as you would for a resume).
- Particularly describe extracurricular activities that show your involvement in the community or in which you held a teaching or coaching role.
- Pay particular attention if there is a section on the form that asks you to comment on how your education and past experience have especially prepared you for the position. Think this through very carefully before you write so that what you say shows a natural progression from past experience to the job you are applying for. If you can, and if they fit naturally, add a few words to demonstrate how the position fits your overall career plan. This can be a particularly difficult section to write so do not be afraid to obtain an opinion of its effectiveness from another person.

ATTENDING AN INTERVIEW

This is the third step in the job application process and the first occasion when you meet a prospective employer (or, more often, the employer's representative) face to face.

PREPARE FOR THE INTERVIEW

The key to a good interview is thorough presentation. If you have prepared yourself well, the interview will most likely run smoothly and you will present yourself confidently.

As soon as you are invited to attend an interview—or, better still, before you are called—start researching facts about the company (or organization, if it is a government establishment). Presumably, you will have done some research before submitting your letter of application. Now you need to identify additional information, such as the number of persons the company employs, specific fields in which it is involved, work for which it is particularly well known, its major products and services, important contracts it has received (news of which has been released to the media), locations of branch offices, and the company's involvement in community activities. Such knowledge can be extremely useful during the interview, because it permits you to ask intelligent questions at appropriate places—questions which indicate to the interviewer that you have done your homework.

You also need to prepare for difficult questions an interviewer may pose to test your readiness for the interview and the sincerity of your application. You may be asked:

Why do you want to join our organization?

How do you think you can contribute to our company?

Why do you want to leave your present employer? (Asked only of persons who are already employed.)

Why did you leave such-and-such a company on such-and-such date? (Asked of persons whose resumes show no explanation for a previous employment termination.)

What do you expect to be doing in five years? Ten years?

What salary do you expect?

If you have not prepared for such questions, and so hesitate before answering, an interviewer may interpret your hesitation to mean you find a question difficult to answer or there are factors you would rather conceal. In either case, you may inadvertently provide an entirely misleading impression of yourself.

An interviewer who asks what salary you expect is partly testing your preparation for the interview and partly assessing how accurately you value yourself. For an undergraduate at a university or college, the question is largely academic: undergraduates compare notes and quickly learn what starting salaries are being offered. But for a person who recently has been or currently is employed, the question is important and must be anticipated. Always know the salary you would like to receive and think you are worth. Avoid quoting a salary range, such as "between 19 and 22 thousand dollars," because it indicates unsureness. Quote a definite figure, such as $21,000, and you will sound much more confident. If you fear that the salary you want to quote may be too high, you can always add the qualification ". . . depending, of course, on the opportunities for advancement and the fringe benefits your company offers."

You should be ready to ask questions during the interview. Just as the interviewer wants to acquire information about you, so should you want to learn things about the company and the opportunities it can offer. Consider what questions you would like answered, jot them onto a small card, and store the card in a pocket or purse. Then, when the interviewer asks "Now, do you have any questions?", you can pull out the card.

Make the entries on your card brief and clearly legible, and keep the list short so you can scan it quickly. Remember, too, that the quality of your questions will demonstrate how carefully you have given thought to the interview.

CREATE A GOOD INITIAL IMPRESSION

Remember that you are being evaluated from the moment you step into the interview room. Consequently

- walk in briskly and cheerfully,
- shake hands firmly, because a limp handshake creates an image of a limp, indefinite applicant,
- repeat the person's name as you are introduced and look him or her directly in the eye, and
- sit when invited to so do, pushing yourself well back in the chair, making yourself comfortable, and avoiding folding your arms across your chest (which psychologically suggests you are resisting questions).

PARTICIPATE THROUGHOUT THE INTERVIEW

An interview normally falls into three fairly easy-to-distinguish parts. The initial part is an exchange of pleasantries between yourself and the interviewer, who wants you to be at ease. To help you adjust to the interview environment, he or she may ask questions on topics you can answer confidently, such as a major news item or something from the hobbies and interests section in your resume. This initial part of the interview normally is short.

In comparison, the middle part of the interview is quite long. During this part the interviewer tries to find out as much as possible about you. He or she will want to hear your opinions and have you demonstrate your knowledge on certain topics. The interviewer will want to control the direction the interview takes but will expect you to develop your answers and to comment on each topic in sufficient depth to establish that you have real knowledge and experience, backed up by well-thought-out opinions.

The closing portion of the interview also is short. The interviewer will ask if you have questions to ask and will discuss details about the company and employment with it. By this stage the interviewer should have a pretty good impression of you, and you should know whether you want to be employed by the company he or she represents.

An effective interviewer will pose questions and subsequent prompts in such a way that you are carried easily from one discussion point to the next and are automatically encouraged to provide comprehensive answers. If, however, you face an inexperienced or inadequately prepared interviewer, the responsibility becomes yours to develop your answers in greater depth than the questions seem to call for.

For example, the interviewer may ask, "How long did you work in a mobile calibration lab?"

You might be tempted to reply "Three years," and then sit back and wait for the next question. You would do much better to reply: "For three years total. The first year and a half I was one of four technicians on the Minneapolis-to-Sioux City circuit. And then for the next year and a half I was the lab supervisor on the Fort Westin-to-Manomonee route."

An answer developed in this depth often provides the prompt (that is, piece of information) from which the interviewer can frame the next question.

Sometimes you will face a single interviewer, while at other times you may face an interview board of two to five people. In a single-interviewer situation you will naturally direct your replies to the interviewer and should make a point of establishing eye contact from time to time. (To maintain continuous eye contact would be uncomfortable for both you and the interviewer.) In a multiple-interviewer situation

- direct most of your questions, and your responses to general questions, to the chairperson (but if an answer is long, occasionally look briefly at and talk momentarily to other board members).
- if a particular board member asks you a specific question, address your response to that person.
- if a board member has been identified as a specialist in a particular discipline, direct questions to that board member if they especially apply to that field.

In certain interviews—often when applicants are being interviewed for a high-stress position—you may be presented with a "stress" question. A stress question is designed to place you in a predicament to which there may be two or even more answers or courses of action that could be taken. You are expected to think *briefly* about the situation presented to you and then to select what you believe is the best answer or course of action. Often you will be challenged and expected to defend the position you have taken.

The secret is not to let yourself be rattled and to defend your answer rationally and reasonably even though the questioner's challenging may seem harsh or unreasonable. Remember that the interviewer is probably more interested in seeing how you cope in the stress situation than in hearing you identify the correct answer.

Here are seven additional factors to consider:

- Use your voice to good effect: make sure everyone can hear you, speak at a moderate speed (thinking out your answers before speaking) and, where appropriate, let your enthusiasm *show*.

- Be ready to ask questions, but have a clear idea of what you want to ask before you pose them. An interviewer will recognize a good question and the clarity of thought behind it.
- If you do not know the answer to a question, say you don't know rather than try bluffing your way through.
- If you do not understand the question, again don't bluff. Either say you do not understand or, if you think you know what the interviewer is driving at, rephrase the question and ask if you have interpreted it correctly. (Never imply that the interviewer posed the question poorly.)
- Use humor with great care. What to you may be extremely funny may not match the interviewer's sense of humor.
- Bring demonstration materials to the interview if you wish (such as a technical proposal or report you authored, or a drawing of a complex circuit you designed) but be aware that you may not have an opportunity to display them. If the topic they support comes up during the interview, introduce them naturally into your response to a question. But remember that the interviewer does not have time to read your work, so the point you are trying to make must be readily identifiable. Never force demonstration materials on an interviewer.
- Do not smoke unless the interviewer also smokes and invites you to do so.

Finally, try to be yourself. Remember that interviewers want to see the kind of person you really are. If you relax and answer questions comfortably and purposefully, they will gain a good impression of you. If you try too hard to be the kind of person you think the interviewers want you to be, or to give the kind of answers you think they want rather than the answers you really believe in, they may detect it and judge you accordingly.

ACCEPTING A JOB OFFER

The telephone rings and the personnel representative you met during your interview tells you that the company is offering you employment at a salary of $xxxxx. You accept the offer! And then she asks when you can start work. (Employers recognize that if you are attending college there will be a waiting period until your course if finished and you have graduated; similarly, an employed engineer or engineering technician has to resign from his or her present position, normally giving either two weeks' or one month's notice.) You quote a starting date to the personnel representative, which she agrees to, and then she says she will confirm the offer in writing. She also asks you to write a letter confirming your acceptance of the position.

The two letters become, in effect, a mild contractual agreement: the employer offers you work under certain conditions, which you agree to. The letters can also prevent any misunderstandings from developing, which can occur if arrangements are made only by telephone. Consequently your acceptance letter should:

- Announce that you are accepting the offer of employment.

- Repeat any important details, such as the agreed salary and starting date.
- Thank the employer for considering you.

The following acceptance letter conforms to this pattern:

> Dear Ms. Tataryn:
>
> I am confirming my telephone acceptance of your May 19 offer of employment as an engineering technician in the Controls Department. I understand that I am to join the company on June 15 and that my salary will be $23,500 annually.
>
> Thank you for considering me for this position. I very much look forward to working for Magnum Electronics.
>
> Sincerely,

Sometimes an applicant may receive two offers of employment at the same time, and will have to decline one. The letter declining employment should follow roughly the same pattern:

- Decline the offer.
- Briefly explain why.
- Thank the employer.

Here is an example:

> Dear Mr. Genser:
>
> I regret I will be unable to accept your kind offer of employment. Since my interview with you I have been offered employment elsewhere and now must honor my commitment to the other company.
>
> Thank you for considering me for this position.
>
> Regards,

Declining a job offer pleasantly and formally in a carefully worded letter like this is insurance for the future: one day you may want to be employed by that employer!

Assignments

PROJECT 10.1: PREPARING A RESUME

You are to prepare a resume describing your background, education, work experience, extracurricular activities, and other interests. Do it in three parts:

Part 1. If you do not already have one, prepare a personal data file (PDR), using 200 × 125 mm (8 × 5 in.) file cards.

Part 2. Write down the following information:

a. The name of a real employer for whom you would like to work at the end of your course.
b. The type of position you would be able to hold with that particular employer.
c. The type of resume that would be most effective to use.

Part 3. Prepare the resume (assume that you will be graduating from your course in two months).

PROJECT 10.2: APPLYING FOR A LOCALLY ADVERTISED POSITION

From your campus student employment center or your local newspaper, identify a company currently advertising a position you could apply for at the end of your course.

Part 1. Write a letter applying for the position (assume that you will be graduating in six weeks). Also assume that you are attaching a resume to your letter. If you are replying to a newspaper advertiser, attach a copy of the advertisement to your letter.

Part 2. Assume that the company you wrote to in Part 1 sends you an application form. Obtain a standard application form from your campus student employment center and complete it as though it is the advertiser's form.

Part 3. Now assume that the company has telephoned and asked you to attend an interview next Tuesday. On a sheet of paper, write down five questions you would ask during the interview. After each question, explain why the question is important and what answer you hope it will elicit from the interviewer.

PROJECT 10.3: REPLYING TO OTHER ADVERTISERS

This project assumes that you are seeking permanent employment at the end of a technical or engineering-oriented educational program. You are to reply to any one of the following advertisements, using your present situation and actual background. If it is still early in your training program, you may update the time and assume that you are now two months before your graduation date. In each case, enclose a resume with your application letter.

ROPER CORPORATION (OHIO DIVISION)
requires a
CHEMICAL TECHNICIAN

to join a project group conducting research and development into the organic polymers associated with the coating industry. The successful applicant will also assist in the development of control techniques for producing automated color tinting. Apply in writing, stating salary expected to:

Phyllis Cairns
Personnel Manager
P. O. Box 1728
Montrose, Ohio 45287

(Advertisement in Montrose Herald, April 17)

ARCHITECTURAL DRAFTSPERSON
required by
a well-established firm of architects and planners

Applicants should have completed a Design and Drafting course at a community college or similar training institution, in which strength of materials and structural design were a curriculum requirement. Experience in preparation of reports and proposals will be considered in selection of successful applicant.
    Write to:

The Corydon Agency
Room 604—300 Main St.
(your city)

(Advertisement of March 28 on College Notice Board)

We require a
CIVIL ENGINEERING TECHNICIAN

with an interest in building construction to prepare estimates, do quantity takeoffs, and assume responsibility for company sales and promotion activities.
    Apply in writing to:
        Personnel Department
        PRECASCON CONCRETE COMPANY
        227 Dryden Avenue

(Advertisement in your local newspaper, February 26)

H. L. WINMAN AND ASSOCIATES
475 Reston Avenue
Cleveland, Ohio 44104

CIVIL ENGINEERING TECHNICIANS

This established firm of consulting engineers needs three Engineering Technicians to assist in the construction supervision of several grade separation structures to be built over a two-year period. Graduation from a recognized Civil Engineering Technology course is a prerequisite. Successful applicants will be hired as term employees. Opportunities are excellent for transfer to permanent employment before the project ends.

Apply in writing to:

Mr. A. Rittman, Head
Special Projects Department

(Advertisement on College Notice Board)

MACRO ENGINEERING INC
invites applications from
ENGINEERS, TECHNOLOGISTS, AND TECHNICIANS
interested in Metrology

We operate a first-class Standards Laboratory that is a calibration center for precision electrical, electronic, and mechanical measuring instruments used in our military and commercial equipment maintenance programs.

Applicants for senior positions should hold a B.S. in Electrical Engineering. Laboratory Technicians should be graduates of a recognized electronics or mechanical technology course who have specialized in precision measurement.

Applications are invited from forthcoming graduates, who will work under the direction of our Standards Engineer. All applicants must be able to start work on July 1.

Apply in writing to:

Mr. F. Stokes
Chief Engineer
Macro Engineering Inc
600 Deepdale Drive
Phoenix, Arizona 85007

(Advertisement in Arizona Republic, March 18)

MONTROSE PAPER COMPANY

Offers excellent opportunities for recent graduates to join an expanding manufacturing organization in the pulp and paper industry.

ENGINEERS AND ENGINEERING ASSISTANTS

Positions are available for mechanical engineers and technicians to assist in the design, installation, and testing of prototype production equipment. Previous experience in a manufacturing plant would be helpful. Innovative ability will be a decided asset.

ELECTRICAL ENGINEERING TECHNICIAN

This person will assist the Plant Engineer in the maintenance of power distribution systems. Applicants should be graduates of a two-year course in Electrical Technology with good knowledge of automatic controls and machine application. Ability to read blueprints and working drawings is essential.

ELECTRONICS ENGINEER OR TECHNICIAN

Two positions are open for electronics specialists who will maintain and troubleshoot microprocessor-controlled production equipment.

COMPUTER SPECIALIST

This position will suit either a graduate of a Computer Technology course or an Electronics Engineering Technician who has specialized in Computer Electronics. Duties will consist of installation, maintenance, and troubleshooting of existing and future computer equipment.

ENVIRONMENT SPECIALISTS

Persons selected will test air pollutants and water effluents from our paper mill and production plant and assess their environmental impact. Applicants should be graduates of a recognized course in the environmental or biological sciences.

Salaries for the above positions will be commensurate with experience and qualifications. Excellent fringe benefit program available. Write in confidence to:

Manager of Industrial Relations
MONTROSE PAPER COMPANY
Montrose, Ohio 45287

NORTH AMERICAN WILDLIFE INSTITUTE
requires
SCIENTISTS AND BIOLOGISTS
for its wildlife preservation projects in Florida,
New Mexico, Manitoba, Quebec, and Alaska.

Applicants should have a degree in the environmental sciences or an appropriate diploma from a junior college or technical institute. Persons anticipating graduation are particularly encouraged to apply.
Write to:

Dr. Willis G. Katronetsky
Chairman, North American Wildlife Institute
Room 1620—385 East 47th Street
New York, N.Y. 10017

(Advertisement in last month's issue of American Wildlife)

ENGINEERING TECHNICIAN

required by manufacturer of
automated car wash equipment

Duties: Assist plant engineer design and test mechanical and electrical components; supervise installations in cities across U.S. and Canada; troubleshoot problems at existing installations.

Applicants must be free to travel extensively. Excellent salary and promotion possibilities.
Apply in writing to:

Personnel Manager
DIAL-A-WASH Incorporated
2020 Waskeka Drive
Montrose, Ohio 45287

(Advertisement in Montrose Herald, April 23)

PRODUCTION TECHNICIAN/DRAFTSPERSON

Duties:

Under the general direction of the Chief Engineer, to be responsible for detailed product design of prototype machines. Also will investigate and recommend improvements to existing product lines.

Qualifications:

Graduate of a recognized Mechanical Engineering degree or diploma course, or a Drafting course with emphasis on Mechanical Drawing. Experience with or knowledge of agricultural equipment would be an asset.

    Please apply to:

        Chief Engineer
        Agricultural Manufacturing Industries
        3720 Harvard Avenue

(Advertisement in your local newspaper, February 26)

CELLFAX
requires recently graduated
ELECTRONICS ENGINEERS AND TECHNOLOGISTS

We are a well-established manufacturer of a wide range of cellular telephones and facsimile equipment.

We plan to augment our sales force by hiring electronics specialists who can assess customers' needs and recommend suitable equipment. If you have strong communication skills, apply in writing (and complete confidence) to:

        Reg Drabik
        Marketing Manager
        Cellfax Inc
        P.O. Box 8807

(Advertisement in last Saturday's local newspaper)

PROJECT 10.4: APPLICATION FOR UNSOLICITED EMPLOYMENT

This project assumes that you are seeking permanent employment at the end of a technical training program but few job openings have been advertised in your field. Write to Macro Engineering Inc in Phoenix, or to H. L. Winman and Associates (for the attention of one of the department heads in Cleveland, or local branch manager Vern Rogers), applying for employment. Use your knowledge of the company and your real background. If it is still early in your training program, you may update the time and assume that it is now two months before your graduation.

PROJECT 10.5: APPLICATION FOR SUMMER EMPLOYMENT

Assume that you are looking for summer employment. Reply to either one of the following advertisements:

REMICK AIRLINES
offers
SUMMER EMPLOYMENT
to a limited number of college students.

Duties: To work as part-time dispatchers, baggage handlers, aircraft cleaners, cafeteria assistants, etc, during the peak summer period (June 15–September 10) at Remick Airlines terminals at:
Fort Wiwchar, Michigan
Weekaskasing Falls, Wisconsin
Lake Lawlong, Nevada

Good pay; ample free time; free transportation to destination and return.
Write, describing previous work experience (if any), to:

Dale Ferguson	or:	Sarah Wiebe
Flight Operations Manager		305-5870 LaFrance Avenue
Room 301		Minneapolis, MN 55404
Montrose Municipal Airport		
Montrose, OH 45286		

(Notice on College Notice Board)

CITY OF MONTROSE, OHIO
TRAFFIC DIVISION

The city of Montrose is carrying out an Origination-Destination (OD) Study throughout June and July, with computation and analysis to be carried out in August.

Approximately 40 students will be required for the OD Study, and 20 for the computation and analysis. For the latter work, some knowledge of statistical analysis will be an advantage.

Students interested in this work are to apply in writing to V. L. Sinjun, Room 310, Civic Center, City of Montrose, Ohio 45287.

(Notice on College Notice Board)

NORTHERN POWER COMPANY
BOX 1760, FAIRBANKS, ALASKA 99706

Applications are invited from students in two-, three-, or four-year technical programs who are seeking three months of well-paid summer employment. The work will include construction of living quarters and associated facilities for mining camps near Prudhoe Bay, Alaska. Terms of employment will be:

13 full weeks—June 10 to September 10
Excellent pay: $475 per week
Transportation: Paid
Accommodation: Paid
Health: Applicants should be in excellent physical condition.

Students interested in this work are to write to Ms. Rheena O'Connell at the above address. Although previous experience in construction work is not essential, knowledge of general construction methods will be considered an asset for some positions.

(From Construction—North, February 20)

PROJECT 10.6: UNSOLICITED SUMMER EMPLOYMENT APPLICATION

Assume that you are looking for summer employment but few summer jobs have been advertised. Write to H. L. Winman and Associates or Macro Engineering Inc, asking for a summer job. Use your knowledge of the companies, plus your actual background, to write an interesting letter.

Address your letter to the attention of Tanys Young in Cleveland or George Dunn in Phoenix. If there is an H. L. Winman and Associates branch in your area, you may address your letter to branch manager Vern Rogers.

CHAPTER 11

The Technique of Technical Writing

This chapter concentrates on a few writing techniques that will enable you to convey information both quickly and efficiently. It considers technical writing from four points of view: how to create the whole document, how to structure paragraphs, how to write individual sentences, and how to use specific words. At the end of the chapter you will find several pages of exercise that test your ability to adopt an effective writing style and, in some cases, establish a suitable tone.

In writing this chapter, I have assumed you are already proficient in grammar or "English" and can recognize and correct basic writing problems. If you need practice in basic writing, I suggest you refer to a textbook such as the *Prentice Hall Handbook for Writers*. You can also refer to the glossary of terms in Chapter 12 for information on how to resolve many of the problems that crop up in technical writing (how to form abbreviations and compound adjectives, how to spell problem words, when to use numerals or spell out numbers in narrative, and so on).

THE WHOLE DOCUMENT

Three factors affect the whole document: the tone you set, the writing style you adopt, and how you arrange your information on the page. Of the three, tone is by far the least tangible: a reader is less likely to be aware of the tone you establish than the writing techniques you use and the arrangement of paragraphs and headings.

TONE

Whether your writing should be formal or informal will depend on the situation and your familiarity with the reader. Formal reports should adopt a formal tone. (Note, however, that a formal tone is neither stiff nor pompous; there is no room for writing that makes readers feel uncomfortable because they are not as knowledgeable as you are.) Business letters are generally less formal, depending on their importance. For example, a management-level letter proposing a joint venture on a major defense project would be formal, whereas letters between engineers discussing mutual technical problems would be informal. A memorandum report can be informal, since normally it is an in-plant document most often written between people who know each other.

Varying levels of tone are evident in the following extracts from three separate H. L. Winman and Associates' documents, all written on the same subject.

1. **Extract from a Memorandum.** John Wood's Material Testing Laboratory has compression-tested samples of concrete for Karen Woodford of the Civil Engineering Department. In his memorandum reporting the test results, John writes:

Informal Tone   I have tested the samples of concrete you took from the sixth floor of Tarryton House and none of them meets the 4800 psi you specified. The first failed at 4070 psi; the second at 3890 psi; and the third at 4050 psi. Do you want me to send these figures over to the architect, or will you be doing it?

2. **Extract from a Letter Report.** Karen Woodford conveys this information to the architect in a brief letter report:

Semiformal Tone   Our tests of three samples taken from the sixth floor of Tarryton House show that the concrete at 52 days still was 800 psi below your specification of 4800 psi. We doubt whether further curing will increase the strength of this concrete more than another 150 psi. We suggest, however, that you examine the design specifications before embarking on an expensive and time-consuming remedy.

3. **Extract from a Formal Report.** The architect rechecked the design specifications and decided that 4200 to 4300 psi still would not satisfy the design requirements. He then requested H. L. Winman and Associates to prepare a formal report that he could present to the general contractor and the concrete supplier. Karen Woodford's report said, in part:

Formal Tone   At the request of the architect we cut three 30 × 15 cm diameter cores from the sixth floor of Tarryton House 52 days after the floor had been poured. These cores were subjected to a standard compression test with the following results (detailed calculations are attached at Appendix A):

Core No.	Location	Failed at:
1	0.46 m W of col 18S	4070 psi
2	0.84 m N of col 22E	3890 psi
3	1.42 m N of col 46E	4050 psi

The average of 4003 psi for the three cores is 797 psi below the design specification of 4800 psi. Since further curing will increase the strength of the concrete by no more than 150 psi, we recommend rejecting this concrete pour.

Although the information conveyed by these three examples is similar, the tone the writer adopts varies in response to each situation.

The sequence in which you write longer reports also affects tone. To set the right tone throughout, write in reverse order, starting with the full development. Writing a report in the order in which it will be read is difficult, if not impossible. An engineering technician who writes the summary before the full development will use too many adjectives and adverbs, big words when shorter words would be more effective, and dull opening statements such as "This report has been written to describe the investigation into defective RL-80 video terminals carried out by H. L. Winman and Associates." He or she will be writing without having established exactly what to say in the full development.

The comments on Craig Derwent's formal report in Chapter 6 explain how Craig set about writing his report in reverse order (see page 168). In brief, they identify the following writing sequence:

Step 1: Assemble and document all the details and technical data. These will become the appendixes to your report.

Step 2: Write the discussion, or full development. Direct it to the type of technical reader who will use or analyze your report in depth.

Step 3: Write the introduction, conclusions, and recommendations. Keep them brief and direct them to a semitechnical reader or person in a supervisory or managerial position.

Step 4: Write the summary. Direct it to a nontechnical reader who has absolutely no knowledge of the project or the contents of your report.

STYLE

Style is affected by the complexity of the subject you are describing and the technical level of the reader(s) to whom you are writing. Consequently you need to "tailor" your writing style to suit each situation, following these guidelines:

1. When presenting low-complexity background information and descriptions of nontechnical or easy-to-understand processes, write in an easygoing style that tells readers they are encountering information that does not require total concentra-

tion. Use slightly longer paragraphs and sentences, and insert a few adjectives and adverbs to color the description and make it more interesting.

2. For important or complex data, use short paragraphs and sentences. Present one item of information at a time. Develop it carefully to make sure it will be fully understood before proceeding to the next item. Use simple words. This punchy style will warn readers that the information demands their full attention.

3. When describing a step-by-step process, start with a narrative-type opening paragraph that introduces the topic and presents any information that the readers should know or would find interesting; then follow it with a series of subparagraphs each describing a separate step, in which
   - they each develop only one item or aspect of the process,
   - they are short, and
   - they are parallel in construction (the importance of parallelism is discussed later in this chapter).

APPEARANCE

In some industrial documents—particularly specifications, technical instructions, and military reports—the paragraphs and subparagraphs are numbered. The simplest paragraph numbering system starts at 1 and numbers the paragraphs consecutively to the end of the document. More complex systems combine numbers, letters, and decimals to allow for subparagraphing. Three possible arrangements are shown in Figure 11-1.

Method A is a numerals-only decimal system, simple and unambiguous, which is often used by the military and for specifications. However, there can be problems when a document has several levels of subparagraphs. If the subparagraphs are indented as full blocks, farther to the right at each sublevel, the extended decimal numbers (e.g. 3.4.1.5) can take up a lot of space and so create very narrow subsubsubparagraphs on the right side of the page. Yet if the sentences of all subparagraphs are run back to the left margin, the level of subparagraphing can become difficult for the reader to identify.

TYPICAL PARAGRAPH NUMBERING SYSTEMS

1.	A.	1.
2.	1.	2.
3.	2.	3.
3.1	(a)	3.1
3.1.1	(b)	3.2
3.1.2	(1)	a)
3.1.2.1	(2)	b)
	B.	(1)
etc., up to	1.	(2)
6 digits	2.	etc.
	etc.	
METHOD A	METHOD B	METHOD C

Figure 11-1. Suggested paragraph and subparagraph numbering systems.

Method B uses a combination of letters and numerals that more readily permit indenting of full subparagraphs. Extremely simple, it is popular with many report writers even though it can be ambiguous (there can be several paragraph 1s, 2s, etc, one set under A, another under B, and so on).

The third method, C, combines the decimal and letter-number arrangements for a system with no ambiguity between main paragraphs and subparagraphs. This is the method chosen by Anna King for H. L. Winman and Associates' reports that carry paragraph numbers, and she suggests that engineers use normal-size paragraphs for the first two levels but only short paragraphs or single sentences for the lower-level subparagraphs:

1. 2.	} main paragraphs
2.1 2.2	} full subparagraphs
a) b)	} short subparagraphs
(1) (2)	} very short subparagraphs (as one sentence only)

This paragraph numbering system combines well with the heading arrangements illustrated in Figure 11-2.

Longer reports need headings to introduce main sections and become dividers between subsections. The three main types of headings you are likely to use are illustrated in Figure 11-2. They are center headings, side headings, and paragraph headings. If you are using a typewriter, then underline the lower-case headings. If you are using word processing and can print boldface, set the headings in boldface and do not underline them.

PARAGRAPHS

The role of the paragraph is complex. It should be able to stand alone but normally is not expected to. It must contribute to the whole document, yet it must not be too obtrusive (except when called on to emphasize a specific point). And it should convey only one idea, although made up of several sentences each containing a separate thought.

Experienced writers construct effective paragraphs almost subconsciously. They adjust length, tone, and emphasis to suit their topic and the atmosphere they want to create, and sometimes even stretch or bend the rules to obtain exactly the right impact. But even they once had to master the techniques of good paragraph writing, although now they let the rhythm of the words guide them far more than the rules.

We are not so fortunate. We have to learn the rules and apply them consciously. Yet we must not let our approach become too pedantic, or become so

MAIN CENTER HEADING: WORD PROCESSING/LASER PRINTER

The main center heading normally is capitalized and printed boldface, but is *not* underlined. Subsidiary center headings may be capitalized in a smaller font or set in boldface lower case letters with the initial letter of each principal word capitalized, but again *not* underlined.

Subparagraphing Without Paragraph Numbering

Many word processing programs and most laser printers allow text to be printed in various typefaces and fonts.

Side Headings

Side headings, as the one immediately preceding this paragraph shows, are usually keystroked in lower case letters with initial letters of principal words capitalized, and printed boldface.

Each paragraph is typed level with the same margin as the side heading. In technical writing, the first line of each paragraph is seldom indented.

Subparagraph Headings and Subparagraphing

Suparagraph headings are indented four or five columns or typewriter spaces. Each subparagraph is typed as a solid indented block, which helps readers *see* the subordination of ideas.

Secondary Subparagraphing

If further subparagraphing is necessary, the headings and subparagraphs are indented a farther four or five columns or typewriter spaces.

Heading Built into the Paragraph. In this lesser-used arrangement, the text continues immediately after the heading. Paragraph headings like these can be used for main paragraphs, subparagraphs, and secondary subparagraphs.

MAIN CENTER HEADING: ELECTRIC TYPEWRITER

This section shows typewritten headings and paragraphs, and also demonstrates how to integrate paragraph numbers into the text.

Subparagraphing With Paragraph Numbering

Typewritten headings set in capital letters normally are not underlined. Headings set in lower case type are always underlined unless printed boldface.

1.   Side Headings

    1.1   Normally, when paragraph numbers are used, side headings are assigned simple paragraph numbers.

    1.2   Subparagraph Headings and Subparagraphing

        If only one paragraph follows a side heading or subparagraph heading, it is not assigned a separate number and is set with its left margin flush (level) with the heading. Further subparagraphing looks like this:

        a)   This would be the first subparagraph.

        b)   This would be the second subparagraph.

        c)   Such subparagraphs should be short (ideally no more than one sentence).

Figure 11-2. Combined system of headings, paragraphs, and paragraph numbers. The top half demonstrates word processing, the lower half shows typewriting.

bound by the rules that we write in a stilted manner that is uninteresting and unrhythmic. Rather, we should consider them as building blocks that help create good writing, not as bars that imprison our creativity.

Good paragraph writing depends on three elements:

Unity
Coherence
Adequate development

These elements cannot stand alone. All three must be present if a paragraph is to be useful to its reader.

UNITY

If a paragraph is to have unity, it must be built entirely around a central idea. This idea is expressed in a topic sentence—often the first sentence—and developed in supporting sentences. In effect, this permits us to construct paragraphs using the pyramid technique, with the topic sentence taking the place of the summary and the supporting sentences representing the full development (see Figure 11-3).

The topic sentence does not always have to be the first sentence in a paragraph; there are even occasions when it does not appear at all and its presence is only implied. But until you are a proficient writer you would be wise to place your topic sentences right up front, where both you and your readers can see them. Then, when you have experience to support your actions, you can experiment and try placing topic sentences in alternative positions.

The following paragraph has the topic sentence right up front. It is strongly unified because its topic is clearly expressed and the supporting sentences develop it fully (the numbers that precede each sentence are for reference):

(1) *Joey, when we began our work with him, was a mechanical boy.* (2) He functioned as if by remote control, run by machines of his own powerfully creative fantasy. (3)

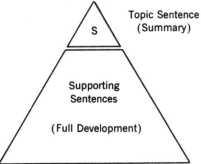

Topic Sentence
(Summary)

Supporting
Sentences

(Full Development)

Figure 11-3. Pyramid technique applied to the paragraph.

Not only did he himself believe that he was a machine but, more remarkably, he created this impression in others. (4) Even while he performed actions that are intrinsically human, they never appeared to be other than machine-started and executed. (5) On the other hand, when the machine was not working we had to concentrate on recollecting his presence, for he seemed not to exist. (6) A human body that functions as if it were a machine and a machine that duplicates human functions are equally fascinating and frightening. (7) Perhaps they are so uncanny because they remind us that the human body can operate without a human spirit, that body can exist without soul. (8) And Joey was a child who had been robbed of his humanity.[1]

This opening paragraph of a technical article demonstrates that technical writing does not have to be dull. Let's analyze it, identify why it has unity, and discover why it holds our interest and makes us want to read on.

The paragraph is summarized in sentence (1), the topic sentence, which sets the scene in nontechnical terms that create interest. Sentence (2) amplifies the topic sentence by using slightly more technical terms and introducing the fact that Joey's mechanical aspects are self-created. Then sentence (3) underscores the strength of Joey's imagination by telling us that he caused adult onlookers to feel that he really was a machine. This is carried further in sentence (4), which relates his machinelike actions to normal activities, and sentence (5), which shows how very real his machinelike qualities were, even in moments of inactivity. Then sentences (6) and (7) introduce the fascination (and an element of apprehension) that Joey's condition caused in persons observing him. Sentence (8) restates the idea expressed in sentence (1), but this time in human as opposed to mechanical terms.

The two examples that follow are paragraphs that do not start with a topic sentence. In this descriptive passage on computers, the topic sentence is at the end of the paragraph:

> The advantages of a computer have been described by Thelma Forbes, who points out that a small business with a manual inventory system can lose up to 10% of its total business volume each year simply because stock levels are improperly predicted and maintained, and records are tardily updated. In comparison, she contends, a computer system anticipates rather than reacts to a company's needs, and so can increase business volume and replace that lost 10% each year. However, a computer system can just as easily become a liability if it is too small, too large, overly expensive, or improperly programmed. *Plainly, for all the benefits a computer can bring us, not just any system will do.*[2]

In this continuing description of Joey, the "Mechanical Boy," the topic sentence is only implied. If it had been stated it would have read something like this: "Joey's machinelike actions were realistic."

[1]From Bruno Bettelheim, "Joey, A 'Mechanical Boy.' " Copyright © March 1959 by *Scientific American,* Inc. All rights reserved.
[2]Brian E. Lundeen, *Evaluation of Computer Systems for the Western Farm Implement Company.* Report No. E-28, Antioch Business Consultants, Foothills, Colorado.

For long periods of time, when his "machinery" was idle, he would sit so quietly that he would disappear from the focus of the most conscientious observation. Yet in the next moment he might be "working" and the center of our captivated attention. Many times a day he would turn himself on and shift noisily through a sequence of higher and higher gears until he "exploded," screaming "Crash, crash!" and hurling items from his ever present apparatus—radio tubes, light bulbs, even motors or, lacking these, any handy breakable object. (Joey had an astonishing knack for snatching bulbs and tubes unobserved.) As soon as the object thrown had shattered, he would cease his screaming and wild jumping and retire to mute, motionless nonexistence.[3]

COHERENCE

Coherence is the ability of a paragraph to hold together as a solid, logical, well-organized block of information. A coherent paragraph is abundantly clear to its readers; they can easily follow the writer's line of reasoning and have no problem in progressing from one sentence to the next.

Most technical people are logical thinkers and should be able to write logical, well-organized paragraphs. But the organization must not be kept a secret; it must be apparent to every reader who encounters their work. Simply summarizing a paragraph in the topic sentence and then following it with a series of supporting sentences does not make a coherent paragraph. The sentences must be arranged in an identifiable order, following a pattern that helps the reader understand what is being said.

This pattern will depend on the topic and the type of document. Paragraphs describing an event or a process most likely will adopt a sequential pattern; those describing a piece of equipment will probably be patterned on the shape of the equipment or the arrangement of its features. Patterns that can be used for typical writing situations are illustrated in Table 11-1.

Narrative Patterns.   You can write narrative-type paragraphs whenever you want to describe a sequence of steps or events. The past-present-future pattern of a progress report or occurrence report, such as Bob Walton's accident report in Figure 4-3, is a typical example. The pattern should be clearly evident, as in two of the following three paragraphs:

A coherent paragraph (in chronological order)   The accident occurred when D. Friesen was checking in at the Remick Airlines counter. He placed the company Polaroid camera on the counter while he completed flight boarding procedure. When the passenger ahead of him lifted a carry-on bag from the counter, its shoulder strap tangled with the carrying strap of the camera and pulled the camera to the floor. D. Friesen examined the camera and discovered a 1½-inch crack across its back. Remick Airlines' representative K. Trane took details of the incident and will be calling you to discuss compensation.

---

[3]Bettleheim, "Joey, 'A Mechanical Boy.' "

Table 11-1 Paragraph Patterns Used in Technical Writing

TOPICS OR SUBJECTS			WRITING PATTERNS	DEFINITIONS AND EXAMPLES
Event Occurrence Situation Process Procedure Method	Generally Narrative		Chronological order	The sequence in which events occurred; e.g. the steps taken to control flooding, how a new product was developed, how an accident happened
			Logical order	Used when chronological order would be too confusing (when describing a multiple-activity process, or several events that occurred concurrently); requires careful analysis to achieve clear narrative
			Cause to effect	The factors (causes) leading up to the result (effect) they produce; e.g., dirty air vents and failure to lubricate bearings cause overheating and eventual failure of equipment
			Evidence to conclusion	The factors, data, or information (evidence) from which a conclusion can be drawn; e.g. how results obtained from a series of tests help identify the source of a problem
			Comparison	A comparison of factors (price, size, weight, convenience) to show the differences between products, processes, or methods
Equipment Scene Appearance Building Location Features	Generally Descriptive	By Shape	Vertical	From top to bottom* of a tall, narrow subject
			Horizontal	From left to right* of a shallow, wide subject
			Diagonal	From corner to corner (for items arranged across a subject)
			Circular	From center to circumference* (for items arranged concentrically) Clockwise* (for items arranged around a subject)
		By Features	In order of: Size	From smallest item to largest item*
			Importance	From most important to least important item*
			Operation	The sequence in which the items are used when the equipment is operated

*Or vice versa.

A much less coherent paragraph (containing the same information but not presented in an identifiable pattern)

The accident occurred when D. Friesen was checking in at the Remick Airlines counter. K. Trane, a Remick Airlines representative, took details of the incident and will be calling you to discuss compensation. The damaged camera received a 1½-inch crack across the back. When the passenger ahead of D. Friesen removed a carry-on bag from the counter, its shoulder strap tangled with the carrying strap of the company Polaroid camera and pulled it to the floor. D. Friesen had placed the camera on the counter while he completed flight boarding procedure.

A coherent paragraph (tracing events from evidence to conclusion)

We noticed a mild shimmy at speeds above 65 mph about 10 days after the new tires had been installed. As a visual check of all four wheels revealed no obvious defects, we rotated the four wheels to different positions on the vehicle. This did not seem to change its point of origin. To pin down the cause, we replaced each wheel in turn with the spare wheel and found that the shimmy disappeared when the spare was in the left front position. We removed the wheel from that position, tested it, and found that it had been incorrectly balanced.

Descriptive Patterns.   You can write descriptive paragraphs to describe scenes, buildings, equipment, and any subject having physical features. The pattern can be defined by the shape of a subject, the order in which parts are operated, the arrangement of parts from smallest to largest, or the importance of the various parts. For example:

Excerpt from paragraph (features in order of importance)

The most important control on the bombardier's panel is the firing button, which when not in use is held in the black retaining clip at the bottom left corner. Next in importance is the fusing switch at the top right of the panel; when in the "OFF" position it prevents the bombs from being dropped live. Two safety switches, one immediately above the firing button retaining clip and the other to the right of the bank of selector switches, prevent the firing button from being withdrawn from its clip unless both are in the "LIVE" (up) position.

Continuity.   A fully coherent paragraph must also have smooth transitions between the sentences. Smooth transitions give a sense of continuity that makes readers feel comfortable. As they finish one sentence, there is a logical bridge to the next. This can be accomplished by using linking words and by referring back to what has already been said. In the example just quoted there is a natural flow from "The most important . . ." in the first sentence to "Next in importance . . ." in the second. The third sentence then refers back to the firing button and so relates the newly introduced safety switches to the previous information. The transitions are equally good in the first paragraph describing damage to a Polaroid camera, each sentence containing a component that is a development from one of the previous sentences. This is not true of the second paragraph, in which each new sentence introduces a new subject with no reference to what has already been said.

ADEQUATE DEVELOPMENT

Paragraph development demands good judgment. You must identify your readers clearly enough so that you can look at each paragraph from their point of view. Only then can you establish whether your supporting sentences amplify the topic sentence in sufficient detail to satisfy their interest.

Simple insertion of additional supporting sentences does not necessarily meet the requirements for adequate development. The supporting sentences must contain just the right amount of pertinent information, all directed to a particular reader. There must never be too little or too much. Too little results in fragmented paragraphs that offer snippets of information which arouse readers' interest but do not satisfy their needs. On the other hand, too much information can lead to long, repetitious paragraphs that annoy readers. Compare the following paragraphs, all describing the result of exploration crews' first venture with machinery across the Peel Plateau east of Alaska, intended for a reader who is interested in the problems of working in the subarctic but who has never seen what the terrain is like.

Inadequate development	Trails left by tractors look like long narrow scars cut in the plateau. Many of them have been there for years. All have been caused by permafrost melting. They will stay like this until the vegetation grows in again.
Adequate development	To the visitor viewing this far northern terrain from the air, the trails left by tractors clearing undergrowth for roads across the plateau look like long, narrow scars. Even those that have been there for as long as 28 years are still clearly defined. All have been caused by melting of the permafrost, which started when the surface moss and vegetation were removed and will continue until the vegetation grows in again—perhaps in another 30 years.
Over-development	To conservationists, whose main interest is the protection of the environment, viewing this far northern terrain from the air is a heart-rending sight. To them, the trails left by tractors clearing undergrowth for roads across the plateau look like long, narrow scars. The tractors were making way for the first roads to be built by man over an area that until now had been trodden only by Inuit indigenous to the area, and the occasional trapper. Some of these trappers had journeyed from Quebec to seek new sources of revenue for their trade. But now, in the very short time span of 28 years, man has defiled the terrain. With his machines he has cut and gouged his way, thoughtlessly creating havoc that will be visible to those that follow for many decades. Those that preceded him for centuries had trodden carefully on the permafrost, leaving no trace of their presence. The new trails, even those that have been there for 28 years, are visible almost as though they had been cut yesterday. And all were caused by melting of the permafrost. . . .

In these examples the descriptive pendulum has swung from one extreme to the other. The first paragraph leaves the reader with questions: What were the tractors doing? How many years? How soon will the vegetation grow in again? The second paragraph develops the topic sentence in just enough detail; it explains why the tractors left semipermanent scars and predicts how long they will remain. The third paragraph, though interesting, is filled with irrelevant information and repetitive statements that detract from the main theme. It might be suitable for a novel but not for a technical report or description.

The first paragraph of the article on Joey, the "Mechanical Boy" also is adequately developed because it provides just sufficient information to support the author's original intent: to introduce the boy. It confines itself to a brief word picture of Joey's problem (but not of Joey himself) and its effect on both the boy and the adults who worked with him. Because its readers would be technical men and women who specialize in other technical fields, the paragraph introduces its subject in general terms that will be understood by and provoke interest in a wide range of readers. The author has not described who Joey is, where he is, and what his surroundings are like. Instead, he has started with a thought-provoking topic sentence and used the whole paragraph to develop the image of the "mechanical" child. To have attempted to introduce names, locations, and similar information would have destroyed the impact of this very effective introductory paragraph.

CORRECT LENGTH

How long should your paragraphs be? The answer is simple: no longer than you need to cover a particular topic for a particular reader. But also keep these points in mind:

1. Variety in paragraph length has a lively visual effect. A series of equal-length paragraphs creates the impression of dullness. If the paragraphs are consistently short, readers will feel cheated because they are not presented with enough information. (Too many short paragraphs create a jackrabbit effect—all starts and stops.) Conversely, a succession of very long paragraphs will make readers feel they are swimming in a sea of information in which they cannot readily identify the main topic; worse, they may subconsciously assume that what you have written is going to be "heavy going," and so start reading your words with a negative bias. Generally, try to limit paragraph length to no more than 10 type-written lines.
2. Readers attach importance to a paragraph that is clearly longer or shorter than those surrounding it. A very short paragraph among a series of generally longer paragraphs attracts attention, just as a longer paragraph among a series of short paragraphs implies that the information it contains is particularly important.
3. Paragraph length needs to be adjusted to suit the complexity of your topic and the technical level of your readers. Generally, complex topics demand short paragraphs containing small portions of information. But their length should be conditioned by your knowledge of specific readers. Even a complex topic can be covered in long paragraphs for persons who are technically knowledgeable.

SENTENCES

Although sentences normally form an integral part of a larger unit—the paragraph—they still must be able to stand alone. While helping to develop the whole paragraph, each has to develop a separate thought. In doing so they play an important part in placing emphasis—in stressing points that are important and playing down those that are not.

Just as experienced writers first had to learn the elements of good paragraph writing, so they also had to learn the elements of good sentence construction. These are:

> Unity
> Coherence
> Emphasis

Unity and coherence perform a similar function in the sentence as they do in a well-written paragraph.

UNITY

Although the comparison is not quite so clearly defined, we can still apply the pyramid technique to the sentence in the same way it is applied to the paragraph and the whole document. In this case the summary is replaced by a primary clause that presents *only one thought*, and the full development by subsidiary clauses and phrases that develop or condition that thought (see Figure 11-4). Thus a unified sentence, as its name implies, presents and develops only a single thought.

Compare these two sentences:

A unified sentence that expresses one main thought   The Amron Building will make an ideal manufacturing plant because of its convenient location, single-level floor, good access roads, and low rent.

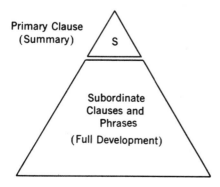

Primary Clause (Summary)   S

Subordinate Clauses and Phrases

(Full Development)

Figure 11-4. Pyramid technique applied to the sentence.

A complicated sentence that tries to express two thoughts	The copier should never have been placed in the general office, where those using it interrupt the work being done by the stenographic pool, which has been consistently understaffed since the beginning of the year.

The first sentence has unity because everything it says relates to only one topic: that the Amron Building will make a good manufacturing plant. The second sentence fails to have unity because readers cannot tell whether they are supposed to be agreeing that the copier should have been placed elsewhere, or sympathizing that there has been a shortage of stenographers. No matter how complex a sentence may be, or how many subordinate clauses and phrases it may have, every clause must either be a development of or actively support only one thought, which is expressed in the primary clause.

COHERENCE

Coherence in the sentence is very similar to coherence in the paragraph. Coherent sentences are continuously clear, even though they may have numerous subordinate clauses, so that the message they convey is apparent throughout. Like the paragraph, they need good continuity, which can be obtained by arranging the clauses in logical sequence, by linking them through direct or indirect reference to the primary clause, and by writing them in the same grammatical form. Parallelism, discussed at the end of this chapter, plays an important role here.

The unified sentence discussing the Amron Building has good coherence because its purpose is continuously clear and each of its subordinate clauses links comfortably back to the lead-in statement (through the phrase "because of its . . ."). The clauses are written in parallel form (they have the same grammatical pattern), which is the best way to carry the reader smoothly from point to point. The first sentence in the following examples lacks coherence because there is no logic to the arrangement or form of the subordinate clauses. Compare it with the second sentence, which despite its length is still coherent because it continuously develops the thought of "late" and "damaged" expressed in the primary clause.

An incoherent sentence	The Amron Building will make an ideal manufacturing plant because of its convenient location, which also should have good access roads, the advantage of its one-level floor, and it commands a low rent. (34 incoherent words)
A coherent sentence	Many of the 61 samples shipped in December either arrived late or were damaged in transit, even though they were shipped one week earlier than usual to avoid the Christmas mail tie-up, and were packed in polyurethane as an extra precaution against rough handling. (45 coherent words)

EMPHASIS

Whereas unity and coherence ensure that a sentence is clear and uncluttered, properly placed emphasis helps readers identify the sentence's important parts. By arranging a sentence effectively, you can help readers attach importance to the whole sentence, to a clause or phrase, or even to a single word.

Emphasis on the Whole Sentence. You can give a sentence more emphasis than sentences that precede and follow it by manipulating its length or by stressing or repeating certain words. You can also imply to the reader that all the clauses within a sentence are equally important (without affecting its relationship with surrounding sentences) by balancing its parts.

Although you should aim for variety in sentence length, there may be occasions when you will want to adjust the length of a particular sentence to give it greater emphasis. Readers will attach importance to a short sentence placed among several long sentences, or to a long sentence among predominantly short ones. They will also detect a sense of urgency in a series of short sentences that carry them quickly from point to point. This technique is used effectively by storytellers:

> The prisoner huddled against the wall, alone in the dark. He listened intently. He could hear the guards, stomping and muttering. Cursing the cold, probably.

In technical writing we have little occasion to write very short sentences, except perhaps to impart urgency to a warning of a potentially dangerous situation:

> Dangerously high voltages are present on exposed terminals. Before opening the doors:
> 1. Set the master control switch to "OFF."
> 2. Hang the red "NO!" flag on the operator's panel.
> Never cheat the interlocks.

Neither must we err in the opposite direction and write overly long sentences that are confusing. The rule that applies to paragraph writing—adjust the length to suit complexity of topic and technical level of reader—applies equally to sentence writing. If, on average, your sentences tend to exceed 22 words, then probably they are too long.

Similarity of shape can signify that all parts of a sentence are of equal importance. Clauses separated by a coordinate conjunction (mainly "and," "or," "but," and sometimes a comma) tell a reader that they have equal emphasis. The following sentences are "balanced" in this way:

Eight test instruments were used for the rehabilitation project, *and* were supplied free of charge by the Dere Instrument Company.

The gas pipeline will be 300 miles shorter than the oil pipeline, *but* will have to cross much more difficult terrain.

The upper knob adjusts the instrument in the vertical plane, *and* the lower knob adjusts it in the horizontal plane.

Emphasis on Part of a Sentence.   While coordination means giving equal weight to all parts of a sentence, subordination means emphasizing a specific part and deemphasizing all other parts. It is effected by placing the most important information in the primary clause and placing less important information in subordinate clauses. In each of the following sentences the main thought is italicized to identify the primary clause:

When the technician momentarily released his grip, *the control slipped out of reach.*

The bridge over the underpass was built on a compacted gravel base, partly to save time, partly to save money, and partly because materials were available on site.

He lost control of the vehicle when the wasp stung him.

When the wasp stung him, *he lost control of the vehicle.*

Emphasis on Specific Words.   Where we place individual words in a sentence has a direct bearing on their emphasis. Readers automatically tend to place emphasis on the first and last words in a sentence. If we place unimportant words in either of these impact bearing positions, they can rob a sentence of its emphasis:

Emphasis misplaced	Such matters as equipment calibration will be handled by the standards laboratory however.
Emphasis restored	Equipment calibration, however, will be handled by the standard laboratory.
Emphasis misplaced	Without exception change the oil every six days at least.
Emphasis restored	Change the oil at least every six days.

The verbs we use have a powerful influence on emphasis. Strong verbs attract the reader's attention, whereas weak verbs tend to divert it. Active verbs are strong because they tell *who did what.* Verbs in the passive voice are weak because they merely pass along information; they describe *what was done by whom.* The following sentences are written using both active and passive verbs. Note that the versions written in the active voice are consistently shorter:

Passive Voice	*Active Voice*
Elapsed time is indicated by a pointer.	A pointer indicates elapsed time.
The project was completed by the installation crew on May 2.	The installation crew completed the project on May 2.
It is suggested that meter readings be recorded hourly.	I suggest you record meter readings hourly.
The samples were passed first to quality control for inspection, and then to the shipping department where they were packed in polyurethane.	Quality control inspected the samples and then the shipping department packed them in polyurethane.

There are occasions when you will have to use the passive voice because you are reporting an event without knowing who took the action, or prefer not to name a person. For instance, you may prefer to write:

The strain gauge was read at 10-minute intervals.

rather than:

Kevin McCaughan read the strain gauge at 10-minute intervals.

Unfortunately, many scientists and engineers still write in the passive voice. (Some have even been advised—wrongly—to shun the active voice because it is too strong and does not fit their "professional" character.) This outdated belief is gradually being eroded.

COMPLETENESS

Every day we see examples of incomplete sentences, on television, in magazines, and especially in advertising. Media writers use them to create a crisp, intentionally choppy effect:

SHEER COMFORT!
25,000 feet high. Wide seats, just like your living room. Tempting meals. Complimentary refreshments. Only on Remick Airlines. Our Business Class. Try us!

But if we do the same in our business letters and reports, our sentences are likely to be read with raised eyebrows.

A common and very easy error to make, particularly when a piece of writing is going well and you do not want to interrupt your enthusiasm, is to inadvertently form a **sentence fragment**. Normally you correct it later, when you are checking what you have written, but sometimes your familiarity with media writing can cause you not to notice it. These are examples of sentence fragments:

- The meeting achieved its objective. Even though three members were absent.
- The staff were allowed to leave at 3 p.m. Seeing the air-conditioning had failed.

The first sentence in each of these examples is complete (it has a subject-verb-object construction), but the two second sentences are incomplete because, to be understood, each depends on information in the first sentence.

A useful way to check whether or not a sentence is complete is to read it aloud entirely on its own. If it contains a complete thought, it will be understood just as it stands. For example, *The meeting achieved its objective* is a complete thought because you do not need additional information to understand it; and so is *The staff were allowed to leave at 3 p.m.* But you cannot say the same when you read these aloud:

- *Even though three members were absent.*
- *Seeing the air-conditioning had failed.*

In most cases a sentence fragment can be corrected by removing the period that separates it from the sentence it depends on, inserting a comma in place of the period, and adding a conjunction or connecting word such as "and," "but," "which," "who," or "because":

- The meeting achieved its objective, even though three members were absent.
- The staff were allowed to leave at 3 p.m. *because* the air-conditioning had failed. (*Seeing* has been changed to *because*.)

Here are two others:

- Staff will have to bring bag lunches or go out for lunch from October 6 to 10. While the lunchroom is being renovated.
    Change the period to a comma.
- All 20-year employees are to be presented with long-service awards. Including three who retired earlier in the year. At the company's annual banquet.
    Change both periods to commas.

In particular, check sentences which start with a word that ends in "—ing" (refer*ring*, answer*ing*, be*ing*) or an expression that ends in "to" (with reference *to*):

- With reference to your letter of June 6. We have considered your request and will be sending you a check.
    Change the period to a comma.

- Referring to the problem of vandalism to employees' automobiles in the parking lot. We will be hiring a security guard to patrol the area from 8 a.m. to 6 p.m., Monday through Friday.

  Although the period could be changed to a comma, a better sentence could be formed by reconstructing the fragment:

- To resolve the problem of vandalism to employees' automobiles in the parking lot, we will be hiring. . . .

A similar sentence error occurs if you link two separate thoughts in a single sentence, joining them with only a comma or even no punctuation. The effect can jar a reader uncomfortably. For example:

- Ms. Solvason has been selected for the word-processing seminar on March 11, she is not eager to attend.

This awkward construction is known as a **run-on sentence**. It can be corrected by

1. replacing the comma with a period, to form two complete sentences:
   . . . on March 11. She is not eager to attend.
2. retaining the comma and following it with "which" or "but":
   . . . on March 11, which she is . . .
   . . . on March 11, but she is. . . .

A run-on sentence with no punctuation is even more noticeable:

- The customer said he never received an invoice I made up a new one.

Here, either a period or a comma and a linking word can be inserted between "invoice" and "I":

   . . . an invoice. I made up . . .
   . . . invoice, so I made up . . .

You will be surprised how easily such simple sentence or punctuation faults can slip by unnoticed (until of course they reach the reader, who immediately spots them!).

WORDS

The right word in the right place at the right moment can greatly influence your readers. A heavy, ponderous word will slow them down; an overused expression will make them doubt your sincerity; a complex word that they do not recognize will annoy them; and a weak or vague word will make them think of you as indefinite. But the right word—short, clear, specific, and necessary—will help them understand your message quickly and easily.

WORDS THAT TELL A STORY

Words should convey images. We have many strong, descriptive words in our individual vocabularies, but most of the time we are too lazy to employ them because the same old routine words spring easily to mind. We write "put" when we would do better to write "position," "insert," "drop," "slide," or any one of numerous descriptive verbs that better describe the action. Compare these examples of vague and descriptive words:

Vague Words	*Descriptive Words*
While the crew was in town they *got* some spare parts.	they *bought* they *purchased* they *borrowed* they *requisitioned*
We have *contacted* the site.	We have *telephoned* . . . We have *visited* . . . We have *written to* . . . We have *spoken to* . . . We have *faxed* . . .
The project will *take a long time*.	will *last four months*. will *require 300 work hours*. will *employ two persons for three weeks*.

Story writers use descriptive words to convey active images to their readers. Because we are concerned with *technical* writing does not mean we should avoid seeking colorful words. One descriptive word that defines size, shape, color, smell, or taste is much more valuable than a dozen words that only generalize. (But a word of warning: reserve most of these colorful, descriptive words for your verbs and nouns rather than for adverbs and adjectives.)

Analogies can offer a useful means for describing an unfamiliar item in terms a nontechnical reader will recognize:

A resistor is a piece of ceramic-covered carbon about the size of a cribbage peg, with a two-inch length of wire protruding from each end.

Specific words tell the reader that you are a definite, purposeful individual. Vague generalities imply that you are unsure of yourself. If you write:

It is considered that a fair percentage of the samples received from one of our suppliers during the preceding month contained a contaminant.

you give your reader four opportunities to wonder whether you really know much about the topic:

1. "It is considered"   Who has voiced this opinion?
2. "a fair percentage."   How many?
3. "one of our suppliers"   Who? One in how many?
4. "contained a contaminant"   What contaminant? In how strong a
                                      concentration?

All these generalities can be avoided in a shorter, more specific sentence:

> We estimate that 60% of the samples received from RamSort Chemicals last June were contaminated with 0.5% to 0.8% mercuric chloride.

This statement tells the reader that you know exactly what you are talking about. As a technical person, you should never create any other impression.

LONG VERSUS SHORT WORDS

Big words create a barrier between writer and reader. Some writers use big words to hide their lack of knowledge; others because they start writing without first defining clearly what they want to say. There are many long scientific words that we have to use in technical writing—we should surround them with short words whenever possible so our writing will not become ponderous and overly complex.

LOW INFORMATION CONTENT EXPRESSIONS

Words and expressions of low information content (LIC) contribute little or nothing to the facts conveyed by a sentence. They are untidy lodgers. Remove them and the sentence appears neater and says just as much. The problem is that practically everyone inserts LIC words into sentences, and we become so accustomed to them that we do not notice how they fill up space without adding any information. For example:

Vague   Pursuant to the client's original suggestion, Mr. Richards was of the
and   opinion that the structure planned for the client would be most suitable
Wordy   for immediate erection on the site until recently occupied by the old
          established costume manufacturer known as Garrick Garments. In
          accordance with the client's anticipated approval of this site, Mr. Rich-
          ards has taken great pains to design a multilevel building that can be
          considered to use the property to an optimum extent.

"Waffle" is the word technical editor Anna King uses to describe such cumbersome writing. By eliminating unnecessary expressions (*pursuant to; of the opinion that; for immediate erection on; in accordance with; can be considered to use; an optimum extent*), Anna cut the original 75 words to a much more effective 42 words:

Clear   Mr. Richards believes the type of building the client wants should be
and   erected on the site previously occupied by Garrick Garments. He has
Direct   assumed that the client will approve this site, and has designed a
multilevel building which fully develops the property.

Table 11-2 contains some of the words and phrases we must try to delete from our writing. They are difficult to identify because they often sound like good but rather vague prose. In the following sentences the LIC words have been italicized; they should be either deleted or replaced, as indicated by the notes in parentheses.

Table 11-2
Examples of Low Information Content (LIC)
Words and Phrases

The LIC words and phrases in this partial list are followed by an expression in parentheses (to illustrate a better way to write the phrase) or by an (X), which means that it should be dropped entirely.

actually (X)
a majority of (most)
a number of (many; several)
as a means of (for; to)
as a result (so)
as necessary (X)
at present (X)
at the rate of (at)
at the same time as (while)
bring to a conclusion (conclude)
by means of (by)
by use of (by)
communicate with (talk to; telephone; write to)
connected together (connected)
contact (talk to; telephone; write to)
due to the fact that (because)
during the course of, during the time that (while)
end result (result)
exhibit a tendency (tend)
for a period of (for)
for the purpose of (for; to)
for this reason (because)
in all probability (probably)
in an area where (where)
in an effort to (to)

in close proximity to (close to; near)
in connection with (about)
in fact, in point of fact (X)
in order to (to)
in such a manner as to (to)
in terms of (in; for)
in the course of (during)
in the direction of (toward)
in the event that (if)
in the form of (as)
in the neighborhood of; in the vicinity of
    (about; approximately; near)
involves the use of (employs; uses)
involves the necessity of (demands;
    requires)
is designed to be (is)
it can be seen that (thus; so)
it is considered desirable (I or we want to)
it will be necessary to (I, you, or we must)
of considerable magnitude (large)
on account of (because)
previous to, prior to (before)
subsequent to (after)
with the aid of (with)
with the result that (so, therefore)

Note: Many of these phrases start and end with words such as *as, at, by for, in, is, it, of, to,* and *with.* This knowledge can help you to identify LIC words and phrases in your writing.

- The control is actuated by *means of* No. 3 valve. (delete)
- Adjust the control *as necessary* to obtain maximum deflection. (delete)
- Tests were run for *a period of* three weeks. (delete)
- If the project drops behind schedule *it will be necessary to* bring in extra help. (*I, you, he, we,* or *they will* bring in extra help)
- By Wednesday we had a backlog of 632 units, *and for this reason* we adopted a two-shift operation. (replace with *so*)
- A new store will be opened *in an area* where market research has *given an indication that there actually is a* need for more retail outlets. (delete both expressions; replace the second one with *indicated the*)

Clichés and hackneyed expressions are similar to LIC words and phrases, except that their presence is more obvious and their effect can be more damaging. Whereas LIC words are intangible in that they impart a sense of vagueness, hackneyed expressions can make a writer sound pompous, garrulous, insincere, or artificial. Referring to oneself as "the writer," starting and ending letters with such overworked phrases as "We are in acknowledgment of" and "please feel free to call me," and using semilegal jargon ("the aforementioned correspondent," "in the matter of," "having reference to") are all typical hackneyed expressions. Further examples are listed in Table 11-3.

Presenting the right word in the correct form is an important aspect of technical writing. Comments on how to abbreviate words, how to form compound adjectives, whether or not you should split an infinitive (generally, you may), and when you should spell out numbers or present them as numerals are contained in a glossary of usage in Chapter 12.

Table 11-3
Typical Clichés and Hackneyed Expressions

and/or	in the foreseeable future
as a matter of fact	in the matter of
as per	last but not least
attached hereto	needless to say
at this point in time	please feel free to
enclosed herewith	pursuant to your request
for your information	regarding the matter of
(as an introductory phrase)	this will acknowledge
if and when	we are pleased to advise
in our opinion	we wish to state
in reference to	with reference to
in short supply	you are hereby advised

PARALLELISM

Parallelism in writing means "similarity of shape." It is applied loosely to whole documents, more firmly to paragraphs, sentences, and lists, and tightly to grammatical form. Readers generally do not notice that parallelism is present in a good piece of writing—they only know that the sentences read smoothly. But they are certainly aware that something is wrong when parallelism is lacking. Good parallelism makes readers feel comfortable, so that even in long or complex sentences they never lose their way. Its ability to help readers through difficult passages makes parallelism particularly applicable to technical writing.

THE GRAMMATICAL ASPECTS

If you keep your verb forms similar throughout a sentence in which all parts are of equal importance (that is, in which the sentence has coordination, or is balanced), you will have taken a major step toward preserving parallelism. For example, if I write "unable to predict" in the early part of a balanced sentence, I should write "able to convince" rather than "successful in convincing" in a latter part:

Parallelism violated	Mr. Johnson was *unable to predict* the job completion date, but was *successful in convincing* management that the job was under control.
Parallelism maintained (A)	Mr. Johnson was *unable to predict* the job completion date, but was *able to convince* management that the job was under control.
More direct alternative (B) (parallelism maintained)	Mr. Johnson *predicted* no completion date, but *convinced* management that the job was under control.

The rhythm of the words is much more evident in the latter two sentences. In sentence (B) particularly, the parallelism has been stressed by converting the still awkward ". . . unable . . . able . . ." of sentence (A) to the positively parallel "predicted . . . convinced . . .".

Two other examples follow:

Parallelism violated	Pete Hansk likes surveying airports and to study new construction techniques.
Parallelism maintained	Pete Hansk likes to survey airports and to study new construction techniques.
	or
	Pete Hansk likes surveying airports and studying new construction techniques.

*Parallelism   His hobbies are developing new software programs and stereo compo-
violated*   nent construction.

*Parallelism   His hobbies are developing new software programs and constructing
maintained*   stereo components.
               or
               His hobbies are new software development and stereo component
               construction.

Parallelism is particularly important when you are joining sentence parts with the coordinating conjunctions "and," "or," and "but" (as in the preceding examples), with a comma, or with correlatives, such as

> either . . . or
> neither . . . nor

Each part of a correlative must be followed by an expression in the same grammatical form. That is, if "either" is followed by a verb, then "or" must also be followed by the same form of verb:

*Parallelism   You may either repair the test set or it may be replaced under the
violated*   warranty agreement.

*Parallelism   You may either *repair* the test set or *replace* it under the warranty
maintained*   agreement.

APPLICATION TO TECHNICAL WRITING

Although parallelism is a useful means for maintaining continuity in general writing, in technical writing parallelism has special application. Parallelism can clarify difficult passages and give rhythm to what otherwise might be dull material. When building sentences that have a series of clauses, you can help the reader see the connection between elements by molding them in the same shape throughout the sentence. There is complete loss of continuity in this sentence because it lacks parallelism:

*Parallelism   In our first list we inadvertently omitted the 7 lathes in room B101, 5
violated*   milling machines in room B117, and from the next room, B118, we also
               forgot to include 16 shapers.

When the sentence is written so that all the item descriptions have a similar shape, the clarity is restored:

*Parallelism   In our first list we inadvertently omitted 7 lathes in room B101, 5 milling
restored*   machines in room B117, and 16 shapers in room B118.

Emphasis can also be achieved by presenting clauses in logical order. If they are arranged climactically (in ascending order of importance), or in order of increasing length, the reader's interest will gain momentum right through the sentence:

> No one ever offered to help the navigator struggle out to the aircraft, a parachute hunched over his shoulder, a roll of plotting charts tucked under his arm, a sextant and an astrocompass dangling from one hand, and his precious navigation bag heavy with instruments, computer, radar charts, and the route plan for the 10-hour flight clenched firmly in the other.

Within the paragraph parallelism has to be applied more subtly. If it is too obvious, the similarity in construction can be dull and repetitive. The verbs often are the key: keep them generally in the same mood and they will help bind the paragraph into one cohesive unit (this is closely tied in with coherence, discussed under "Paragraphs"). This paragraph has good parallelism:

> The effects of sound are difficult to measure. What to some people is simply background noise, to others may be ear-shattering, peace-destroying drumming. The roar of a jet engine, the squeal of tires, the clatter of machinery, the hiss of air-conditioning, the chatter of children, and even the repetitive squeak of an unoiled door hinge, can seriously affect them and create a distinct feeling of uneasiness.

Similarity of shape is most obvious in the third sentence, with its rhythmic

> *roar* of a jet engine
> *squeal* of tires
> *clatter* of machinery
> *hiss* of air-conditioning
> *chatter* of children
> *squeak* of an unoiled door hinge

Here we have words that paint strong images. They not only have the same grammatical form but also are bound together because they relate to the same sense: hearing.

Of course, parallelism can be of particular effect in descriptive writing like this, where the author wants to convey exciting images. In technical writing the subjects are more mundane, giving us less opportunity for imaginative imagery. Nevertheless, you should try to use parallelism to carry your reader smoothly through your paragraphs, as in this description of part of a surveyor's transit:

> Two sets of clamps and tangent screws are used to adjust the leveling head. The upper clamp fastens the upper and lower plates together, while the upper tangent screw permits a small differential movement between them. The lower clamp fastens the lower plate to the socket, while the lower tangent screw turns the plate

through a small angle. When the upper and lower plates are clamped together they can be moved freely as a unit; but when both the upper and lower clamps are tightened the plates cannot be moved in any plane.

APPLICATION TO SUBPARAGRAPHING

Subparagraphing seldom occurs in literature but is used frequently in technical writing to separate events or steps, describe an operation or procedure, or list parts or components. Subparagraphing always demands good parallelism.

In the following paragraph, a series of tests has been divided into subparagraphs (for clarity, only the initial words of each test are shown here):

> Three tests were conducted to isolate the fault:
> * In the first test a matrix was imposed upon the video screen and. . . .
> * The second test consisted of voltage measurements taken at. . . .
> * A continuity tester was connected to the unit for test 3 and. . . .

The last subparagraph is not parallel with the first two. To be parallel it must adopt the same approach as the others (it should first mention the test number and then say what was done):

> For the third test, a continuity tester was connected to the unit and. . . .

To be truly parallel the subparagraphs should start in *exactly* the same way:

> In the first test a matrix was. . . .
> In the second test a series of. . . .
> In the third test a continuity tester was. . . .

But this would be too repetitive. The slight variations in the original version make the parallelism more palatable.

If more than three tests have to be described, a different approach is necessary. To continue with "A fourth test showed . . . ," "For the fifth test . . . ," and so on would be dull and unimaginative. A better method is to insert a number or a bullet in front of each paragraph:

> Seven tests were conducted to isolate the fault.
> 1. A matrix was imposed upon the video screen and. . . .
> 2. Voltage measurements were taken at. . . .
> 3. A continuity tester was connected to the unit and. . . .

Parallelism has been retained, and we now have a more emphatic tone that lends itself to technical reporting.

A third version employs active verbs together with parallelism to build a strong, emphatic description that can be written in either the first or third person:

In laboratory tests conducted to isolate the fault we:
1. *Imposed* a matrix upon the video screen and. . . .
2. *Measured* voltage at. . . .
3. *Connected* a continuity tester to the unit and. . . .

To change from the first person to the third person, only the lead-in sentence has to be rewritten:

In tests conducted to isolate the fault, the laboratory:
1. *Imposed.* . . .
2. *Measured.* . . .
3. *Connected.* . . .

Assignments

Exercise 11.1.   The following sentences and short passages lack compactness, simplicity, or clearness, and particularly contain low information content (LIC) expressions. Improve them by deleting unnecessary words, or by partial or complete rewriting.

1. The production line supervisor installed a microwave oven in the rest area in the vicinity of line station 22.
2. It was installed in an effort to increase the morale of the assemblers working at night.
3. The microwave tower is located in close proximity to television station KMON.
4. A power outage disrupted work for a period of 3 hours and 22 minutes.
5. If you experience further problems, please feel free to call on me at any time.
6. The *4Tell* software will be initiated as a means of preventing slippage of the schedule.
7. An accounting error resulted in an operating loss of considerable magnitude last year.
8. For your information, it was decided by the shareholders that an amount of $250,000 should be assigned to a fund for the purpose of future plant expansion.
9. Check for grain temperature with the use of a 45 inch long probe inserted vertically downward into the depths of the grain.
10. It is with considerable concern that we have noticed a large number of discrepancies between the projected and actual expenditures for 1991.
11. The Freeling Lake Mine survey project has, at this point in time, reached the point of being two days ahead of schedule.

12. Any attempt to operate the engine in excess of 1500 rpm is likely to result in and be the cause of accelerated bearing failure.
13. The drawings attached hereto are in reference to the May 12 proposed tentative plan for modification of the MicroStat in accordance with directive TP-7.
14. If it is your intention to effect repairs to your model 2120 printer, it would be in your best interests first to assure the availability of parts for this older type of model before starting work.
15. The end result was a 22% decrease in acidity following the introduction of Limasol Plus into the solution.
16. The Plant Safety Committee members were in agreement that, under the requirements of last year's hazardous materials legislation, WHMIS training must be in the final stages of completion no later than September 1.
17. Our statement No. 28201 is enclosed, in the amount of $687.56, in reference to maintenance and repairs performed on the electronic control system used in connection with your high speed printing press.
18. It is our estimation that the cost for the maintenance of the Arbutus control system for a period of three years (July 1, 1992 to June 30, 1995) will be in the region of $220,000 to $260,000.
19. Between a speed of 66 and 72 mph the wheel at the left front exhibits a tendency to shimmy.
20. With the completion of this sentence you will bring to a conclusion this exercise on the identification of LIC expressions.

Exercise 11.2. The following sentences offer choices between words that sound similar or are frequently misused. Select the correct word in each case.

1. The test documentation (appears/seems) to be complete.
2. Signals from space probe 811 traveled 86 billion miles (farther/further) than signals from space probe 260.
3. Three attempts were made to (elicit/illicit) a response from station K-2.
4. Before clamping the scale onto its base, (orient/orientate) the 0° point on the scale so that it is opposite the "N" mark.
5. Our vehicle was (stationary/stationery) when truck T48106 skidded into it.
6. The inspection team calibrated 38 of the 47 test instruments between May 10 (and/to) 16. The (balance/remainder) were to be calibrated on May 24 and 25.
7. Fuel consumption is (affected/effected) by driving speed.
8. The quality control inspection team recommended instituting a (preventive/preventative) maintenance program to reduce system outages.
9. We could not complete the modifications on schedule (as/for/because) the monitor had not yet been shipped to the site.
10. Although the report documented the investigation results in full detail, its executive summary (implied/inferred) there were hidden costs that were not readily evident.
11. When compared (to/with) the previous period, the (amount/number) of products rejected by the quality control inspectors decreased from 237 to 189.
12. Under current human rights legislation, employers have to be particularly (discreet/discrete) when making personal enquiries about potential employees.

13. Analyses of the data (is/are) being delayed until all the data (has/have) been received.

14. The inspection team will submit (its/it's/their) final report on October 26.

15. Improved maintenance in 1991 resulted in 27 (fewer/less) service interruptions than in 1990.

16. When the temperature within the cabinet rises above 28.5°C, the blower motor cuts in automatically and operates (continually/continuously) until the cabinet temperature drops to 25°C.

17. Before accepting each job lot of steel from the foundry, samples are tested mechanically and electronically to ensure the steel is (free from/free of) flaws.

18. The chief engineer agreed that the agenda (was/were) too comprehensive for a one-hour meeting.

19. When the secretary telephoned to say that Mr. Humphries was too ill to travel, the program committee had to find an (alternate/alternative) keynote speaker for the conference.

20. The (principal/principle) reason for including this exercise in *Technically—Write!* is to draw your attention to the glossary of technical usage.

Exercise 11.3. Rewrite the following sentences to make them more emphatic (in many cases, change them from the passive to active voice). Where necessary, create a "doer" if one is not identified.

1. The minutes of the April 20 project meeting were recorded by technician Dan Wolski.

2. It was recommended by the quality control inspector that production lots No. 214 and 307 be reworked.

3. Yesterday's power failure was caused by an explosion at Winfield power station.

4. The 57 personal computers at Multiple Industries will be linked by a local area network (LAN) and central file server, which has been designed by Vancout Business Systems.

5. In a determined effort to meet the December 5 contract delivery date, an evening shift was approved by management, with its operation to be implemented from October 28.

6. Due to poor visibility and drifting snow caused by a blizzard on January 15, the highway to Montrose international airport was closed by State Police from 4 p.m. to 9 a.m. the following day.

7. Although delays in the delivery of spare parts had been identified by Cal Poloskiew in two separate letters to the supplier on September 10 and 28, no action was taken by the supplier until he had received four more complaints from site engineers in October.

8. When the *4Tell* software was evaluated, it was discovered by Sheila Fieldstone that the program was only 80% compatible with the operating system used by the company for its mainframe computer.

9. In a memo dated February 10 from Pierre St. Germaine, Fern Wilshareen was asked to recommend an engineering technologist from her department to attend the Engineering Technology Symposium in Dallas on April 3 and 4 as the company's representative.

10. In an announcement made by David N. Courtland, president of Vancourt Business Systems, shareholders were told that the feasibility of establishing a small

manufacturing facility on the British island of Guernsey, off the coast of France, was being considered by the Board as a means of establishing a foothold for the company in Europe.

Exercise 11.4. Improve the parallelism in the following sentences.

1. Getting approval to purchase a replacement control unit will not be a problem, but there will be difficulty in finding a local supplier who can deliver it by June 16.
2. If you decide to order three Nabushi 310 portable computers, the price will be $1995 each, but the price will be $2250 each if you order only one or two.
3. When the Montrose branch office closed, two staff members were transferred to the Syracuse office and a generous severance package was arranged for five who were laid off.
4. A survey was carried out to define property boundaries and as a means for settling the dispute between property owners.
5. We selected the Nabushi 700 laser jet printer because it is fast, moderately priced, readily available, and its scalable fonts can be used directly with our word-processing software.
6. The accident delayed installation work for three days and was instrumental in increasing operating costs by 4.6%.
7. A consultant from Superior Computers Inc will hold three in-house seminars on how to install version 4.8 of the 4-7-9 software and procedures for using it with our operating system.
8. A power outage occurred on December 23 because of
   • ice accumulation on overhead cables,
   • a transformer breakdown at Point Lacroix,
   • the system also became overloaded locally, when residents switched on their Christmas lights, and
   • neighboring states experienced exceedingly low temperatures that created an excessive demand.
9. As soon as you check into your hotel in Atlanta, call branch manager Shirley Kahn. Inform her of your arrival, making arrangements to hold a department head meeting the following morning.
10. The following procedure is to be followed for submitting expense accounts when on protracted field assignments:
    (1) They are to be submitted weekly.
    (2) Complete them on Monday for the preceding week.
    (3) They should be completed in quadruplicate.
    (4) Mail copies 1, 2, and 3 to the head office, retaining copy 4 for your files.
    (5) Be sure to mail them no later than noon on Tuesday.

Exercise 11.5. Abbreviate the terms shown in italics in the following sentences. In some cases you will also have to express numerals in the proper form. (Rules for forming abbreviations appear in Guidelines 2 and 6 of Chapter 12, and rules for writing numbers in narrative are in Guideline 4.)

1. The sound level recorded at the south end of the assembly area was 58.6 *decibels*, which was .8 *decibels* higher than the level recorded at the north end of the area.

2. 8000 of the 68656 cylinders for the Norland contract were manufactured with a 22.5 *millimeter inside diameter*.

3. A 24-*inch diameter* culvert, *approximately* 330 *yards* long, will connect the proposed development to the main drainage channel.

4. Rotate the antenna in a *counterclockwise* direction until the needle on the dial dips to between the .6 and 1.4 markings.

5. Vibration became severe at speeds above 54 *miles per hour*, which forced us to drive more slowly than anticipated. However, this enabled us to achieve a fuel consumption of 21.7 *miles per gallon*.

6. Radio station KMON's two operating frequencies are 88.1 *kilohertz* and 103.6 *megahertz*.

7. A radar station on a platform 360 *feet above mean sea level* can "see" an aircraft flying at thirty thousand *feet* up to a range of 420 *miles (average)*.

8. Paragraph three on page seven of the proposal quoted an additional six hundred and fifty seven dollars ($657.00), tax *included*, for repairing the damage to the roof caused by hurricane Hazel.

9. One degree of latitude is the equivalent of sixty *nautical miles*, so that each minute of latitude equals one *nautical mile*, or 6,080 *feet*.

10. The new furnace produces 56000 *British thermal units*, which is seven *per cent* more than the unit it replaces.

11. Comparing temperatures is not difficult if you remember that zero *degrees Celsius* is the same as thirty-two *degrees Fahrenheit*, twenty-two *degrees Celsius* is equal to seventy-two *degrees Fahrenheit*, and that both *temperatures* are the same at minus forty *degrees*.

12. Tank *serial number* 2821 holds one thousand *liters* of oil, which is equivalent to 264 *United States gallons* or 220 *Imperial gallons*.

Exercise 11.6. Correct any of the following sentences that are not complete or have been punctuated improperly.

1. The hard disk crashed at 4:15 p.m. Before I had time to copy today's work onto a diskette.

2. Seventy-eight percent of the staff participated in this year's blood donor campaign on November 12 and 13, a record.

3. A 34-minute break in the power supply at 10:28 a.m. on May 20 interrupted the continuity of that day's low temperature tests.

4. With reference to your letter of September 8, and our subsequent meeting on September 15 during which you outlined the problems you have experienced with your model 3 Microstat. I have conveyed your dissatisfaction to the manufacturer, who has agreed both to repair your Microstat free of charge and provide you with a replacement while the work is being done.

5. Progressive corrosion inside the pipes has reduced liquid flow by 21% since 1983. A condition which, if not corrected, could cause system shutdown in less than 12 months.

6. Work on the Feldstet contract was completed on February 16 three days ahead of the February 19 scheduled completion date, a cause for celebration.

7. In last month's report I estimated that we could complete repairs to the antenna towers and cables by June 8, unfortunately gale force winds on June 2 tore down

two towers and pulled 31 cable anchors out of the soil, this caused 11 additional days work and has set back the repair completion date to June 19 (if we work through the weekend, that is) otherwise it will be June 21.

8. When the digital exchange was installed at Multiple Industries, eight lines were left unused for anticipated staff expansion. Also for providing dedicated lines for a planned facsimile transmission network between branches.

9. The overhead steam pipe ruptured at 10:10 a.m., fortunately the office was empty as everyone had gone down to the cafeteria for coffee break. Although damage was effected to the computer equipment and oak furniture.

10. Before submitting requests for purchasing new or replacement equipment, ensure that
    • manufacturer details are complete,
    • price quotations are attached.
    • The appropriate specifications are quoted, and
    • An alternative supplier is listed.
    • The divisional manager approves the request.

Exercise 11.7. Select the correctly spelled words among the choices offered.

1. The (computer/computor) is supplied with a built-in 99-year (calendar/calender).
2. A (coarse/course)-grained (aggregate/agreggate) is used as a base before pouring the concrete.
3. Profits in the (forth/fourth) quarter increased by (forty/fourty) percent.
4. The interest on the loan is not a (deductable/deductible) expense.
5. The (affluent/effluent) produced by the paper mill is (enviromentally/environmentaly/environmentally) sound.
6. The preface to a book or report is sometimes called a (forward/foreward/foreword).
7. The (cite/site) is (inaccessable/inaccessible) except by helicopter.
8. Version 6.0 of the *4Tell* software (supercedes/supersedes) version 5.5.
9. The tests show that the materials have (similar/similiar) properties.
10. To the young business owner seeking a cash flow loan, the (colatteral/collatteral/collateral) demanded by the bank seemed (exhorbitant/exorbitant).
11. When Multiple Industries bought all the outstanding shares of Torrance Electronics, the latter company became a (wholely/wholly) owned subsidiary.
12. Silica gel is a (desiccant/dessiccant/dessicant), or drying agent, that is packed with electronic equipment before shipment.
13. Software designers who have (entepreneurial/entrepeneurial/entrepreneurial) drive do not (necessarily/neccessarily) have good management expertise.
14. After (lengthy/lengthly) deliberation, the division manager admitted that Ken Wynne's innovative design was indeed (ingenious/ingenuous).
15. When we have (accumulated/accummulated) all the results from product tests, we will (prescribe/proscribe) definitive purchase specifications.
16. An (auxiliary/auxilliary) heater cuts in when temperature drops below 3°C.
17. The (eigth/eighth) test demonstrated that the process is (feasable/feasible).

18. The incandescent lamps have been replaced with (flourescent/fluorescent/fluourescent) lamps.
19. The sales manager was (embarassed/embarrassed/embarrased) that customers were being (harassed/harrassed/harrased) by overly zealous sales staff.
20. Well? How many words did you (mispell/misspell)?

Exercise 11.8. Rewrite the following letters and memorandums to improve their effectiveness.

1. Dear Mr. Reimer

   It is with the sincerest regret that Vancourt Computers Inc has to inform you that there will be an unfortunate delay of about three weeks in filling your esteemed order. (Ref. your P.O. 2863 dated April 29) Due to measures entirely beyond our control the ship carrying a shipment of 1100 portable (or laptop) model 7000 computers from Nabuchi Electronics in Taiwan developed engine trouble in mid-Pacific and had to limp back to its home port (in fact it had to be towed the last 120 nautical miles!) Your 3 computers are among the 1100. The shipping company has recently informed us that the ship will sail on June 6 and arrive in San Francisco on June 19. We will air express your 3 units as soon as they have cleared customs and absorb the additional expense ourselves. We hope that this will help you understand our position.

   Yours very truly,
   Vern Karpov

2. To: Greg Haugen, Site 14
   From: Chris Halliday, Central Stores
   Date: June 11, 19xx

   In response to your request of May 29, the necessary action has been taken (on June 9.) The manometer you requisitioned (Req 3730) has been shipped to Site 14. In the event that it has not been delivered to the site within ten days, a "loss report" (attached hereto) should be completed and forwarded to this office marked for my personal attention.

3. Customer Service Manager
   Emerald Air Express

   To whomever it may concern:

   I am writing with reference to a shipment one of your drivers delivered to me on August 17. This particular shipment was an envelope containing eight papers which I needed urgently. For your records, your weighbill number 7284 06292 36 is in reference to this shipment. The documents originated in Toronto, Canada, and the package was picked up by one of your drivers and delivered to your Toronto office on August 13. The shipping envelope was marked "Next Morning Delivery."

   Your driver—here, who delivered the package to me—insisted that the shipment had been sent C.O.D., and I had to pay $48.76 for it before he would hand over the envelope. Only after the driver had gone did I discover that the shipment had been traveling for 4 days!!!

   What I want to know is this: Why was I charged such an outrageous sum for documents that were time-sensitive and which, by the time it reached me, was of

no value? I look forward to receiving your check at your earliest convenience.
I remain (annoyed),
Peter LeMay

4. Dear Mr. Shasta

   I am in receipt of your letter of Nov 17, you'll be glad to hear the problem you outlined is under consideration. A defective chip has been discovered. In the output stage. Replacements are hard to get, I phoned around but no one has one locally. So I've ordered one directly from the manufacturer—Mansell Microprocessors—and asked them to ship it air express. I'll telephone you immediately it comes in. But it won't be for three weeks, too many are back-ordered, they can't ship it before Dec 12.

   Sincerely,

   T. L. Pedersen

5. Dear Ms. Sorchan

   Inspection of your residential lot at 2127 Victoria Street. Survey shows your neighbor's fence 3.7 in. inside your property (your neighbor to south, that is, at #2123). Fence to the north is okay: its yours and its 1.2 in. inside your property. Does your neighbor at #2123 know about this? You have basis for watertight legal action if that's the rout you want to take. Our survey invoice #236 enclosed.

   Regards,

   Wilton Candrow

6. To: Annette Lesk
   From: Mark Hoylan
   Date: 09/07/92
   Subject: Progress at MMW

   As you are well aware, Ken Poitras and I have been at Morriss Machine Works for two and a half weeks now, where we are shoring up the flooring for the NCR machine to be installed this week, and have had to build a 10 ft extention along the width of the north wall.

   All this work is now complete and we should've been heading back to the office by now except theres a problem: the NCR machine arrived today and instead of being installed its sitting outside under a tarp. Why? Because no one seems to of calculated that a 42 inch wide machine (which is it's narrowest dimension) can't be greased through a 36 inch wide door!

   So . . . Mr. Grindelbauer who is the Machine Works manager (actually, I think he's the owner, too) has asked Ken and yours truly to stay behind and tear out part of the wall to make the door wider (which we're doing now) so the machine can be put in tomorrow, and then for us to rebuild the wall and reinstall the door which I reckon will take two extra days. I tried to explain to him there would be an extra charge for doing all this as its not in the contract, and he said for you to call him. Will you do that? Thanks.

   Mark

   P.S. I reckon the additional time will be 32 hours and there will be two additional nights accommodation and per diem, plus some extra lumber and wallboard at cost which will probably come in at about $135.

   PPS We'll be back in the office on the tenth. OK?

Exercise 11.9.   Rewrite this one-paragraph notice to make it clearer, more personal, and more likely to encourage readers to do as it requests.

PROCEDURE RE EXPENSE CLAIMS

Expense claims must be handed to the Accounts Section before 10:30 a.m. on Wednesday for payment on Friday. Personnel failing to hand in their forms at the proper time may do so at any time until 4:30 p.m. on Thursday but must wait until Monday for payment. Under no account will a late claim be paid in the same week that it was filed. Claims handed in after Thursday will be processed with the following week's claims and will be paid on the next Friday.

CHAPTER 12

Glossary of Technical Usage

A standard glossary of usage contains rules for combining words into compound terms, for forming abbreviations, for capitalizing, and for spelling unusual or difficult words. This glossary also offers suggestions for handling many of the technical terms peculiar to industry. Hence, it is oriented toward the technical rather than the literary writer.

The glossary is preceded by six guidelines for specific aspects of technical writing:

Guideline 1: Combining Words into Compound Terms
Guideline 2: Abbreviating Technical and Nontechnical Terms
Guideline 3: Capitalization and Punctuation
Guideline 4: Writing Numbers in Narrative
Guideline 5: Summary of Numerical Prefixes and Symbols
Guideline 6: Introduction to Metric Units

The entries in the glossary are arranged alphabetically. Among them are words that are most likely to be misused or misspelled, such as:

- Words that are similar and frequently confused with one another; e.g. *imply* and *infer*; *diplex* and *duplex*; *principal* and *principle*.
- Common minor errors of grammar, such as *comprised of* (should be *comprises*), *most unique* (*unique* should not be compared), *liaise* (an unnatural verb formed from *liaison*).
- Words that are particularly prone to misspelling; e.g. *desiccant, oriented, immitance*.
- Words for which there may be more than one "correct" spelling; e.g. *programer* or *programmer*.

Where two spellings of a word are in general use (e.g. *symposiums* and *symposia*), both are entered in the glossary and a preference is shown for one of them.

Definitions have been included when they will help you select the correct word for a given purpose, or to differentiate between similar words having different meanings. These definitions are intentionally brief and are intended only as a guide; for more comprehensive definitions, consult an authoritative dictionary (I recommend *Webster's New Collegiate Dictionary*).

All entries in the glossary are in lower case letters. Capital letters are used where capitals are recommended for a specific word, phrase, or abbreviation. Similarly, periods have been eliminated except where they form part of a specific entry. For example, the abbreviation for "inch" is *in.*, and the period that follows it has been inserted intentionally to distinguish it from the word "in".

Finally, think of this glossary as a guide rather than a collection of hard and fast rules. Our language is continually changing, so that what was fashionable yesterday may seem pedantic today and a cliché tomorrow. I expect that in some cases your views will differ from mine. Where they do, I hope that the comments and suggestions I offer will help you to choose the right expression, word, or abbreviation, and that you will be able to do so both consistently and logically.

GUIDELINES

GUIDELINE 1: COMBINING WORDS INTO COMPOUND TERMS

One of the biggest problems for technical writers is knowing whether multiword expressions should be compounded fully, joined by hyphens, or allowed to stand as two or more separate words. For example, should you write:

> cross check, cross-check or crosscheck?
> counter clockwise, counter-clockwise, or counterclockwise?
> change over, change-over, or changeover?

The tendency today is to compound a multiword expression into a single term. But this bare statement cannot be applied as a general rule because there are too many variations, some of which appear in the glossary.

Most multiword expressions are compound adjectives. When two words combine to form an adjective they are either joined by a hyphen or compounded to form one word. They are usually joined by a hyphen if they are formed from a noun-adjective expression:

Adjective + *Noun*	*As a Compound Adjective*
vacuum tube	vacuum-tube voltmeter
cathode ray	cathode-ray tube
high frequency	high-frequency oscillator

But when one of the combining words is a verb, they often combine into a one-word adjective. Under these conditions they normally will compound into a single-word noun:

Two Words	*As a Noun*	*As an Adjective*
lock out	lockout	lockout voltage
shake down	shakedown	shakedown test
cross over	crossover	crossover network

Three or more words that combine to form an adjective in most cases are joined by hyphens. For example, *lock test pulse* becomes *lock-test-pulse generator*. Occasionally, however, they are compounded into a single term, as in *counterelectromotive force*. Specific examples are in the glossary.

Obviously, these "rules" cannot be taken at full face value because there are occasions when they do not apply. A useful guide for doubtful combinations is the *Government Printing Office Style Manual*,[1] which contains a list of over 19,000 compound terms.

GUIDELINE 2: ABBREVIATING TECHNICAL AND NONTECHNICAL TERMS

You may abbreviate any term you like, and in any form you like, providing you indicate clearly to the reader how you intend to abbreviate it. This can be done by stating the term in full, then showing the abbreviation in parentheses to indicate that from now on you intend to use the abbreviation. Here is an example:

> Always spell out single digit numbers (sdn). The only time sdn are not spelled out is when they are being inserted in a column of figures.

When forming abbreviations of your own, take care not to form a new abbreviation when a standard one already exists. For example, if you did not know that there is a commonly accepted abbreviation for pound (weight), you might be

---

[1]*United States Government Printing Office Manual*, Superintendent of Documents, Washington, D.C., 20402.

tempted to use *pd*. This would not sit well with your readers, who might resent replacement of their old friend *lb*.

There are three basic rules that you should observe when forming abbreviations:

1. *Use lower case letters*, unless the abbreviation is formed from a person's name:

centimeter	—cm
kilogram	—kg
approximately	—approx
decibel	—dB (the *B* represents *Bell*, for Alexander Graham Bell)

2. *Omit all periods*, unless the abbreviation forms another word:

horsepower	—hp
cubic centimeter	—cm^3
cathode ray tube	—crt
foot/feet	—ft
inch	—in.
singular	—sing.

3. *Write plural abbreviations in the same form as the singular abbreviation:*

inches	—in.
pounds	—lb
kilograms	—kg
hours	—h *or* hr

There are, however, exceptions which have grown as part of our language. Through continued use these unnatural abbreviations have been generally accepted as the correct form. A few examples follow:

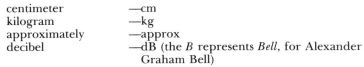

for example	—e.g.
(exempli gratia)	
that is	—i.e.
(id est)	
morning	—a.m.
(ante meridiem)	
afternoon	—p.m.
(post meridiem)	
inside diameter	—ID
number(s)	—No.

(There is a slowly growing trend to write these as eg and ie)

Other unusual abbreviations exist in each technical discipline. Although the glossary offers a reasonably comprehensive list of standard abbreviations and some technical abbreviations, for specific technical terms you may need to refer to a list of standard abbreviations compiled by one of the technical societies in your discipline.

GUIDELINE 3: CAPITALIZATION AND PUNCTUATION

Lower case letters should be used as much as possible in technical writing. Too many capital letters cause an untidy appearance and reduce the effectiveness of capital letters inserted for emphasis. The glossary therefore recommends lower case letters except (1) where usage has resulted in general adoption of a capitalized form (as in No., GCA, and the Brinell and Rockwell hardness tests), (2) for proper nouns, and (3) for expressions coined from proper nouns which have not yet been accepted as common terms. (Note, for example, that although we still use capital letters when referring to Doppler's principle, we write of doppler radar in lower case letters.)

The tendency today is to underpunctuate technical writing. This does not mean you may omit punctuation with a "when in doubt, leave it out" attitude. But you need not punctuate every clause and subclause. The intent is to obtain smooth reading; the criterion is to make the message clear. Hence, you should insert punctuation where it is needed rather than rigorously divide the information into tight little formal compartments.

GUIDELINE 4: WRITING NUMBERS IN NARRATIVE

The conventions that dictate whether a number should be written out or expressed in figures differ between ordinary writing and technical writing. In technical writing you are much more likely to express numbers in figures.

The following rules are intended mainly as a guide. They will apply most of the time, but there will be occasions when you will have to make a decision between two rules that conflict. Your decision should then be based on three criteria:

- Which method will be most readable?
- Which method will be simplest to type?
- Which method did I use previously under similar circumstances?

Good judgment and a desire to be consistent will help you select the best method each time.

The basic rule for writing numbers in technical narrative is:

- Spell out single-digit numbers (one to nine inclusive).
- Use figures for multiple-digit numbers (10 and above).

Exceptions to this rule are:

Always Use Figures:

1. When writing specific technical information, such as test results, dimensions, tolerances, temperatures, statistics, and quotations from tabular data.

2. When writing any number that precedes a unit of measurement: 3 in.; 7 lb; 121.5 MHz.

3. When writing a series of both large and small numbers in one passage: During the week ending May 27 we tested 7 transmitters, 49 receivers, and 38 power supplies.

4. When referring to section, chapter, page, figure (illustration), and table numbers: Chapter 7; Figure 4.

5. For numbers that contain fractions or decimals: 7¼; 7.25.

6. For percentages: 3% gain; 11% sales tax.

7. For years, dates, and times: At 3 p.m. on January 9, 1993.

8. For sums of money: $2000; $28.50; $20; 27 cents or $0.27 (preferred).

9. For ages of persons.

Always Spell Out:

1. Round numbers that are generalizations: about five hundred; approximately four thousand.

2. Fractions that stand alone: repairs were made in less than three quarters of an hour.

3. Numbers that start a sentence. (Better still, rewrite the sentence so that the number is not at the beginning.)

Additional rules are:

Spell out one of the numbers when two numbers are written consecutively and are not separated by punctuation: 36 fifty-watt amplifiers or thirty-six 50-watt amplifiers. (Generally, spell out whichever number will result in the simplest or shortest expression.)

Insert a zero before the decimal point of numbers less than one: 0.75; 0.0037.

Use decimals rather than fractions (they are easier to type), except when writing numbers that are customarily written as fractions.

Insert commas in large numbers containing five or more digits: 1,275,000; 27,291; 4056. (Insert a comma in four-digit numbers only when they appear as part of a column of numbers.)

Write numbers that denote position in a sequence as 1st, 2nd, 3rd, 4th, . . . 31st, . . . 42nd, . . . 103rd, . . . 124th, . . .

GUIDELINE 5: SUMMARY OF NUMERICAL PREFIXES AND SYMBOLS

The following table summarizes the numerical prefixes and abbreviations used in the glossary and conforms to the requirements for writing metric units.

Multiple/ Submultiple	Prefix	Symbol	Multiple/ Submultiple	Prefix	Symbol
10^{18}	exa	E	10^{-1}	deci	d
10^{15}	peta	P	10^{-2}	centi	c
10^{12}	tera	T	10^{-3}	milli	m
10^{9}	giga	G	10^{-6}	micro	μ
10^{6}	mega	M	10^{-9}	nano	n
10^{3}	kilo	k	10^{-12}	pico	p
10^{2}	hecto	h	10^{-15}	femto	f
10	deca	da	10^{-18}	atto	a

GUIDELINE 6: INTRODUCTION TO METRIC UNITS

For this fourth edition, the glossary includes terms and symbols within the International System of Units (SI). The trend toward worldwide adoption of metric units of measurement means that for some time both the basic inch/pound system and the metric (SI) system will be in use concurrently. The terms and symbols introduced here are those you are most likely to encounter in your technical reading or may want to use in your technical writing.

The acronym "SI" is used in all languages to represent the name "Système International d'Unités." Both the acronym and the name were adopted for universal usage in 1960 by the eleventh Conférence Générale des Poids et Mesures (CGPM), which is the international authority on metrication. Since then, some of the metric terms the conference established have crept into our language. For example, *Hertz*, the unit of frequency measurement, was first introduced as a replacement for *cycles per second* in the early 1960s; now it is used universally, both in the technical world and by the general public. Other terms already gaining recognition are:

	non-SI	SI
Temperature:	degrees Fahrenheit	degrees Celsius
Length:	miles, yards, feet, inches	kilometers, meters, millimeters
Weight:	tons, pounds, ounces	kilograms, grams, milligrams
Liquid volume:	gallons, quarts	kiloliters, liters

The glossary defines the basic SI units and shows how they should be written, either in full or abbreviated as a symbol, and their multiples and submultiples. It does not attempt to explain how the units are derived, nor does it include many of the less common units. The individual glossary entries are supplemented by the following guidelines, many of which are equally applicable to non-SI terms and units:

Guidelines.   SI symbols must be written, typed, or printed

1. in upright type, even if the surrounding type slopes or is in italic letters.
2. in lower case letters, except when the name of the unit is derived from a person's name (e.g. the symbol **F** for *farad* is derived from *Faraday*), in which case the first letter of the symbol is capitalized (e.g. **Wb** for *weber*).
3. with a space between the last numeral and the first letter of the symbol: **355 V, 27 km** (not 355V, 27km).
4. with no "s" added to a plural: **1 g, 236 kg**.
5. with no period after the symbol, unless it forms the last word in a sentence.
6. with no space between the multiple or submultiple symbol and the SI symbol: **3.6 kg, 150 mm, 960 kHz**.
7. with a solidus (oblique stroke:/) to represent the word "per": **cm/s** (centimeters per second). Only one solidus should be used in each expression.
8. with a dot at midletter height (·) to represent that the symbols are multiplied: **lm · s** (lumen seconds). (If your computer will not type a dot at midletter height, insert an asterisk in its place.)
9. always as a symbol, when a number is used with the SI unit (e.g. ". . . the tank holds 400 L"), but spelled out when no number is used with the unit (e.g. ". . . capacity is measured in liters [*not:* ". . . capacity is measured in L"]).

Spelling: "-er" or "-re"?   An anomaly exists concerning the spelling of **liter** and **meter**. SI stipulates the **-re** spelling (**litre** and **metre**), but in the United States we have traditionally used the **-er** spelling. I have continued to use the more familiar traditional spelling in *Technically—Write!*, but should caution you that the spelling may change as metric units of measurement become more widely established in the United States.

THE GLOSSARY

General abbreviations used throughout the glossary are:

abbr — abbreviate(d); abbreviation
adj — adjective
def — definition
lc — lower case
pl — plural
pref — prefer, preferred, preference
rec — recommend(ed)
SI — International System of Units

A

a; an use *an* before words that begin with a silent *h* or a vowel; use *a* when the *h* is sounded or if the vowel is sounded as *w* or *y; an hour* but *a hotel, an opinion* but *a European*

aberration

ab initio def: from the beginning

above- as a prefix, *above-* combines erratically: *aboveboard, above-cited, aboveground, above-mentioned*

abrasion

abscess

abscissa

absence

absolute abbr: **abs**

absorb(ent); adsorb(ent) *absorb* means to swallow up completely (as a sponge absorbs moisture); *adsorb* means to hold on the surface, as if by adhesion

abut; abutted; abutting; abutment

accelerate; accelerator; accelerometer

accept; except *accept* means to receive (normally willingly): *he accepted the company's offer of employment; except* generally means exclude: *the night crew completed all the repairs except rewiring of the control panel*

access; accessed; accessible

accessory; accessories abbr: **accy**

accidental(ly)

accommodate; accommodation

account abbr: **acct**

accumulate; accumulator

achieve means to conclude successfully, usually after considerable effort; avoid using *achieve* when the intended meaning is simply to reach or to get

acknowledgment

acquaint; acquaintance; acquainted; acquainting

acquiesce def: agree to

acquire; acquisition

acre; acreage

across not *accross*

actually omit this word: it is seldom necessary in technical writing

actuator

A.D. def: Anno Domini

adapt; adept; adopt *adapt* means to adjust to; *adept* means clever, proficient; *adopt* means to acquire and use

adapter; adaptor *adapter* pref

adaption; adaptation *adaption* pref

addendum pl: *addenda*

adhere to never use *adhere by*

ad hoc def: set up for one occasion

adjective (compound) two or more words that combine to form an adjective are either joined by a hyphen or compounded into a single word; see Guideline 1; abbr: *adj*

adsorb(ent) see **absorb**

advantageous

advice; advise use *advice* as a noun and *advise* as a verb: *the engineer's advice was sound; the technician advised the driver to take an alternative route;* spell: **adviser, advisable**

aerate

aerial see **antenna**

aero- a prefix meaning of the air; it combines to form one word; *aerodynamics, aeronautical;* in some instances it has been replaced by *air: airplane, aircraft*

aesthetic alternate spelling of **esthetic**

affect; effect *affect* is used only as a verb, never as a noun; it means to produce an effect upon or to influence (*the potential difference affects the transit time*); *effect* can be used either as a verb or as a noun; as a verb it means to cause or to accomplish (*to effect a change*); as a noun it means the consequences or result of an occurrence (as in *the detrimental effect upon the environment*), or it refers to property, such as *personal effects*

affidavit

aforementioned; aforesaid   avoid using these ambiguous expressions

after-   as a prefix, usually combines to form one word: *afteracceleration, afterburner, afterglow, afterheat, afterimage;* but: *after-hours*

agenda   although plural, *agenda* is generally treated as singular: *the agenda is complete*

aggravate   the correct definition of *aggravate* is to increase or intensify (worsen) a situation; try not to use it when the meaning is *annoy*

aggregate

aging; ageing   *aging* pref

agree to; agree with   to be correct, you should *agree to* a suggestion or proposal, but *agree with* another person

air-   as a prefix, normally combines to form one word; *airborne, airfield, airflow;* exceptions: *air-condition(ed) (er) (ing), air-cool(ed) (ing)*

air horsepower   abbr: **ahp**

airline; air line   an *airline* provides aviation services; an *air line* is a line or pipe that carries air

alkali; alkaline   pl: *alkalis* (pref) or *alkalies*

allege; alleged; alleging

allot; allotted; allotment

all ready; already   *all ready* means that all (everyone or everything) is ready; *already* means by this time: *the samples are all ready to be tested; the samples have already been tested*

all right   def: everything is satisfactory; never use *alright*

all together; altogether   *all together* means all collectively, as a group; *altogether* means completely, entirely: *the samples have been gathered all together, ready for testing; the samples are altogether useless*

allude; elude   *allude* means refer to; *elude* means avoid

almost   never contract *almost* to *most*; it is correct to write *most of the software has been tested*, but wrong to write *the software is most ready*

alphanumeric   def: in alphabetical, then numerical sequence

alternate; alternative   *alternate(ly)* means by turns: *the inspector alternated among the four construction sites; alternative(ly)* offers a choice between only two things: *the alternative was to return the samples;* however, it is becoming common for *alternative* to refer to more than two, as in *there are three alternatives*

alternating current   abbr: **AC**

alternator

altitude   abbr: **alt**

a.m.   def: before noon (*ante meridiem*)

amateur

ambient   abbr: **amb**

ambiguous; ambiguity

American Wire Gage   abbr: **AWG**

among; between   use *among* when referring to three or more items; use *between* when referring to only two; avoid using *amongst*

amount; number   use *amount* to refer to a general quantity: *the amount of time taken as sick leave has decreased;* use *number* to refer to items that can be counted: *the number of applicants to be interviewed was reduced to six*

ampere(s)   abbr: **A** (pref) or **amp;** other abbr: **kA, mA, μA, nA, pA, A/m** (ampere per minute)

ampere-hour(s)   abbr: **Ah** (pref) or **amp-hr** (more common)

amplitude modulation   abbr: **AM**

an   see **a**

anaemic; anemic   *anemic* pref

anaesthetic; anesthetic   *anesthetic* pref

analog; analogous

analyze(r)

AND-gate

and/or   avoid using this term; in most cases it can be replaced by either *and* or *or*

angstrom   abbr: **Å**

anion   def: negative ion

anneal; annealed; annealing

annihilate; annihilated; annihilation

antarctic   see **arctic**

ante-   a prefix that means before; combines to form one word: *antecedent, anteroom*

ante meridiem   def: before noon; abbr: **a.m.;** can also be written as *antemeridian*, but never as *antimeridian*

antenna   the proper plural in the technical sense in *antennas; antennae* should be limited to zoology; *antenna* has generally replaced the obsolescent *aerial*

anti-   a prefix meaning opposite or contradictory to; generally combines to form one word: *antiaircraft, antiastigmatism, anticapacitance, anticoincidence, antisymmetric;* if combining word starts with *i* or is a proper noun, insert a hyphen: *anti-icing, anti-American*

antimeridian   def: the opposite meridian (of longitude); e.g. the antimeridian of 96°C 30′W is 83° 30′E

anxious   although *anxious* really implies anxiety, current usage permits it to be used when the meaning is simply keen or eager

anybody; any body   *anybody* means any person; *any body* means any object: *anybody can attend; discard the batch if you find any body containing foreign matter*

anyone; any one   *anyone* means any person; *any one* means any single item: *you may take anyone with you; you may take any one of the samples*

anyway; any way   *anyway* means in any case or in any event; *any way* means in any manner; *the results may not be as good as you expect, but we want to see them anyway; the work may be done in any way you wish*

apparatus; apparatuses

apparent; apparently

appear(s); seem(s)   use *appears* to describe a condition that can be seen: *the equipment appears to be new;* use *seems* to describe a condition that cannot be seen: *the computer seems to be fast*

appendix   def: the part of a report that contains supporting data; pl: *appendixes* (pref) or *appendices*

appreciate   the pref meaning for *appreciate* is "to value," as in *we appreciate your help;* it is sometimes used (wrongly) as a synonym for *understand*

approximate(ly)   abbr: **approx;** but *about* is a better word

arbitrary

arc; arced; arcing

architect; architecture

arctic   capitalize when referring to a specific area: *beyond the Arctic Circle;* otherwise use lc letters: *in the arctic;* never omit the first *c*

area   the SI unit for area is the *hectare* (abbr: **ha**)

areal   def: having area

around   def: on all sides, surrounding, encircling; avoid confusing with *round*

arrester; arrestor   *arrester* pref

artwork

as   avoid using when the intended meaning is *since* or *because;* to write *he could not open his desk as he left his keys at home* is incorrect (replace *as* with *because*)

as per   avoid using this hackneyed expression, except in specifications

asphalt

assembly; assemblies   abbr: **assy**

assure   means to state with confidence that something has been or will be made certain; it is sometimes confused with *ensure* and *insure,* which it does not replace; see **ensure**

as well as   avoid using when the meaning is *and*

asymmetric; asymmetrical

asynchronous

athletic   not *atheletic*

atmosphere   abbr: **atm**

atomic weight   abbr: **at. wt**

attenuator

atto   def: 10^{-18}; abbr: **a**

audible; audibility

audio frequency   abbr: **af** (pref) or **a-f**

audiovisual   one word

audit; auditor

aural   def: that which is heard; avoid

confusing with *oral*, which means that which is spoken

author; writer   avoid referring to yourself as *the author* or *the writer;* use *I, me,* or *my*

authoritative

auto-   a prefix meaning self; combines to form one word: *autoalarm, autoconduction, autoionization, autoloading, autotransformer*

automatic frequency control   abbr: **afc** (pref) or **AFC**

automatic volume control   abbr: **avc** (pref) or **AVC**

auxiliary   abbr: **aux**

average   see **mean**

avocation   def: an interest or hobby; avoid confusing with *vocation*

ax; axe   **ax** pref; pl: *axes*

axis   the plural also is *axes*

azimuth   abbr: **az**

B

back-   as a prefix normally combines into one word: *backboard, backdate(d), backlog*

balance: remainder   use *balance* to describe a state of equilibrium (as in *discontinuous permafrost is frozen soil delicately balanced between the frozen and unfrozen state*), or as an accounting term; use *remainder* when the meaning is the rest of: *the remainder of the shipment will be delivered next week*

balk(ed)

ball bearing

bandwidth

bare; bear   *bare* means barren or exposed; *bear* means to withstand or to carry (or a wild animal)

barometer   abbr: **bar.**

barrel; barreled; barreling   the abbr of *barrel(s)* is **bbl**

barretter

barring   def: preventing, excepting

bases   this is the plural of both *base* and *basis*

basically

B.C.   def: Before Christ

because; for   use *because* when the clause it introduces identifies the cause of a result:

he could not open his desk because he left his keys at home; use *for* when the clause introduces something less tangible: *he failed to complete the project on schedule, for reasons he preferred not to divulge*

becquerel   def: a unit of activity of radionuclides (SI): abbr: **Bq;** other abbr: **PBq, TBq, GBq, kBq;** in SI, the *becquerel* replaces the *curie*

benefit; benefited; benefiting

beside, besides   *beside* means alongside, at the side of; *besides* means as well as

between   see *among*

bi-   a prefix meaning two or twice; combines to form one word: *biangular, bidirectional, bifilar, bilateral, bimetallic, bizonal*

biannual(ly); biennial(ly)   *biannual(ly)* means twice a year; *biennial(ly)* means every two years

bias; biased; biases; biasing

billion   def: 10^9

billion electron volts   although the pref abbr is **GeV,** *beV* and *bev* are more commonly used

Bill of Materials   abbr: **BOM**

bimonthly   def: every two months

binary

binaural

bioelectronics

bionics   def: application of biological techniques to electronic design

birdseye   (view)

biweekly   def: every two weeks

blow-   as a prefix combines to form one word: *blowhole, blowoff, blowout*

blueprint

blur; blurred; blurring; blurry

board feet   abbr: **fbm** (derived from *feet board measure*)

boiling point   abbr: **bp**

bonus; bonuses

borderline

brakedrum; brake lining; brakeshoe

brake horsepower; brake horsepower-hour   abbr: **bhp, bhp-hr**

brand-new

break-   when used as a prefix to form a

compound noun or adj, *break* combines into one word: *breakaway, breakdown, breakup;* in the verb form it retains its single-word identity: *it was time to break up the meeting*

bridging

Brinell hardness number   abbr: **Bhn**

British thermal unit   abbr: **Btu**

buoy; buoyant

bureaucracy; bureaucrat

bur(r)   *burr* pref

buses; bused; busing, bus bar

business; businesslike; businessperson   avoid using *businessman* or *businesswoman* unless referring to a specific male or female person

by-   as a prefix, *by-* normally combines to form one word: *bylaw, byline, bypass, byproduct*

byte   abbr: **kbyte** and **Mbyte** (pref), or **kb** and **Mb**

C

calendar; calender; colander   a *calendar* is the arrangement of the days in a year; *calender* is the finish on paper or cloth; a *colander* is a sieve

caliber

calk

cal(l)iper   *caliper* pref

calorie   abbr: **cal**

calorimeter; colorimeter   a *calorimeter* measures quantity of heat; *a colorimeter* measures color

cancel(l)ed; cancel(l)ing   *canceled, canceling* pref, but always *cancellation*

candela   def: unit of luminous intensity (replaces *candle*); abbr: **cd**; recommended abbr for candela per square foot and square meter are **cd/ft^2** and **cd/m^2**

candlepower; candlehour(s)   abrr: **cp, c-hr**

candoluminescence

cannot   one word pref; avoid using *can't* in technical writing

canvas; canvass   *canvas* is a coarse cloth used for tents; *canvass* means to solicit

capacitor

capacity for   never use *capacity to* or *capacity of*

capillary

capital letters   abbr: **caps.**; use capital letters as little as possible (see Guideline 3)

car-   as a prefix normally combines to form one word: *carload, carlot, carpool, carwash*

carburet(t)or   *carburetor* pref; a third, seldom used spelling is *carburetter;* also: **carburetion**

carcino-   as a prefix combines to form one word

case-   as a prefix normally combines to form one word: *casebook, caseharden, casework(er);* exceptions: *case history, case study*

cassette

caster; castor   use *caster* when the meaning is to swivel freely, and *castor* when referring to castor oil, etc.

catalog; cataloged; cataloging

catalyst; catalytic

category; categories; categorical

cathode-ray tube   abbr: **crt** (pref) or **CRT** (commonly used)

cation   def: positive ion

-ceed; -cede; -sede   only one word ends in *-sede: supersede;* only three words end in *-ceed; exceed, proceed, succeed;* all others end in *-cede:* e.g. *precede, concede*

centerline   abbr: **₵** (pref) or **CL**

Celsius   abbr: **C;** see **temperature**

census

center-to-center   abbr: **c-c**

centi-   def: 10^{-2}; as a prefix combines to form one word: *centiampere, centigram;* abbr: **c;** other abbr:

centigram	**cg**
centiliter	**cL**
centimeter	**cm**
centimeter-gram-second	**cgs**
centimeter per second	**cm/s**
square centimeter	**cm^2**

centigrade   abbr: **C;** in SI, **centigrade** has been replaced by **Celsius;** see **temperature**

centri-   a prefix meaning center; combines to form one word: *centrifugal, centripetal*

chairperson   avoid using *chairman* or *chairwoman*

chamfer

changeable; changeover

channel; channeled; channeling

chapter   abbr: **chap.**

chargeable

chassis   both singular and plural are spelled the same

check-   as a prefix combines to form one word: *checklist, checkpoint, checkup* (noun or adjective)

checksum   def: a term used in computer technology

chisel; chiseled; chiseling

chrominance

cipher

circuit; circuitous; circuit breaker

cite   def: to quote; see **site**

climate   avoid confusing *climate* with *weather; climate* is the average type of weather, determined over a number of years, experienced at a particular place; *weather* is the state of the atmospheric conditions at a specific place at a specific time

clockwise(turn)   abbr: **CW**

co-   as a prefix, *co-* generally means jointly or together; it usually combines to form one word: *coexist, coequal, cooperate, coordinate, coplanar* (*co-worker* is an exception); it is also used as the abbr for *complement of* (an arc or angle): *codeclination, colatitude*

coalesce; coalescent

coarse; course   *coarse* means rough in texture or of poor quality; *course* implies movement or passage of time: *a coarse granular material; the technical writing course*

coaxial   abbr: **coax.**

coefficient   abbr: **coef**

coerce; coercion

collaborate; collaborator   avoid writing *collaborate together* (delete *together*)

collapsible

collateral

cologarithm   abbr: **colog**

colon   when a colon is inserted in the middle of a sentence to introduce an example or short statement, the first word following

the colon is not capitalized; a colon rather than a semicolon should be used at the end of a sentence to introduce subparagraphs that follow; a hyphen should not be inserted after the colon

colorimeter   see **calorimeter**

column   abbr: **col.**

combustible

comma   a comma normally need not be used immediately before *and, but,* and *or,* but may be inserted if to do so will increase understanding or avoid ambiguity; also see Guideline 3.

commence   in technical writing, replace *commence* with the more direct *begin* or *start*

commit; commitment; committed; committing

committee

communicate   it is vague to write *I communicated the results to the client;* use a clearer verb: *I mailed . . . , wrote . . . , faxed . . . , telephoned*

compare; comparable; comparison; comparative   use *compared to* when suggesting a general likeness; use *compared with* when making a definite comparison

compatible; compatibility

complement; compliment   *complement* means the balance required to make up a full quantity or a complete set; to *compliment* means to praise; *in a right angle, the complement of 60°C is 30°; Mr. Perchanski complimented Janet Rudman for writing a good report*

composed of; comprising; consists of   all three terms mean "made up of" (specific items); if any one of these terms is followed by a list of items, it implies that the list is complete; if the list is not complete, the term should be replaced by *includes* or *including*

compound terms   two or more words that combine to form a compound term are joined by a hyphen or are written as one word, depending on accepted usage and whether they form a verb, noun, or adjective; the trend is toward one-word compounds; see Guideline 1.

comprise; comprised; comprising   to write

comprised of is incorrect, because the verb *comprise* includes the preposition *of*

concur; concurred; concurrent; concurring

condenser

conductor

confer; conferred; conferring; conference; conferee

conform use *conform to* when the meaning is to abide by; use *conform with* when the meaning is to agree with

conscience

conscious

consensus means a general agreement of opinion; hence to write *concensus of opinion* is incorrect; e.g. write: *the consensus was that a further series of tests would be necessary*

consistent with never *consistent of*

consists of; consisting of see **composed of**

contact *contact* should not be used as a verb when *write, visit, speak, fax,* or *telephone* better describes the action to be taken

contend; contention; contentious

continual; continuous *continual(ly)* means happens frequently but not all the time: *the generator is continually being overloaded* (is frequently overloaded); *continous(ly)* means goes on and on without stopping: *the noise level is continuously at or above 100 dB* (it never drops below 100dB)

continue(d) abbr: **cont**

continuous wave abbr: **cw**

contra- as a prefix normally combines into a single word

contrast when used as a verb, *contrast* is followed by *with*; when used as a noun, it may be followed by either *to* or *with* (*with* pref)

control; controlled; controlling; controller

conversant with never *conversant of*

converter; convertible

conveyor

cooperate

coordinate; coordinator

copyright not *copywrite*

corollary

correlate

correspond *to correspond to* suggests a

resemblance; *to correspond with* means to communicate in writing

corroborate

corrode; corrodible; corrosive

cosecant abbr: **csc** (pref) or **cosec**

cosine abbr: **cos**

cotangent abbr: **cot**

coulomb def: a quantity of electricity, electric charge (SI); abbr: **C**; other abbr: **kC, mC, μC, nC, pC, C/m^2**

council; counsel a *council* is a group of people; a *counsel* is a lawyer; *to counsel* is to give advice; also: **counseled; counseling; counselor**

counter- a prefix meaning opposite or reciprocal; combines to form one word: *counteract, counterbalance, counterflow, counterweight*

counterclockwise(turn) abbr: **CCW**

counterelectromotive force abbr: **cemf;** also known as *back emf*

counts per minute abbr: **cpm**

course see **coarse**

criteria; criterion the singular is *criterion*, the plural is *criteria*: e.g. *one criterion; seven criteria;* there is a trend (which is *not* recommended) to use *criteria* as singular and *criterions* as plural

criticism; criticize, critique

cross- as a prefix, combines erratically: *cross-check, crosshatch, crosstalk, cross-purpose, cross section*

cross-refer(ence) abbr: **x-ref**

cryogenic

cryptic

crystal abbr: **xtal**

crystalline; crystallize

cubic abbr: **cu** or ^{3}; other abbr:

cubic centimeter(s)	**cm^3** (pref); **cc**
cubic decimeter(s)	**dm^3**
cubic foot (feet)	**ft^3** (pref); **cu ft**
cubic feet per minute	**cfm** (pref); **ft^3/min**
cubic feet per second	**cfs** (pref); **ft^3/sec**
cubic inch(es)	**in.^{3}** (pref); **cu in.**
cubic meter(s)	**m^3**
cubic millimeter(s)	**mm^3**
cubic yard(s)	**yd^3** (pref); **cu yd**

curie abbr: **Ci;** other abbr: **mCi, μCi;** in SI the *curie* is replaced by the *becquerel*

curriculum pl: *curriculums* (pref) or *curricula*

cursor

cycles per minute abbr: **cpm**

cycles per second abbr: **cps;** although still occasionally used, this term has been replaced by **hertz**

cylinder; cylindrical abbr: **cyl**

D

dais def: a platform

daraf def: the unit of elastance

data def: gathered facts; although *data* is plural (derived from the singular *datum*, which is rarely used), it is more often used as a singular noun: *when all the data has been received, the analysis will begin*

dateline

date(s) avoid vague statements such as "last month" and "next year" because they soon become indefinite; write as a specific date, using day (in numerals), month (spelled out), and year (in numerals): *January 27, 1992* or *27 January 1992* (the latter form has no punctuation); to abbreviate, reduce month to first three letters and year to last two digits: *Jan 27, 92* or *27 Jan 92;* for SI, use numerals only, in this order: year, month, day: *1992 01 27*

day- as a prefix, generally combines to form one word: *daybook, daylight, daytime, daywork*

days days of the week are capitalized: *Monday, Tuesday*

de- a prefix that generally combines to form one word: *deaccentuate, deactivate, decentralize, decode, deemphasize, deenergize, deice, derate, destagger;* an exception is *de-ionize*

dead- as a prefix combines erratically: *deadbeat, dead center, dead end, deadline, deadweight, deadwood*

debug; debugged; debugging

decelerate def: to slow down; never use *deaccelerate*

decibel abbr: **dB;** the abbr for decibel referred to 1 mW is **dBm**

decimals for values less than unity (one), place a zero before the decimal point: *0.17, 0.0017*

decimate def: to reduce by one tenth; can also mean to destroy much of

decimeter abbr: **dm**

declination abbr: **dec**

deductible

defective; deficient *defective* means unserviceable or damaged (generally lacking in quality); *deficient* means lacking in quantity (it is derived from *deficit*), and in the military sense incomplete: *a short circuit resulted in a defective transmitter; the installation was completed on schedule except for a deficient rotary coupler which will not be delivered until June 10*

defend; defendant; defense

defer; deferred; deferring; deferrable; deference

definite; definitive *definite* means exact, precise; *definitive* means conclusive, fully evolved; e.g. *a definite price* is a firm price; *a definitive statement* concerns a topic that has been thoroughly considered and evaluated

degree(s) abbr: **deg** (pref in narrative) or ° (following numerals); see **temperature**

demarcation

demi- a little-used prefix meaning half (generally replaced by *semi-*); combines to form one word: *demivolt*

demonstrate; demonstrator; demonstrable not *demonstratable*

depend; dependence; dependent; dependable

deprecate; depreciate *deprecate* means to disapprove of; *depreciate* means to reduce the value of: *the use of "as per" in technical writing is deprecated; the vehicles depreciated by 50% the first year and 20% the second year*

depth

desiccant; desiccate(d)

desirable

desktop the abbr for **desktop publishing** is **dtp** (pref) or **DTP**

despite def: in spite of; *despite* should never be followed by the word *of*

deter; deterred; deterrence; deterrent; deterring

deteriorate

develop not *develope*

device; devise the noun is *device*, the verb is

devise: a unique device; he devised a new software program

dext(e)rous   *dexterous* pref

diagram; diagramed; diagraming; diagramatic

dial; dialed; dialing

dialog(ue)   *dialogue* pref

diaphragm

diazo

didn't   never use this contraction in technical writing; use **did not**

dielectric

diesel; diesel-electric

dietitian

differ   use *differ from* to demonstrate a difference; use *differ with* to describe a difference of opinion

different   *different from* is preferred; *different to* is sometimes used; never use *different than*

diffraction

diffusion; diffusible

dilemma   means to be faced with a choice between two unhappy alternatives; should not be used as a synonym for *difficulty*

diplex; duplex   *diplex operation* means the simultaneous transmission of two signals using a single feature, e.g. an antenna; *duplex operation* means that both ends can transmit and receive simultaneously

direct current   abbr: **dc**

directly   def: immediately; do not use when the meaning is a soon as

disassemble; dissemble   *disassemble* means to take apart; *dissemble* means to conceal facts or put on a false appearance

disassociate   see *dissociate*

disc; disk   both are correct and commonly used; **disk** pref

discernible

discreet; discrete   *discreet* means prudent or discerning: *his answer was discreet; discrete* means individually distinctive and separate: *discrete channels; discretion* is formed from *discreet*, not from *discrete*

disinterested; uninterested   *disinterested* means unbiased, impartial; *uninterested* means not interested

dispatch; despatch   *dispatch* pref

disseminate

dissimilar

dissipate

dissociate; disassociate   *dissociate* pref

distil(l)   *distil* pref; but always *distilled, distillate, distillation*

distribute; distributor

don't; doesn't   such contractions should not appear in technical writing

donut; doughnut   for electronics/nucleonics, use *donut*

doppler   capitalize only when referring to the Doppler principle

double-   as a prefix combines erratically: *double-barrelled, doublecheck, double-duty, double entry, doublefaced*

down-   as a prefix combines into one word: *downgrade, downrange, downtime, downwind*

dozen   abbr: **doz**

drafting; draftsperson   avoid writing *draftsman* or *draftswoman*

drawing(s)   abbr: **dwg**

drier; dryer   the adjective is always *drier;* the pref noun is *dryer: this material is drier; place the others back in the dryer;* also: **dryly**

drop; droppable; dropped; dropping

due to   an overused expression; *because of* pref

duo-   a prefix meaning two; combines to form one word: *duocone, duodiode, duophase*

duplex   see **diplex**

duplicator

dutiable; duty-free

E

each   abbr: **ea**

east   capitalize only if *east* is part of the name of a specific location: *East Africa;* otherwise use lc letters: *the east coast of the U.S.;* abbr: **E;** the abbr for *east-west* (control, movement) is **E-W;** *eastbound* and *eastward* are written as one word

eccentric; eccentricity

echo; echoes

economic; economical   use *economical* to

describe economy (of funds, effort, time); use *economic* when writing about economics: *an economical operation* (it did not cost much); *an economic disaster* (it will have a major effect on the economy)

effect see **affect**

efficacy; efficiency *efficacy* means effectiveness, ability to do the job intended; *efficiency* is a measurement of capability, the ratio of work done to energy expended; *we hired a consultant to assess the efficacy of our training methods; the powerhouse is to have a high-efficiency boiler*

e.g. def: *for example;* avoid confusing with **i.e.;** may also be abbr **eg**

eighth

electric(al) if in doubt, use *electric;* generally, *electric* means produces or carries electricity, whereas *electrical* means related to the generation or carrying of electricity; abbr: **elec**

electro- a prefix generally meaning pertaining to electricity; it normally combines to form one word: *electroacoustic, electroanalysis, electrodeposition, electromechanical, electroplate;* if the combining word starts with *o,* insert a hyphen: *electro-optics, electro-osmosis*

electromagnetic units abbr: **emu**

electromotive force abbr: **emf**

electronic(s) use *electronic* as an adjective, *electronics* as a noun: *electronic countermeasures; your career in electronics*

electron volt(s) abbr: **eV** (pref) or **ev**

electrostatic units abbr: **esu**

elevation abbr: **el**

elicit; illicit *elicit* means to obtain or identify; *illicit* means illegal

eligible; eligibility

ellipse

embarrass; embarrassed; embarrassing; embarrassment

embedded

emigrate; immigrate *emigrate* means to go away from; *immigrate* means to come into

emit, emitter, emittance; emission; emissivity

enamel; enameled; enameling

encase; incase *encase* pref

encipher

enclose; inclose *enclose* pref; *inclose* is used mainly as a legal term

endorse, indorse *endorse* pref

enforce not *inforce*

engineer; engineered; engineering

enquire; inquire *inquire* pref

enrol; enroll both are correct, but *enroll* pref; universal usage prefers *ll* for **enrolled** and **enrolling;** a single *l* may be used for *enrolment,* but **enrollment** is more common

en route def: on the road, on the way; never use *on route*

ensure; insure; assure use *ensure* (or *insure*) when the meaning is to make certain of: *use the new oscilloscope to ensure accurate calibration;* use *insure* when the meaning is to protect against financial loss: *we insured all our drivers;* use *assure* when the meaning is to state with confidence that something has been or will be made certain: *he assured the meeting that production would increase by 8%*

entrepreneur; entrepreneurial

entrust; intrust *entrust* pref

envelop; envelope *envelop* is a verb which means to surround or cover completely; *envelope* is a noun that means a wrapper or a covering

environment; environmental; environmentally

EPROM def: abbr for erasable programmable read-only memory; can also be abbr as **eprom**

equal *equality* and *equalize* have only one *l; equally* always has *ll; equaled* and *equaling* preferably have only one *l,* but sometimes are seen with *ll*

equi- a prefix that means equality; combines to form one word: *equiphase, equipotential, equisignal*

equilibrium; equilibriums

equip; equipped; equipping, equipment

equivalent abbr: **equiv**

erase; erasable

errata although *errata* is plural (from the singular *erratum,* which is seldom used), it can be used as a singular or plural noun: both *the errata are ready* and *the errata is complete* are acceptable

erratic

erroneous

escalator

especially; specially *specially* pref when it refers to an adjective, as in *a specially trained crew; especially* should introduce a phrase, as in *they were well trained, especially the computer technicians*

esthetic; aesthetic *esthetic* pref

et al def: and others

et cetera abbr: **etc**; def: and so forth, and so on; use with care in technical writing: *etc* can create an impression of vagueness or unsureness; *the transmitters, etc, were tested* is much less definite than either *the transmitter, modulator, and power supply were tested* or, if to restate all the equipment is too repetitious, *the transmitting equipment was tested*

everybody; every body *everybody* means every person, or all the persons; *every body* means every single body: *everybody was present; every body was examined for gunpowder scars*

everyone; every one *everyone* means every person, or all the persons; *every one* means every single item: *everyone is insured; every one had to be tested in a saline solution*

exa def: 10^{18}; abbr: **E**

exaggerate

exceed

excel; excelled; excellent; excelling

except def: to exclude; see **accept**

exhaust

exhibit; exhibitor

exhort

exorbitant

expedite; expediter (pref) or **expeditor**

explicit; implicit *explicit* means clearly stated, exact; *implicit* means implied (the meaning has to be inferred from the words): *the supervisor gave explicit instructions* (they were clear); *that the manager was angry was implicit in the words he used*

extemporaneous

extracurricular

extraordinary

extremely high frequency abbr: **ehf**

F

face- as a prefix normally combines to form one word: *facedown, facelift, faceplate;* exceptions: *face-saver, face-saving*

Fahrenheit abbr: **F**; see **temperature**

fail-safe

fallout one word in the noun form

familiarize; familiarization

farad def: a unit of electric capacitance; abbr: **F**; other abbr: **μF, nF, pF**

farfetched; far-out; far-reaching; farseeing; farsighted

farther; further *farther* means greater distance: *he traveled farther than the other technicians; further* means a continuation of (as an adjective) or to advance (as a verb): *the promotion was a further step in her career plan*, and *to further his education, he took a part-time course in industrial drafting*

fascinate; fascination

fasten; fastener

faultfinder; faultfinding

feasible; feasibility

February

feet; foot abbr: **ft**; other abbr:

feet board measure (board feet)	**fbm**
feet per minute	**fpm**
feet per second	**fps**
foot-candle(s)	**fc** (pref); **ft-c**
foot-pound(s)	**fp** (pref); **ft-lb**
foot-pound-second (system)	**fps** system

femto def: 10^{-15}; abbr: **f**; other abbr:

femtoampere(s)	**fA**
femtovolt(s)	**fV**

ferri- a prefix meaning contains iron in the ferric state; combines to form one word: *ferricyanide, ferrimagnetic*

ferro- a prefix meaning contains iron in the ferrous state; combines to form one word: *ferroelectric, ferromagnetic, ferrometer*

ferrule; ferule   a *ferrule* is a metal cap or lid; a *ferule* is a ruler

fewer; less   use *fewer* to refer to items that can be counted: *fewer technicians than we predicted have been assigned to the project*; use *less* to refer to general quantities: *there was less water available than predicted*

fiber; fibrous; Fiberglas   *Fiberglas* is a trade name

field-   as a prefix normally does not combine into a single word or hyphenated form: *field glasses, field test, field trip;* but: *fieldwork(er)*

figure numbers   in text, spell out the word *Figure* in full, or abbr it to **Fig.**: *the circuit diagram in Figure 26* and *for details, see Fig. 7;* use the abbreviated form beneath an illustration; always use numerals for the figure number

final; finally; finalize

fire-   as a prefix combines eratically: *firearm fire alarm, firebreak, fire drill, fire escape, fire extinguisher, firefighter, firepower, fireproof, fire sale, fire wall*

first   to write *the first two . . .* (or three, etc) is better than to write *the two first . . . ;* never use *firstly;* as a prefix, *first-* combines erratically: *first-class, firsthand, first-rate*

fix   in technical usage, *fix* means to firm up or establish as a permanent fact; avoid using it when the meaning is to repair

flameout; flameproof

flammable   def: easily ignited; see **inflammable**

flexible

flight   usually combines to form two words: *flight control, flight deck, flight plan*

flip-flop

flotation   this is the correct spelling for describing an item that floats: *flotation gear*

flow chart

fluid   abbr: **fl**; the abbr for fluid ounces is **fl oz**

fluorescence; fluorescent

fluorine; fluoridation

focus; focused; focusing; focuses   pl: *focuses* (pref) or *foci*

foot-   as a prefix normally combines into one word: *footbridge, footcandle, foothold, footnote, footwork*; also see **feet**

for   see **because**

forceful; forcible   use *forceful* to describe a person's character; use *forcible* to describe physical force

fore-   def: that which goes before; as a prefix normally combines into one word: *foreclose, forefront, foreground, foreknowledge, foremost, foresee, forestall, forethought, forewarn*

forecast   this spelling applies to both present and past tenses

forego; forgo   *forego* and *foregoing* mean to go before; *forgo* means to go without

foreman   avoid using in a general sense, except when referring to a person specifically, as in *John Hayward, the foreman;* never use *forewoman* (a better choice is *supervisor*)

foresee

foreword; forward   a *foreword* is a preface or preamble to a book; *forward* means onward: *the scope is defined in the foreword to the book; he requested that we bring the meeting date forward*

for example   abbr: **e.g.** (pref) or **eg**

former; first   use *former* to refer to the first of only two things; use *first* if there are more than two

formula   pl: *formulas* (pref) or *formulae*

forty   def: 40; it is not spelled *fourty*

fourth   def: 4th; it is not spelled forth

fractions   when writing fractions that are less than unity, spell them out in descriptive narrative but use figures for technical details: *by the end of the heat run, nine tenths of the installation had been completed; a flat case 15.4 mm square by 0.5 mm deep;* use decimals rather than fractions, except when a quantity is normally stated as a fraction (such as ³/₈ *in. plywood*)

free-   as a prefix normally combines into one word: *freehand, freehold, freestanding, freeway, freewheel*

free from   use *free from* rather than *free of:* *he is free from prejudice*

free on board   abbr: **fob** (pref), **f.o.b.** (commonly used), or **FOB**

frequency   abbr: **freq**

frequency modulation   abbr: **FM**

fulfil(l)   the pref spelling is *fulfill, fulfilled, fulfilling,* and *fulfillment; fulfil* and *fulfilment* can also be spelled with a single *l*

funnel; funneled; funneling

further   see **farther**

fuse   as a verb, means join together or weld; as a noun, means a circuit protection device

fuselage

fuze   def: a detonation initiation device

G

gage; gauge   both spellings are correct; gage is recommended because it is less likely to be misspelled; *gaging* and *gauging* do not retain the *e*

gallon   gallons differ between U.S. (231 in.^{3}; 3.785 dm^3) and Britain (277.42 in.^{3}; 4.546 dm^3); abbr: **gal;** other abbr:

gallons per day	**gpd**
gallons per hour	**gph**
gallons per minute	**gpm**
gallons per second	**gps**

gang; ganged; ganging

gas; gases; gassed; gassing; gaseous; gassy

gauge   see **gage**

gearbox; gearshift

geiger (counter)

gelatin(e)   *gelatin* pref

geo-   a prefix meaning of the earth; combines to form one word: *geocentric, geodesic, geomagnetic, geophysics*

giga   def: 10^9; abbr: **G;** other abbr:

gigabecquerel(s)	**GBq**
gigahertz	**GHz**
gigajoule(s)	**GJ**
gigaohm(s)	**GΩ; Gohm**
gigapascal(s)	**GPa**
gigavolt(s)	**GV**

gimbal

glue; gluing; gluey

glycerin(e)   *glycerin* pref

gotten   never use this expression in technical writing; use *have got* or simply *have*

government   capitalize when referring to a specific government either directly or by implication; use lc if the meaning is government generally: *the U.S. Government; the Government specifications; no government would sanction such restrictions*

gram   abbr: **g;** abbr for gram-calorie is **g-cal**

grammar; grammatical(ly)

grateful   not *greatful*

gray   def; absorbed dose of ionizing radiation (SI); abbr: **Gy;** other abbr: **mGy, μGy;** in SI, the *gray* replaces the *rad*

Greenwich mean time   abbr: **GMT**

grill(e)   when the meaning is a loudspeaker covering, or a grating, *grille* pref

ground (electrical)   abbr: **gnd**

ground crew; ground floor

guage   wrongly spelled; the correct spelling is *gauge* or *gage* (pref)

guarantee   never *guaranty*

guesstimate

guideline; guidelines

gyroscope   abbr: **gyro**

H

half; halved; halves; halving   as a prefix, *half* combines erratically; some common compounds are: *half-hour(ly), half-life, half-monthly, halftone, half-wave;* for others, consult your dictionary

hand-   as a prefix normally combines to form one word: *handbill, handbook, handful, handfuls, handhold, handpicked, handset, handshake*

hangar; hanger   a *hangar* is a large building for housing aircraft; a *hanger* is a supporting bracket

hard-   as a prefix normally combines into one word: *hardbound, hardhanded, hardhat, hardware;* exceptions: *hard-earned, hard-hitting*

haversine   abbr: **hav**

H-beam

head-   as a prefix normally combines into one word: *headfirst, headquarters, headset, headstart, headway*

heat-   as a prefix, *heat* combines erratically; some typical compounds are: *heat-resistant,* heat-run, heatsink, heat-treat; for others, consult your dictionary

heavy-duty

hectare def: a large unit of area, used in surveying and agriculture; in SI, *hectare* replaces *acre;* abbr: **ha**

height (not *heighth*) abbr: **ht**; also **heighten; heightfinder; heightfinding**

helix pl: *helices* (pref) or *helixes*

hemi- a prefix meaning half; combines to form one word: *hemisphere, hemitropic*

henry def: a unit of inductance; abbr: **H**; other abbr: **mH, μH, nH, pH**

here- whenever possible avoid using *here-* words that sound like legal terms, such as *hereby, herein, hereinafter, hereof;* they make a writer sound pompous; as a prefix, *here-* combines to form one word

hertz def: a unit of frequency measurement (similar to *cycles per second,* which it replaces); abbr: **Hz**; other abbr; **THz, GHz, MHz, kHz**

heterodyne

heterogeneous; homogeneous *heterogeneous* means of the opposite kind; *homogeneous* means of the same kind

high- as a prefix either combines into one word or the two words are joined by a hyphen: *highhanded, highlight, high-power* (adj), *high-priced, high-speed*

high frequency abbr: **hf**

high-pressure (as an adjective) abbr: **h-p**

high voltage abbr: **hv** (pref) or **HV**

hinge; hinged; hinging

homogeneous see **heterogeneous**

horizontal abbr: **hor**

horsepower abbr: **hp**; the abbr for horsepower-hour is **hp-hr**

hour(s) abbr: **hr** or **h** (SI)

hundred abbr: **C**

hundredweight def: 112 lb; abbr: **cwt**

hybrid

hydro- a prefix meaning of water; combines to form one word: *hydroacoustic, hydroelectric, hydromagnetic, hydrometer*

hyper- a prefix meaning over; combines to form one word: *hyperacidity, hypercritical*

hyperbola the plural is *hyperbolas* (pref) or *hyperbolae*

hyperbole def: an exaggerated statement

hyperbolic cosine, sine, tangent abbr: **cosh, sinh, tanh**

hyphen in compound terms you may omit hyphens unless they need to be inserted to avoid ambiguity or to conform to accepted usage e.g. *preemptive* is preferred without a hyphen, but *photo-offset* and *re-cover* (when the meaning is *to cover again*) both require one; refer to individual entries and Guideline 1

hypothesis pl: *hypotheses*

I

I-beam

ibid def: Latin abbr *ibidem*, meaning in the same place; used in footnoting, but becoming obsolete

ID card

i.e. def: *that is;* avoid confusing with **e.g.;** may also be abbr **ie**

if and when aviod using this expression; use either *if* or *when*

ignition abbr: **ign**

ill- as a prefix combines into a hyphened expression: *ill-advised, ill-defined, ill-timed*

illegible

im- see **in-**

imbalance this term should be restricted for use in accounting and medical terminology; use *unbalance* in other technical fields

immalleable

immaterial

immeasurable

immigrate see **emigrate**

immittance

immovable

impasse

impel; impelled; impelling; impeller

imperceptible

impermeable

impinge; impinging

imply; infer speakers and writers can *imply*

something; listeners and readers *infer* from what they hear or read: *in his closing remarks, Mr. Smith implied that further studies were in order; the technician inferred from the report that his work was better than expected*

impracticable; impractical *impracticable* means not feasible; *impractical* means not practical; a less-preferred alternative for impractical is *unpractical*

in; into *in* is a passive word; *into* implies action: *ride in the car; step into the car*

in-: im-; un- all three prefixes mean not; all combine to form one word; *ineligible, impossible, unintelligible;* if you are not sure whether you should use in-, im-, or un-, use *not*

inaccessible

inaccuracy

inadmissible

inadvertent

inadvisable, unadvisable *inadvisable* pref

inasmuch as a better word is *since*

inaudible

incalculable; not *incalculatable*

incandescence; incandescent

incase *encase* pref

inch(es) abbr: **in.;** other abbr:
inches per second **ips** (pref); **in./s**
inch-pound(s) **in.-lb**

incidentally in most cases the word *incidentally* is unnecessary

inclose use **enclose**

includes; including abbr: **incl;** when followed by a list of items, *includes* implies that the list is not complete; if the list is complete, use *comprises* or *consists of*

incomparable

incompatible

incumbent def: the person currently holding a particular position

incur; incurred; incurring

index pl: *indexes* pref, except in mathematics (where *indices* is common)

indicated horsepower abbr: **ihp;** the abbr for indicated horsepower-hour is **ihp-hr**

indifferent to never use *indifferent of*

indiscreet; indiscrete *indiscreet* means

imprudent; *indiscrete* means not divided into separate parts

indispensable

indorse *endorse* pref

industrywide

ineligible

inequitable

inessential; unessential both are correct; *unessential* pref

inexhaustible

inexplicable

infallible

infer; inferred; inferring; inference also see **imply**

inflammable def: easily ignited (derived from *inflame*); **flammable** is a better word: it prevents readers from mistakenly thinking the *in* of *inflammable* means not

inflexible

infrared

ingenious; ingenuous *ingenious* means clever, innovative; *ingenuous* means innocent, naive; *ingenuity* is a noun derived from *ingenious*

in-house

inoculate

inoperable not *inoperatable*

inquire; enquire *inquire* pref; also *inquiry*

insanitary; unsanitary both are correct; *insanitary* pref

inseparable

inside diameter abbr: **ID**

in situ def: in the normal position

insofar as

instal(l) *install, installed, installer, installing, installation,* and *installment* pref; a single *l* is acceptable for *instal and instalment*

instantaneous

instrument

insure the pref def is to protect against financial loss; can also mean make certain of; see **ensure**

integer; integral; integrate; integrator

intelligence quotient abbr: **IQ**

intelligible

inter-   a prefix meaning among or between; normally combines to form one word: *interact, intercarrier, interdigital, interface, intermodulation, interoffice*

intermediate-pressure   (as an adjective) abbr: **i-p**

intermittent

internal   abbr: **int**

interrupt

into   see **in**

intra-   a prefix meaning within; normally combines to form one word: *intranuclear;* if combining word starts with *a,* insert a hyphen: *intra-atomic*

intractable

intrigue; intrigued; intriguing

intrust   *entrust* pref

IQ

irrational

irregardless   never use this expression; use *regardless*

irrelevant   frequently misspelled as *irrevelant*

irreversible

iso-   a prefix meaning the same, of equal size; normally combines to form one word: *isoelectronic, isometric, isotropic;* if combining word starts with *o,* insert a hyphen: *iso-octane*

its; it's   *its* means belonging to; *it's* is an abbr for it is: *the transmitter and its modulator; if the fault is not in the remote equipment, then it's most likely in master control;* in technical writing *it's* should seldom be used: replace with *it is*

J

jobholder; job hopper; job seeker; job lot

joule   def: a unit of energy, work, or quantity of heat (SI); abbr: **J**; other abbr: **TJ, GJ, MJ, kJ, mJ, J/m³, J/K** (joule(s) per kelvin), **J/kg, J/mol** (joule(s) per mole)

journey; journeys

judg(e)ment   *judgment* pref

judicial; judicious   *judicial* means related to the law; *judicious* means sensible, discerning

juxtaposition

K

kelvin   def: the SI unit for thermodynamic temperature; abbr: **K**

key-   as a prefix normally combines to form one word: *keyboard, keypunch, keying, keystroke;* but *key word*

kilo   def: 10^3; abbr: **k;** other abbr:

kiloampere(s)	**kA**
kilobecquerel(s)	**kBq**
kilobyte(s)	**kbyte** (pref) or **kb**
kilocalorie(s)	**kcal**
kilocoulomb(s)	**kC**
kilogram(s), (see **kilogram**)	**kg**
kilohertz	**kHz**
kilohm(s)	**kΩ; kohm**
kilojoule(s)	**kJ**
kiloliter(s)	**kL**
kilometer(s)	**km**
kilometers per hour	**km/h**
kilomole(s)	**kmol**
kilonewton(s)	**kN**
kilopascal(s)	**kPa**
kilosecond(s)	**ks** (pref); **ksec**
kilosiemens	**kS**
kilovolt(s)	**kV**
kilovolt-ampere(s)	**kVA**
kilovolt-ampere(s), reactive	**KVAr**
kilowatt(s)	**kW**
kilowatthour(s)	**kWh** (pref); **kw-hr**

kilogram   def: the SI unit for mass; abbr: **kg**; other typical abbr: **Mg, g, mg, μg;** also:

kilogram-calorie(s)	**kg-cl**
kilogram(s) per meter	**kg/m**
kilogram(s) per square meter	**kg/m²**
kilogram(s) per cubic meter	**kg/m³**
kilogram meter(s) per second	**kg · m/s**

knockout   as noun or adj, one word

knot   abbr: **kn**

knowledge; knowledgeable

L

label(l)ed; label(l)ing   single *l* pref

laboratory   abbr: **lab**

laborsaving

lacquer

lambert abbr: **L**; use the abbr **L** with care: it is also the SI abbr for *liter*

lampholder

last, latest, latter *last* means final; *latest* means most recent; *latter* refers to the second of only two things (if more than two, use *last*); it is better to write *the last two* (or *three*, etc) than *the two last*

lath; lathe a *lath* is a strip of wood; a *lathe* is a machine

latitude abbr: **lat** or Φ

lay- as a prefix generally combines to form one word (as noun or adj): *layoff, layout, layover*

least common multiple abbr: **lcm**

left-hand(ed) abbr: **LH**

lend; loan use *lend* as a verb, *loan* as a noun; to write or say *"loan* me you textbook" is wrong, but *"lend* me your textbook" is correct

length the SI unit of lenth is the *meter* (or **metre** in Europe), expressed in multiples and submultiples of *kilometers* (**km**), *meters* (**m**), and *millimeters* (**mm**)

lengthy not *lengthly;* also: **lengthening, lengthwise**

less see **fewer**

letter- as a prefix combines erratically: *letterhead, letter-perfect, letter writer*

letter of intent; letter of transmittal pl: *letters of intent, letters of transmittal*

level; leveled; leveling

liable to means under obligation to; avoid using as a synonym for *apt to* or *likely to*

liaison *liaison* is a noun; it is sometimes used uncomfortably as a verb: *liaise*

libel; libeled; libeling; libelous

licence; license *license* pref: *licence* is sometimes used as a noun

light- as a prefix *light-* generally combines to form one word: *lightface* (type), *lightweight;* but *light-year;* the past tense is *lighted*

lightening; lightning *lightening* means to make lighter; *lightning* is an atmospheric discharge of electricity

linear abbr: **lin;** the abbr for lineal foot is **lin ft**

lines of communication not *line of communications*

liquefy; liquefaction

liquid abbr: **liq**

liter; litre the SI spelling is **litre,** but in U.S. **liter** is more common; abbr: **L**; other abbr: **kL, mL, μL**; the abbr for *liter(s) per day/hour/minute/second* are **L/d, L/h, L/m, L/s**

loan see **lend**

loath; loathe *loath* means reluctant; *loathe* means to dislike intensely

lock- as a prefix combines into a single word: *locknut, lockout, locksmith, lockstep, lockup, lockwasher*

locus pl: *loci*

logarithm abbr: (common) **log**; (natural) **ln**

logbook

logistic(s) use *logistic* as an adjective, *logistics* as a noun; *logistic control; the logistics of the move*

long- as a prefix normally combines into a single word or is hyphenated: *long-distance* (adj), *longhand, longplaying, long-term, long-winded;* but *long shot*

longitude abbr: **long** or λ

looseleaf

lose; loose *lose* is a verb that refers to a loss; *loose* is an adjective or a noun that means free or not secured; *three loose nuts caused us to lose a wheel*

louver

low frequency abbr: **lf**

low-pressure (as an adjective) abbr: **l-p**

lubricate; lubrication abbr: **lub**

lumen def: a unit of luminous flux (SI); abbr: **lm**; other abbr:

lumen-hour(s)	$lm \cdot h$ (pref); **lm-hr**
lumens per square foot	lm/ft^2
lumens per square meter	lm/m^2
lumens per watt	**lm/W**
lumen-second(s)	$lm \cdot s$
microlumen(s)	μlm
millilumen(s)	**mlm**

luminance; luminescence; luminosity; luminous

lux def: a unit of illuminance (SI); abbr: **lx**; other abbr: **klx**

M

Mach

macro- a prefix meaning very large; combines to form one word: *macroscopic, macroview*

magneto pl: *magnetos;* as a prefix, normally combines to form one word: *magnetoelectronics, magnetohydrodynamics, magnetostriction;* if combining word starts with *o* or *io,* insert a hyphen: *magneto-optics, magneto-ionization*

magneton; magnetron a *magneton* is a unit of magnetic moment; a *magnetron* is a vacuum tube controlled by an external magnetic field

maintain; maintained; maintenance

majority use *majority* mainly to refer to a number, as in *a majority of 27;* avoid using it as a synonym for many or most; e.g. do not write *the majority of technicians* when the intended meaning is *most*

make- as a prefix normally combines to form one word: *makeshift, makeup*

malfunction

malleable

man to avoid sexist connotations, replace *man* as follows:

for:	write:
man(ned), (ning)	staff(ed), (ing)
man-hour(s)	work-hour(s)
manpower	labor

manage; managed; managing; manageable

maneuver; maneuvered; maneuvering; maneuverable

manufacturer abbr: **mfr**

marketplace

marshal; marshaled; marshaling; marshaler

mass see **kilogram**

material; materiel *material* is the substance or goods out of which an item is made; when used in the plural, it describes items of a like kind, such as *writing materials; materiel* are all the equipment and supplies necessary to support a project or undertaking (a term commonly used in military operational support)

matrix pl: *matrices*

maximum pl: *maximums* (pref) or *maxima;*

abbr: **max;** like *minimize, maximize* can be used as a verb

maybe; may be *maybe* means "perhaps": *maybe there is a second supplier;* the verb form *may be* means "perhaps it will be" or "possibly there is": e.g. *there may be a second supplier*

mean; median the *mean* is the average of a number of quantities; the *median* is the mid-point of a sequence of numbers; e.g. in the sequence of five numbers 1, 2, 3, 7, 8, the mean is 4.2 and the median is 3

mean effective pressure abbr: **mep**

mean sea level *abbr:* **msl** (pref) or **MSL**

mediocre

medium when *medium* is used to mean substances, liquids, materials, or communication or advertising, the plural is *media;* in all other senses the plural is *mediums*

mega def: 10^6; abbr: **M;** other abbr:

megabyte(s)	**Mbyte** (pref) or **Mb**
megacoulomb(s)	**MC**
megaelectronvolt(s)	**MeV**
megahertz	**MHz**
megajoule(s)	**MJ**
meganewton(s)	**MN**
megapascal(s)	**MPa**
megavolt(s)	**MV**
megawatt(s)	**MW**
megohm(s)	**MΩ; Mohm**

memorandum pl: *memorandums* (pref) or *memoranda;* abbr: **memo** (singular) or **memos** (plural)

merit; merited; meriting

metal a single *l* is pref for *metaled* and *metaling* (although *ll* also is acceptable); *metallic* and *metallurgy* always have *ll*

meteorology; metrology *meteorology* pertains to the weather; *metrology* pertains to weights, measures, and calibration

meter; metre def: metric unit of length; the SI spelling is *metre,* but in U.S. *meter* is more common; abbr: **m;** other typical abbr:

square meter(s)	$\mathbf{m^2}$
cubic meter(s)	$\mathbf{m^3}$
meters per second	**m/s**
newton meter(s)	$\mathbf{N \cdot m}$
newtons per square meter	$\mathbf{N/m^2}$

kilogram(s) per cubic meter   **kg/m^3**

micro   def: 10^{-6}; abbr: μ (pref) or **u;** other abbr:

microampere(s)	μ**A**
microcoulomb(s)	μ**C**
microfarad(s)	μ**F**
microgram(s)	μ**g**
microgray(s)	μ**Gy**
microhenry(s)	μ**H**
microhm(s)	$\mu\Omega$; μ**ohm**
microlumen(s)	μ**lm**
micromho(s)	μ**mho**
micrometer(s)	μ**m**
micromole(s)	μ**mol**
micronewton(s)	μ**N**
micropascal(s)	μ**Pa**
microsecond(s)	μ**s** (pref); μ**sec**
microsiemens	μ**S**
microtesla(s)	μ**T**
microvolt(s)	μ**V**
microwatt(s)	μ**W**

micro-   as a prefix meaning very small, normally combines to form one word: *microameter, micrometer, microorganism, microprocessor, microswitch, microview, microwave;* the term *micromicro-* (10^{-12}) has been replaced by **pico**

microphone   abbr: **MIC** (pref) or **mike**

mid-   a prefix that means in the middle of; generally combines into one word: *midday, midpoint, midweek;* if used with a proper noun, insert a hyphen: *mid-Atlantic*

mile   the word *mile* is generally understood to mean a statute mile of 5280 ft (1609 m), so the statement *I drove 326 miles* implies statute miles; when referring to the *nautical mile* (6080 ft; 1853 m), always identify it as such: *the flight distance was 4210 nautical miles (or 4210 nmi);* abbr:

statute mile(s)	**mi**
nautical mile(s)	**nmi** (pref); **n.m.**
miles per gallon	**mpg**
miles per hour	**mph**

mileage; milage   *mileage* pref

milli   def: 10^{-3}; abbr: **m,** other abbr:

milliampere(s)	**mA**
millicoulomb(s)	**mC**
millicurie(s)	**mCi**
millifarad(s)	**mF**
milligram(s)	**mg**

milligray(s)	**mGy**
millihenry(s)	**mH**
millijoule(s)	**mJ**
millikelvin(s)	**mK**
milliliter(s)	**mL**
millilumen(s)	**mlm**
millimeter(s)	**mm**
millimho(s)	**mmho**
millimole(s)	**mmol**
milliohm(s)	**mΩ; mohm**
millinewton(s)	**mN**
millipascal(s)	**mPa**
milliroentgen(s)	**mR**
millisecond(s)	**ms** (pref); **msec**
millisiemens	**mS**
millitesla(s)	**mT**
millivolt(s)	**mV**
milliwatt(s)	**mW**
milliweber(s)	**mWb**

milli-   as a prefix, combines to form one word: *milliammeter, milligram, millimicron*

millibar   def: a unit of pressure (=100 Pa); abbr: **mbar**

mini-   as a prefix combines to form one word: *minicomputer, minireport*

miniature; miniaturization

minimum   pl: *minimums* (pref) or *minima;* abbr: **min**

minority   use mainly to refer to a number, as in *a minority by 2;* avoid using it as a synonyn for several or a few; to write *a minority of the technicians* is incorrect when the intended meaning is *a few technicians*

minuscule   not *miniscule;* def: very small

minute   abbr:

time	**min**
angular measure	$'$

mis-   a prefix meaning wrong(ly) or bad(ly); combines to form one word: *misalign, misfired, mismatched, misshapen*

miscellaneous

miscible

misspelled   not *mispelled*

miter; mitered; mitering

mnemonic

model; modeled; modeling; modeler

mold

mole   def: the SI unit for amount of

substance; abbr: **mol;** other abbr: **kmol, nmol, μmol, mol/m**

momentary; momentarily both mean *for a moment,* not *in a moment*

money- as a prefix normally combines to form one word: *moneymaking, moneysaving*

mono- a prefix meaning one or single; combines to form one word: *monopulse, monorail, monoscope*

monotonous

months the months of the year are always capitalized: *January, February,* etc; if abbr, use only the first three letters: *Jan, Feb,* etc; the abbr for *month* is **mo**

moral; morale often confused; *moral* refers to personal strength of character, the ability to differentiate between right and wrong; *morale* means the general contentedness or happiness of a person or group of people

mortise; mortice *mortise* pref

mosaic

most never use as a short form for *almost;* to say *most everyone is here* is incorrect

movable; moveable *movable* pref

Mr.; Ms. address men as *Mr.* and women as *Ms.;* use *Miss* or *Mrs.* only if you know the person prefers to be so addressed

multi a prefix meaning many; combines to form one word: *multiaddress, multicavity, multielectrode, multistage*

municipal; municipality

N

NAND-gate

nano def: 10^{-9}; abbr: **n;** other abbr:

nanoampere(s)	**nA**
nanocoulomb(s)	**nC**
nanofarad(s)	**nF**
nanohenry(s)	**nH**
nanometer(s)	**nm**
nanosecond(s)	**ns** (pref); **nsec**
nanotesla(s)	**nT**
nanovolt(s)	**nV**
nanowatt(s)	**nW**

naphtha(lene)

nationwide

nautical mile def: 6080 ft (1853 m); abbr: **nmi** (pref) or **n.m.;** see **mile**

navigate; navigator; navigable

NB means note well, and is the abbr for *nota bene;* it's more common to use the word *note*

NC abbr for *normally closed* (contacts)

nebula pl: *nebulas* (pref) or *nebulae*

negative abbr: **neg**

negligible

nevertheless

newton def: a unit of force (SI); abbr: **N;** other abbr: **MN, kN, mN, μN, N · m** (newton meter), **N/m** (newtons per meter)

next write *the next two* (or *next three,* etc) rather than *the two next* (etc)

nickel

night never use *nite;* write *nighttime* as one word

nineteen; ninety; ninth all three are frequently misspelled

NO abbr for *normally open* (contacts)

No. abbr for **number**

noise-cancel(l)ing single *l* pref

nomenclature

nomogram; nomograph *nomogram* pref

non- as a prefix meaning not or negative, normally combines to form one word: *nonconductor, nondirectional, nonnegotiable, nonlinear, nonstop;* if combining word is a proper noun, insert a hyphen: *non-American;* avoid forming a new word with *non-* when a similar word that serves the same purpose already exists (i.e. you should not form *nonaudible* because *inaudible* already exists)

none when the meaning is "not one," treat as singular; when the meaning is "not any," treat as plural: *none* (not one) *was satisfactory; none* (not any) *of the receivers were repaired*

no one two words

NOR-gate

norm def: the average or normal (situation or condition)

normalize

normally closed; normally open (contacts) abbr: **NC, NO**

normal to   def: at right angles to

north   abbr: **N**; other abbr:
northeast   **NE**
northwest   **NW**
north-south   **N-S** (control, movement)
northbound and *northward* are written as one word; for rule on capitalization, see **east**

notable

not applicable   abbr: **N/A**

note well   abbr: **NB** (derived from *nota bene*), but *note* is more common

NOT-gate

notice; noticeable; notification

not to exceed   an overworked phrase that should be used only in specifications; in all other cases use *not more than*

nth   (harmonic, etc)

nuclear   frequently misspelled

nucleus   the plural is *nuclei* (pref) or *nucleuses*

null

number   although *no.* would appear to be the most logical abbr for number (and is pref), **No.** is much more common (the symbol # is no longer used as an abbr for number); the abbr. *no.* or *No.* must always be followed by a quantity in numerals: it is incorrect to write *we have received a No. of shipments;* for the difference in usage between *amount* and *number*, see **amount**

numbers (in narrative)   as a general rule, spell out up to and including nine, and use numerals for 10 and above; for specific rules, see Guideline 4

O

oblique; obliquity

oblivious   def: unaware or forgetful; although *oblivious* should be followed by *of*, it is becoming common practice to follow it with *to;* e.g. *she was oblivious of the disturbance* is correct; *she was oblivious to the disturbance*, although acceptable, is less pref

obsolete; obsolescent

obstacle

obtain; secure   use *obtain* when the meaning is simply to get; use *secure* when the meaning is to make safe or to take possession of (possibly after some

difficulty): *we obtained four additional samples; we secured space in the prime display area*

occasional; occasionally

occur; occurred; occurrence; occurring

o'clock   avoid using; see **time**

of   avoid using in place of *have;* write *we should have measured*, not *we should of measured*

off-   as a prefix either combines into one word, or a hyphen is inserted: *offset, off-center(ed), off-scale, off-the-shelf*

off of   an awkward construction; omit the word *of*

ohm   def: a unit of electric resistance; abbr: Ω or **ohm**; other abbr: **GΩ, Gohm, MΩ, Mohm, kΩ, kohm, mΩ, mohm, μQ, uohm, ohm-cm;** *ohmmeter* has two *m's*

oilfield; oil-filled

OK; okay   these are slang expressions which should never appear in technical writing

omit; omitted; omission

omni-   a prefix meaning all or in all ways; combines to form one word: *omnibearing, omnidirectional, omnirange*

one-   as a prefix mostly combines with a hyphen: *one-piece, one-sided, one-to-one, one-way;* but *oneself* and *onetime*

on; onto   *on* means positioned generally; *onto* implies action or movement: *the report is on Mr. Cord's desk; the speaker stepped onto the platform*

once-over

onward(s)   *onward* pref

opaque; opacity

op cit   def: Latin abbr for *opere citato*, meaning in the work cited; used in footnoting but now obsolete

operate; operator; operable   not *operatable*

optimum   pl: *optima* (pref), and sometimes *optimums;* also: **optimal**

oral   def: spoken; avoid confusing with *aural*

orbit; orbital; orbited; orbiting

organize; organizer; organization

OR-gate

orientation   this is the noun; the verb form is *orient, oriented, orienting*

orifice

origin; original; originally

oscillate

oscilloscope   slang abbr: **scope**

ounce(s)   abbr: **oz**; other abbr:
ounce-foot   **oz-ft**
ounce-inch   **oz-in.**

out-   as a prefix normally combines to form one word: *outbreak, outcome, outdistance;* when *out-* is followed by *of*, insert hyphens if used as a compound adjective (as in *an out-of-date list*), but treat as separate words when used in place of a noun (as in *the printing schedule is out of phase*)

outside diameter   abbr: **OD**

outward(s)   *outward* pref

over-   as a prefix meaning above or beyond, normally combines to form one word: *overbunching, overcurrent, overdriven, overexcited, overrun;* avoid using as a synonym for *more than*, particularly when referring to quantities: *more than 17 were serviceable* is better than *over 17 were serviceable*

overage   means either too many or too old

overall   an overworked word; as an adjective it often gives unnecessary additional emphasis (as in *overall impression*) and should be deleted; avoid using as a synonym for *altogether, average, general*, or *total*

oxidize   *oxidation* is better than *oxidization*

oxyacetylene

P

pacemaker; pacesetter

page; pages   abbr: **p; pp**

paid   not *payed*, when the meaning is to spend

pair(s)   abbr: **pr**

pamphlet

panel; paneled; paneling

paper-   as a prefix mostly combines to form one word: *paperback, paperbound, paperwork;* but *paper-covered, paper-thin*

parabola; parabolas; parabolic; paraboloid

paragraph(s)   abbr: **para**

parallax

parallel; paralleled; paralleling; parallelism;

parallelogram   both *parallel to* and *parallel with* are correct

paralysis; paralyses (pl); **paralyze**

parameter; perimeter   *parameter* means a guideline; *perimeter* means a border or edge

paraphernalia

paraplegic

paraprofessional

parenthesis   this is the singular form; pl: *parentheses*

particles   frequently misspelled

partly; partially   use *partly* when the meaning is "a part" or "in part"; use *partially* when the meaning is "to a certain extent," or when preference or bias is implied

parts per million   abbr: **ppm**

part-time

pascal   def: a unit of pressure (SI); abbr: **Pa**; other abbr: **GPa, MPa, kPa, mPa, μPa, pPa, Pa · s** (pascal second)

pass-   as a prefix normally combines to form one word: *passbook, passkey, passport, password*

passed; past   as a general rule, used *passed* as a verb and *past* as an adjective or noun: *the test equipment has been passed by quality control; past experience has demonstrated a tendency to fail at low temperature; in the past . . .*

pay-   as a prefix normally combines to form one word: *paycheck, payload, payroll;* but *pay day*

pencil; penciled; penciling

pendulum   pl: *pendulums*

penultimate   def: the next to last

people; persons   *people* pref: *all the people were present;* use *persons* to refer only to small numbers of people: *one person was interviewed; seven people failed the test*

per   in technical writing it is acceptable to use *per* to mean either by, a, or an, as in *per diem* (by the day) and *miles per hour;* avoid using *as per* in all writing

per cent; percent   *percent* is pref and is replacing *per cent;* abbr: **%**; use **%** only after numerals: *42%;* use *percent* after a spelled-out number: *about forty percent;*

avoid using the expression *a percentage of* as a synonym for *a part of* or *a small part;* also: **percentage** and **percentile**

perceptible

permeable; permeameter; permeance

permissible

permit; permitted; permitting; permittivity

perpendicular   abbr: **perp**

persevere; perseverance

persistent; persistence; persistency

personal; personnel   *personal* means concerning one person; *personnel* means the members of a group, or the staff: *a personal affair;* the *personnel in the powerhouse*

peta   def: 10^{15}; abbr: **P**; other abbr: **PBq** (petabecquerel)

pharmacy; pharmacist; pharmaceutical

phase   in the nonelectric sense, *phase* means a stage of transmission or development; it should not be used as a synonym for *aspect;* it is used correctly in *the second phase called for a detailed cost breakdown*

phase-in; phaseout   but use two words for the verb forms: *to phase in; to phase out*

phenolic

phenomenon   pl: *phenomena*

photo-   as a prefix, normally combines to form one word: *photoelectric, photogrammetry, photoionization, photomultiplier;* if combining word starts with *o*, insert a hyphen: *photo-offset*

pico   def: 10^{-12}; abbr: **p**; other abbr:

picoampere(s)	**pA**
picocoulomb(s)	**pC**
picofarad(s)	**pF**
picohenry(s)	**pH**
picosecond(s)	**ps** (pref); **psec**
picowatt(s)	**pW**

piecemeal; piecework

piezoelectric; piezo-oscillator

pilot; piloted; piloting

pint   abbr: **pt**

pipeline

plagiarism   def: to copy without acknowledging the original source

plateau   pl: *plateaus* (pref) or *plateaux*

plug; plugged; plugging

plumbbob; plumb line

p.m.   def: after noon (post meridiem)

pneumatic

polarize; polarizing; polarization

policyholder

poly-   a prefix meaning many; combines to form one word: *polydirectional, polyethylene, polyphase*

polyvinyl chloride   abbr: **pvc**

positive   abbr: **pos**

post-   a prefix meaning after or behind; combines to form one word: *postacceleration, postgraduate, postpaid;* but *post office*

post meridiem   def: after noon; abbr: **p.m.;** can also be written as *postmeridian* (less pref)

potentiometer   abbr: **pot.**

pound(s) (weight)   abbr: **lb**; other abbr:

pound-foot	**lb-ft**
pound-inch(es)	**lb-in.**
pounds per square foot	**psf** (pref); **lb/ft^2**
pounds per square inch	**psi** (pref); **lb/in.^{2}**
pounds per square inch, absolute	**psia**

power factor   abbr: **pf** or spell out

powerhouse; power line; powerpack

practicable; practical   these words have similar meanings but different applications that sometimes are hard to differentiate; *practicable* means feasible to do: *it was difficult to find a practicable solution* (one that could reasonably be implemented); *practical* means handy, suitable, able to be carried out in practice; *a practical solution would be to combine the two departments*

practice

pre-   a prefix meaning before or prior; normally combines to form one word: *preamplifier, predetermined, preemphasis, preignite, preset;* if combining word is a proper noun, insert a hyphen: *pre-Roman*

precede; proceed   *precede* means go before; *proceed* generally means carry on or continue: *a brief business meeting preceded the dinner* (the meeting occurred first); *after dinner, we proceeded with the annual presentation of awards;* see **proceed**

precedence; precedent   *precedence* means priority (of position, time, etc): *the pressure test has precedence* (it must be done first); *a precedent* is an example that is or will be followed by others: *we may set a precedent if we grant his request* (others will expect similar treatment)

précis

predominate; predominant; predominantly

prefer; preferred; preference; preferable   avoid overstating *preferable*, as in *more preferable* and *highly preferable*

prescribe; proscribe   *prescribe* means to state as a rule or requirement; *proscribe* means to deny permission or forbid

presently   use *presently* only to mean soon or shortly; never use it to mean *now* (use *at present* instead)

pressure-sensitive

prestigious

pretense; pretence   *pretense* pref

preventive; preventative   *preventive* pref

previous   def: earlier, that which went before; avoid writing *previous to* (use *before*); see **prior**

principal; principle   as a noun, *principal* means (1) the first one in importance, the leader; or (2) a sum of money on which interest is paid: *one of the firm's principals is Martin Dawes; the invested principal of $10,000 earned $950 in interest last year;* as an adjective, *principal* means most important or chief: *the principal reason for selecting the Arrow microprocessor was its low capital cost; principle* means a strong guiding rule, a code of conduct, a fundamental or primary source (of information, etc): *his principles prevented him from taking advantage of the error*

printout

prior; previous   use only as adjectives meaning earlier: *he had a prior appointment,* or *a previous commitment prevented Mr. Perchanski from attending the meeting;* write *before* rather than *prior to* or *previous to*

privilege

proceed; proceeding; procedure   use *proceed to* when the meaning is to start something new; use *proceed with* when the meaning is to continue something that was started previously

producible

program; program(m)ed; program(m)ing; program(m)er   use of single and double *m* varies widely; *mm* pref

prohibit   use *prohibit from;* never *prohibit to*

prominent; prominence

promissory   (note)

proofread

propel; propelled; propelling; propellant (noun); **propellant** (adjective); **propeller**

prophecy; prophesy   use *prophecy* only as a noun, *prophesy* only as a verb

proportion   avoid writing *a proportion of* or *a large proportion of* when *some, many,* or a specific quantity would be simpler and more direct

proposition   in its proper sense, *proposition* means a suggestion put forward for argument; it should not be used as a synonym for *plan, project,* or *proposal*

pro rata   def: assign proportionally; sometimes used in the verb form as *prorate: I want you to prorate the cost over two years*

prospectus; prospectuses

protein

proved; proven   use *proven* only as an adjective or in the legal sense; otherwise use *proved: he has been proven guilty; he proved his case*

psycho-   as a prefix normally combines to form one word: *psychoanalysis, psychopathic, psychosis;* if combining word starts with *o,* insert a hyphen: *psycho-organic*

purge; purging

pursuant to   avoid using this wordy expression

Q

quality control   abbr: **QC**

quantity; quantitative   the abbr of quantity is **qty**

quart   abbr: **qt**

quasi-   a prefix meaning seemingly or almost; insert a hyphen between the prefix and the combining word: *quasi-active, quasi-bistable, quasi-linear*

question mark   insert a question mark after a direct question: *how many booklets will you require?;* omit the question mark when the question posed is really a demand; *may I have your decision by noon on Monday*

questionnaire

quick-   as a prefix normally combines with a hyphen: *quick-acting, quick-freeze, quick-tempered;* exceptions; *quicklime; quicksilver*

quiescent; quiescence

quorum

R

rack-mounted

racon   def: a radar beacon

radian   def: a unit of angular measurement; abbr: **rad**

radiator

radio-   as a prefix, combines to form one word: *radioactive, radiobiology, radioisotope, radioluminescence;* if combining word starts with *o*, omit one of the *o*'s: *radiology, radiopaque;* in other instances *radio* may be either combined or treated as a separate word, depending on accepted usage; typical examples are *radio compass, radio countermeasures, radio direction-finder, radio frequency* (as a noun), *radio-frequency* (as an adjective), *radio range, radiosonde, radiotelephone*

radio frequency   abbr: **rf**

radius   pl: *radii* (pref) or *radiuses*

radix   pl: *radices* (pref) or *radixes*

rain-   as a prefix normally combines to form one word: *raincoat, rainproof, rainwear;* exception: *rain check*

rally; rallied; rallying

RAM   def: random access memory

range; ranging; rangefinder; range marker

rare; rarely; rarity; rarefy; rarefaction

ratemeter

ratio; ratios

rational; rationale   *rational* means reasonable, clear-sighted: *John had a rational explanation*

for the error; rationale means an underlying reason: *Tricia explained the company's rationale for diversifying the product line*

re   def: a Latin word meaning in the case of; avoid using *re* in technical writing, particularly as an abbr for *regarding, concerning, with reference to*

re-   a prefix meaning to do again, to repeat; normally combines to form one word: *reactivate, rediscover, reemphasize, reentrant, reignition, rerun, reset;* if the compound term forms an existing word that has a different meaning, insert a hyphen to identify it as a compound term, as in *re-cover* (to cover again)

reaction   use *reaction* to describe chemical or mechanical processes, not as a synonym for *opinion* or *impression*

reactive kilovolt-ampere; reactive voltampere   see **kilo** or **volt**

readability

readout   (noun and adj form)

realize; realization

recede

receive; receiver; receiving; receivable

rechargeable

recipe; receipt   often confused; *recipe* means cooking instructions; *receipt* means a written record that something has been received

recommend; recommendation

reconcile; reconcilable

reconnaissance

recover; re-cover   *recover* means to get back, to regain; *re-cover* means to cover again

recur; recurred; recurring; recurrence   these are the correct spellings; never write *reoccur* (etc)

recycle; recyclable

reducible

reenforce; reinforce   *reenforce* means to enforce again; *reinforce* means to strengthen; *Rick Davis reenforced his original instructions by circulating a second memorandum; the Artmo Building required 34,750 tons of reinforced concrete*

refer; referred; referring; referral; referee; reference

referendum   pl: *referendums* (pref) or *referenda* (less common)

reiterate   def: to say again

relaid; relayed   *relaid* means laid again, like a carpet; *relayed* means to send on, as a message would be relayed from one person to another

remit; remitted; remitting; remitter; remittance

remodel; remodeled; remodeling

removable

remuneration   *pay, wage,* or *salary* is a better word; often misspelled as *renumeration*

rent-a-car

reoccur(rence)   never use; see **recur**

repairable; reparable   both words mean in need of repair and capable of being repaired; *reparable* also implies that the cost to repair the item has been taken into account and it is economically worthwhile to effect repairs

repellant; repellent   use *repellant* as a noun, *repellent* as an adjective; **repeller**

replaceable

reproducible

rescind

reservoir

reset; resetting; resettability

resin; rosin   these words have become almost synonymous, with a preference for *resin;* use *resin* to describe a gluey substance used in adhesives, and *rosin* to describe a solder flux-core

respective(ly)   this overworked word is not needed in sentences that differentiate between two or more items; e.g. it should be deleted from a sentence such as: *pins 4, 5 and 7 are marked R, S, and V respectively*

resumé   the correct spelling is *résumé* (with two accents), but the single accent or no accent (*resume*) has become standard

retrieve; retrieval

retro-   a prefix meaning to take place before, or backward; normally combines to form one word: *retroactive, retrofit, retrogression;* if combining word starts with *o,* insert a hyphen: *retro-operative*

reverse; reverser; reversal; reversible

revolutions per minute; revolutions per second   abbr: **rpm; rps**

rheostat

rhombus   pl: *rhombuses* (pref) or *rhombi*

rhythm; rhythmic; rhythmically

ricochet; richocheted; richocheting

right-hand(ed)   abbr: **RH**

rivet; riveted; riveter; riveting

road-   as a prefix normally combines to form one word: *roadblock, roadmap, roadside*

roentgen   abbr: **R**

role; roll   a *role* is a person's function or the part that he or she plays (in an organization, project, or play); a *roll,* as a technical noun, is a cylinder; as a verb, it means to rotate; *the technicians' role was to make the samples roll toward the magnet*

ROM   def: read only memory

root mean square   abbr: **rms**

rosin   see **resin**

rotate; rotator; rotatable; rotary

round   def: circular; avoid confusing with **around**

ruggedize

rustproof; rust-resistant

S

salable; saleable   *salable* pref

salvageable

same   avoid using *same* as a pronoun; to write *we have repaired your receiver and tested same* is awkward; instead, write *we have repaired and tested your receiver*

sapphire

satellite

saturate; saturation; saturable

save; savable

sawtooth; saw-toothed

scalar; scaler   *scalar* is a quantity that has magnitude only; *scaler* is a measuring device

scarce; scarcity

sceptic(al); skeptic(al)   *skeptic(al)* pref

schedule

schematic   although really an adjective (as in

schematic diagram), in technical terminology *schematic* can be used as a noun (meaning *a schematic drawing*)

science; scientific(ally); scientist

scissors as a plural word, write *the scissors are;* in the singular form, write *the pair of scissors is*

screwdriver; screw-driven

seamweld

seasonal; seasonable *seasonal* means affected by or dependent on the season; *seasonable* means appropriate or suited to the time of year: *a seasonal activity; seasonable weather*

seasons the seasons are not capitalized: *spring, summer, autumn* or *fall, winter*

secant abbr: **sec**

secede; secession

second as a prefix, *second-* combines erratically: *second-class, second-guess, secondhand, second-rate, second sight;* the abbr for *second* (time) is **sec**, and for *second* (angular measure) it is ″

secure see **obtain**

-sede *supersede* is the only word to end with *-sede;* others end with *-cede* or *-ceed*

seem(s) see **appear(s)**

self- insert a hyphen when used as a prefix to form a compound term: *self-absorption, self-bias, self-excited, self-locking, self-resetting;* but there are exceptions: *selfless, selfsame*

semi- a prefix meaning half; normally combines to form one word: *semiactive, semiannually* (every six months), *semiconductor, semimonthly* (half-monthly), *semiremote, semiweekly* (half weekly); if combining word starts with *i*, insert a hyphen: *semi-idle, semi-immersed*

separate; separable; separator; separation all are frequently misspelled

sequence; sequential

serial number abbr: **ser no.** or **S/N**

series-parallel

serrated

serviceable

serviceperson avoid using *serviceman* or *servicewoman*

servo- as a prefix, combines to form one word: *servoamplifier, servocontrol, servosystem;*

as a noun, *servo* is an abbr for *servomotor* or *servomechanism*

sewage; sewerage *sewage* is waste matter; *sewerage* is the drainage system that carries away the waste matter

shall *shall* is rarely used in technical writing (*will* is pref), except in specifications when its use implies that the specified action is mandatory

short- as a prefix, may combine with a hyphen, as in *short-circuit, short-form* (report), *short-lived, short-term;* in some cases it may combine into one word, as in *shorthand* (writing), *shorthanded, shortcoming, shortsighted*

shrivel(l)ed; shrivel(l)ing single *l* pref

sic a Latin word which means a quotation has been copied exactly, even though there was an error in the original; e.g., *the report stated: "Our participation will be an issential (sic) requirement."*

siemens def: a unit of electric conductance (SI); abbr: **S**; other abbr: **kS, mS, µS**

sight def: the ability to see; see **site**

signal; signaled; signaling; signaler

signal-to-noise (ratio)

silhouette

silverplate; silver-plate use *silverplate* as a noun or adjective, *silver-plate* as a verb

similar not *similiar*

sine abbr: **sin**

singe; singeing the *e* must be retained to avoid confusion with *singing*

singlehanded

siphon not *syphon*

sirup; syrup *syrup* pref

site, sight, cite three words that often are misspelled; a *site* is a location: *the construction site; sight* implies the ability to see: *mud up to the axles became a familiar sight; cite* means quote: *I cite the May 17 progress report as a typical example of good writing*

siz(e)able *sizable* pref

skeptic(al); sceptic(al) *skeptic(al)* pref

skil(l)ful *skillful* pref; note that general usage dictates that *ll* is pref here, contrary to the preference in most *l* and *ll* situations in this glossary

slip   usually combines into a single word: *slippage, slipshod, slipstream;* but *slip ring(s)*

smolder

solder

solely

someone; some one   *someone* is correct when the meaning is any one person; *some one* is seldom used

some time; sometimes   *some time* means an indefinite time: *some time ago; sometimes* means occasionally: *he sometimes works until after midnight*

sound   combines irregularly: *sound-absorbent, sound-absorbing, sound-powered, soundproof, sound track, sound wave*

south   abbr: **S;** other abbr:
southeast   **SE**
southwest   **SW**
southbound and *southward* are written as one word; for rule on capitalization, see **east**

space-   as a prefix normally combines to form one word: *spacecraft, spaceflight*

spare(s)   can be used as a noun meaning spare part(s)

specially   see **especially**

specific gravity   abbr: **sp gr**

specific heat   abbr: **sp ht**

spectro-   as a prefix, combines to form one word: *spectrometer, spectroscope;* if combining word starts with *o*, omit one *o: spectrology*

spectrum   pl: *spectra*

spiral; spiraled; spiraling

split infinitive   to split an infinitive is to insert an adverb between the word *to* and a verb: *to really insist* is a split infinitive; although grammarians used to claim that you should never split an infinitive, they now suggest you may do so if rewriting would result in an awkward construction, ambiguity, or extensive rewriting

spotweld

square   abbr: **sq** or ^{2}; other abbr:
square foot   **ft^2** (pref); **sq**
square inch   **in.^{2}** (pref); **sq in.**
square meter   **m^2**
square centimeter   **cm^2**

square millimeter   **mm^2**
curies per square meter   **Ci/m^2**
milliwatts per square meter   **mW/m^2**

standby; standoff; standstill   all combine into one word when used as a noun or an adjective

standing-wave ratio   abbr: **swr**

state-of-the-art

stationary; stationery   *stationary* means not moving: *the vehicle was stationary when the accident occurred; stationery* refers to writing materials: *the main item in the October stationery requisition was an order for one thousand writing pads*

statute mile   def: 5280 ft (1609 m); see **mile**

statutory

stencil; stenciled; stenciling

stereo-   as a prefix, combines to form one word: *stereometric, stereoscopic; stereo* can be used alone as a noun meaning multichannel system

stimulus   pl: *stimuli*

stock; stockholder; stocklist; stock market; stockpile

stop-   as a prefix usually combines to form one word: *stopgap, stopnut, stopover* (when used as a noun or adjective); *stop payment, stop watch*

stoppage

strato-   a prefix that combines to form one word: *stratocumulus, stratosphere*

stratum   pl: *strata*

structural

stylus   pl: *styluses* (pref) or *styli*

sub-   a prefix generally meaning below, beneath, under; combines to form one word: *subassembly, subcarrier, subcommittee, subnormal, subpoint*

subparagraph   abbr: **subpara;** abbr for *subsubparagraph* is **subsubpara**

subpoena; subpoenaed   *subpena* also used but less pref

subtle; subtlely/subtly   *subtly* pref

succinct

sufficient   in technical writing, *enough* is a better word than *sufficient*

sulfur; sulphur *sulfur* pref; as a prefix, *sulf-* combines to form one word: *sulfanilamide*

summarize

super a prefix meaning greater or over; combines to form one word: *superabundant, superconductivity, superregeneration*

superhigh frequency abbr: **shf**

superimpose; superpose *superimpose* means to place or impose one thing generally on top of another; *superpose* means to lay or place exactly on top of, so as to be coincident with

supersede see **-sede**

supra- a prefix meaning above; normally combines to form one word: *supramolecular;* if combining word starts with *a,* insert a hyphen: *supra-auditory*

surfeit def: to have more than enough

surveyor; surveillance

susceptible

switch- *switchboard, switchbox, switchgear*

swivel; swiveled; swiveling

syllabus pl: *syllabuses* (pref) or *syllabi*

symmetry; symmetrical

sympathize

symposium pl: *symposia* (pref) or *symposiums*

synchro- as a prefix combines to form one word: *synchromesh, synchronize, synchroscope; synchro* can also be used alone as a noun meaning synchronous motor

synonymous use *synonymous with,* not *synonymous to*

synopsis pl: *synopses*

synthesis pl: *syntheses*

synthetic

syphon *siphon* pref

syringe

syrup; syrupy

systemwide

T

tablespoon use *tablespoonfuls* rather than *tablespoonsful;* abbr: **tbsp**

tail- as a prefix normally combines to form one word: *tailboard, tailless, tailwind;* but *tail end, tail fin*

take- *takeoff; takeover; takeup;* as nouns and adjectives these terms all combine into a single word

tangent abbr: **tan**

tangible

tape deck

taxable; tax-exempt; taxpayer

teamwork

teaspoon use *teaspoonfuls* rather than *teaspoonsful;* abbr: **tsp**

technician

tele- a prefix that means at a distance; combines to form one word: *teleammeter, telemetry, telephony, teletype(writer)*

telecom; telecon *telecom* is the abbr for *telecommunication(s); telecon* is the abbr for *telephone conversation*

television abbr: **TV**

temperature abbr: **temp**; combinations are *temperature-compensating* and *temperature-controlled;* when recording temperatures, the abbr for *degree* (deg or °) may be omitted; *an operating temperature of 85C; the water boils at 100C or 212F;* the pref (SI) unit for temperature is the degree Celsius (**°C**)

tempered

template; templet both spellings are correct; *template* pref

temporary; temporarily

tenfold

tensile strength abbr: **ts**

tentative; tentatively

tenuous

tera def: 10^{12}; abbr: **T;** other abbr:

terabecquerel(s)	**TBq**
terahertz	**THz**
terajoule(s)	**TJ**
terawatt(s)	**TW**

terminus pl: *termini* (pref) or *terminuses;* note that this plural is contrary to the pref plurals for most *-us* endings

tesla def: a unit of magnetic flux density, magnetic inductance (SI); abbr: **T;** other abbr: **mT, μT, nT**

that is abbr: **i.e.** (pref) or **ie**

their; there; they're the first two of these words are frequently misspelled, more through carelessness than as an outright error; *their* is a possessive meaning belonging to them: *the staff took their holidays earlier than normal; there* means in that place: *there were 18 desks in the room,* or *put it there; they're* is a contraction of *they are* and should not appear in technical or business writing

there-   as a prefix combines to form one word: *thereafter, thereby, therein, thereupon*

therefor(e)   *therefore* pref

thermo-   a prefix generally meaning heat; combines to form one word: *thermoammeter, thermocouple, thermoelectric, thermoplastic*

thermodynamic temperature   the SI unit is the kelvin (abbr: **K**), expressed in degrees Celsius (°C)

thesis   pl: *theses*

thousand   abbr: **k**
thousand foot-pound(s)   **kip-ft**
thousand pound(s)   **kip**

three-   when used as a prefix, a hyphen normally is inserted between the combining words: *three-dimensional, three-phase, three-ply, three-wire;* exceptions are *threefold* and *threesome*

threshold

through   never use *thru*

tieing; tying   *tying* pref

timber; timbre   *timber* is wood; *timbre* means tonal quality

time   always write time in numerals, if possible using the 24-hour clock: *08:17* or *8:17 a.m., 15:30* or *3:30 p.m.;* 24-hour times may be written as *20:45* (pref), *20:45 hr,* or *20:45 hours;* never use the term "o'clock" in technical writing: write *15:00* or *3 p.m.* rather than *3 o'clock*

time-   typical combinations are *time base, time-card, time clock, time constant, time-consuming, time lag, timesaving, time-table, time-wasting*

tinplate; tin-plate   use *tinplate* as a noun, *tin-plate* as a verb or adjective

to; too; two   frequently misspelled, most often through carelessness; *to* is a preposition that can mean in the direction of, against, before, or until; *too* means as well; *two* is the quantity "2"

today; tonight; tomorrow   never use *tonite*

tolerance   abbr: **tol**

ton; tonne   the U.S. ton is 2000 lb and is known as a *short ton;* the British ton is 2240 lb and is known as a *long ton;* the metric ton is 1000 kg (2204.6 lb) and is known as a *tonne* (abbr: **t**); other terms: *tonmile* and *tonnage*

toolbox; toolmaker; toolroom

top-   *top-heavy; top-loaded; top-up*

torque; torqued; torquing

total; totaled; totaling

toward(s)   *toward* pref

traceable

trade-   as a prefix combines most often into a single word: *trademark, tradeoff;* but *trade-in, trade name,* and *trade show*

trans-   a prefix meaning over, across, or through; it normally combines to form one word: *transadmittance, transcontinental, transship;* if combining word is a proper noun, insert a hyphen: *trans-Canada* (an exception is *transatlantic,* which through common usage has dropped the capital A and combined into one word); *transonic* has only one *s*

transceiver   def: a transmitter-receiver

transfer; transferred; transferring; transferable; transference

transmit; transmitted; transmitting; transmittal; transmitter; transmission

transverse; traverse   *transverse* means to lie across; *traverse* means to track horizontally

travel; traveled; traveling

tri-   a prefix meaning three or every third; combines to form one word: *triangulation, tricolor, trilateral, tristimulus, triweekly*

triple-   all compounds are hyphenated: *triple-acting, triple-spaced*

trouble-free; troubleshoot(ing)

truncated

tune; tunable; tuneup   (noun or adjective)

tunnel; tunneled; tunneling

turbo-   a prefix meaning turbine-powered; combines to form one word: *turboelectric, turboprop*

turbulence; turbulent

turn- *turnaround* and *turnover* form one word when used as adjectives or nouns; *turnstile* and *turntable* always form one word; *turns-ratio* is hyphenated

two- when used as a prefix to form a compound term, a hyphen normally is inserted: *two-address, two-phase, two-ply, two-position, two-wire;* an exception is *twofold*

type- as a prefix normally combines into one word: *typeface, typeset(ting), typewriter, typewriting*

U

ultimatum pl: *ultimatums* (pref) or *ultimata* (seldom used)

ultra- a prefix meaning exceedingly; normally combines to form one word: *ultrasonic, ultraviolet;* if combining word starts with *a,* insert a hyphen: *ultra-audible, ultra-audion*

ultrahigh frequency abbr: **uhf**

un- a prefix that generally means not or negative; normally combines to form one word: *uncontrolled, undamped, unethical, unnecessary;* if combining word is a proper noun, insert a hyphen: *un-American;* if term combines to form an existing word that has a different meaning, insert a hyphen: *un-ionized* (meaning not ionized); if uncertain whether to use *un-, in-,* or *im-,* try using *not*

unadvisable; inadvisable both are correct; *inadvisable* pref

unbalance; imbalance for technical writing, *unbalance* pref

unbias(s)ed *unbiased* pref

under- a prefix meaning below or lower; combines to form one word: *underbunching, undercurrent, underexposed, underrated, undershoot, undersigned*

underage means a shortage or deficit, or too young

unequal(l)ed *unequaled* pref; see **equal**

unessential; inessential *unessential* pref

unforeseen; unforeseeable

uni- a prefix meaning single or one only; combines to form one word: *uniaxial, unidirectional, unifilar, univalent*

uninterested def: not interested; avoid confusing with *disinterested*

unionized; un-ionized *unionized* refers to a group of people who belong to a union; *un-ionized* means not ionized

unique def: the one and only, without equal, incomparable; use with great care and never in any sense where a comparison is implied; you cannot write *this is the most unique design;* rewrite as *this design is unique,* or (if a comparison must be made) *this is the most unusual design*

unmistakable

unnavigable

unparalleled

unpractical *impractical* pref

unsanitary *insanitary* pref

unserviceable abbr: **u/s**

unstable but *instability* is better than *unstability*

untraceable

up- as a prefix combines to form one word: *update, upend, upgrade, uprange, upswing*

upper case def: capital letters; abbr: **uc**

uppermost

up-to-date

use; usable; usage; using; useful

utilize avoid using *utilizes* when *uses* or *employs* would be a better word

V

vacuum

valance; valence *valance* means a cover over a drapery track; *valence* is an electronic or nucleonic term, as in *valence electron*

valve-grind(ing)

vari- as a prefix meaning varied, combines into one word: *varicolored, variform*

variance write *at variance with;* never write *at variance from*

varimeter; varmeter def: a meter for measuring reactive power; *varimeter* pref

vehicle; vehicular

vender; vendor *vendor* pref

versed sine abbr: **vers**

versus def: against; abbr: **vs**

vertex   def: top; pl: *vertexes* (pref) or *vertices;* avoid confusing with *vortex*

very high frequency   abbr: **vhf**

vice versa   def: in reverse order

video-   as a prefix normally combines to form one word: *videocast, videocassette, videotape;* the abbr for videocassette recorder is **VCR** or **vcr**

video frequency   abbr: **vf**

viewfinder; viewpoint

vise   def: a clamp; in U.S. spelled *vise;* elsewhere: *vice*

visor; vizor   *visor* pref

viz   def: namely; this term is seldom used in technical writing

vocation; avocation   *vocation* is a trade or calling; *avocation* means an interest or hobby

voice-over; voiceprint

volt   def: electric potential or potential difference; abbr: **V;** other abbr: **MV, kV, mV, μV, nV;** also:

volt-ampere(s)	**VA**
volt-ampere(s), reactive	**VAr**
volts, alternating current	**Vac**
volts, direct current	**Vdc**
volts, direct current, working	**Vdcw**
volts per meter	**V/m**

volt-   combines into one word: *voltammeter, voltohmyst*

volume   abbr: **vol**

vortex   def: spiral; pl: *vortexes* (pref) or *vortices;* avoid confusingwith *vertex*

VU-meter

W

waive; waiver; waver   *waive* and *waiver* mean to forgo one's claim or give up one's right; *waver* means to hesitate, to be irresolute

walkie-talkie

war-   as a prefix, combines to form one word: *warfare, wartime*

warranty

waste, wastage

water-   combines irregularly: *watercool(ed); water cooler; waterflow; water level; waterline; waterproof; water-soluble; watertight*

watt   def: a unit of power, or radiant flux (SI); abbr: **W;** other abbr: **TW, GW, MW, kW, mW, μW, nW, pW, W/m²;** the abbr for *watt-hour(s)* is **Wh** (pref) or **W-hr;** as a prefix, *watt-* forms *watthourmeter* and *wattmeter*

wave-   normally combines to form one word: *waveband, waveform, wavefront, waveguide, wavemeter, waveshape;* exceptions are *wave angle* and *wave-swept*

wavelength   abbr: λ

waver   see **waiver**

wear and tear   *no* hyphens

weather   use only as a noun; never write *weather conditions;* avoid confusing with *climate* and *whether*

weatherproof

weber   def: a unit of magnetic flux (SI); abbr: **Wb;** other abbr: **mWb**

Wednesday   often misspelled

week(s)   abbr: **wk**

weekend

weight   abbr: **wt**

well-   as a prefix normally combines with a hyphen: *well-adjusted, well-defined, well-timed*

west   abbr: **W;** *westbound* and *westward* are written as one word; for rule on capitalization, see **east**

where-   as a prefix, combines to form one word: *whereas, wherefore, wherein;* when combining word starts with *e*, omit one *e: wherever*

whether; weather   *whether* means if; *weather* has to do with rain, snow, sunshine, etc

while; whilst   *while* pref

whoever

wholly   not *wholely*

wide; width   abbr: **wd**

wideband; widespread

wirecutter(s); wire-cutting; wirewound

withheld; withhold

word processor; word processing   abbr: **WP;** as an adj, insert a hyphen: *the word-processing software*

words per minute   abbr: **wpm**

work-   as a prefix usually combines to form

one word: *workbench, workflow, workload, workshop;* but *work force* and *work station*

working volts, dc   abbr: **Vdcw**

worldwide

wrap; wrapped; wrapping; wraparound

writeoff; writeup   both combine into one word when used as noun or adjective

writer   see **author**

writing   only one *t*

X

x-   *x-axis; X-band; x-particle; x-radiation; x-ray*

Xerox

X-Y recorder

Y

y-   *Y-antenna; y-axis; Y-connected; Y-network; Y-signal*

yard(s)   abbr: **yd**

yardstick

year(s)   abbr: **yr;** typical combinations are *year-end* and *year-round*

your; you're   *your* means belonging to or originating from you: *I have examined your prototype analyzer; you're* is a contraction of *you are* and should not appear in technical or business writing

Z

Z-axis

zero   pl: *zeros* (pref) or *zeroes;* typical combinations are *zero-access, zero-adjust, zero-beat, zero-hour, zero level, zero-set, zero reader*

zip code

zoology; zoological

Index

MARKING CONTROL CHART

This control chart will show you which aspects of your writing need attention and, as time progresses, whether you have successfully corrected your most predominant faults. As each assignment is returned to you, count up the errors indicated as marginal notations by your instructor and enter them on the chart. For instance, if on assignment 1 your instructor enters "F" and "U" once, and "S" three times, in the margin, you are being told you have used the wrong format (F), your work is untidy (U), and you have three spelling errors (S). In column 1 of the chart enter "1" in the squares opposite F and U, and "3" in the square opposite "S".

ASSIGNMENT NUMBER

1	2	3	4	5	6	7	8	9	10	11	12	13	14	15

A — Awkward construction

B — Brevity overdone; too few details

C — Continuity weak; paragraph lacks coherence/unity

D — Development inadequate; support your argument

E — Error! Check your facts, data, information

F — Format incorrect

G — Grammar fault

H — Heavy going; dull; uninteresting

I — Illogical or irrelevant (correct or omit)

J — Jumpy—too many short sentences, reads like primary reader

K — King-size paragraph or sentence (shorten it)

L — Low Information Content words or phrase (delete)

M — Missing words or information

N — No! Never do this; never use slang, contractions, unexplained abbreviations, etc

O — Organization poor

P — Punctuation error, or punctuation missing

Q — Query: what does this mean? not understood; can't read your writing

R — Repetition

S — Spelling error

T — Tone wrong

U — Untidy, messy, or careless work (improve "presentation")

V — Vague; ambiguous; not clear enough

W — Wishy-washy; weak argument; unconvincing

X — X-out (delete, omit) this unnecessary statement

Y — Yak! Yak! Yak!—too wordy; too many generalities

Z — Lacks continuity; needs better transitions

— Numbers wrongly presented

// — Use parallel construction